Lecture Notes in Computer Science 2582

Edited by G. Goos, J. Hartmanis, and J. van Leeuwen

Lecture Notes in Computer Science 2562

Springer
Berlin
Heidelberg
New York
Hong Kong
London
Milan
Paris
Tokyo

Leopoldo Bertossi   Gyula O.H. Katona
Klaus-Dieter Schewe   Bernhard Thalheim (Eds.)

Semantics
in Databases

Second International Workshop
Dagstuhl Castle, Germany, January 7-12, 2001
Revised Papers

Springer

Series Editors

Gerhard Goos, Karlsruhe University, Germany
Juris Hartmanis, Cornell University, NY, USA
Jan van Leeuwen, Utrecht University, The Netherlands

Volume Editors

Leopoldo Bertossi
Carleton University, School of Computer Science
Herzberg Building, 1125 Colonel By Drive, Ottawa, Canada K1S 5B6
E-mail: bertossi@scs.carleton.ca

Gyula O.H. Katona
Hungarian Academy of Sciences, Alfréd Rényi Institute of Mathematics
P.O. Box 127, 1364 Budapest, Hungary
E-mail: ohkatona@renyi.hu

Klaus-Dieter Schewe
Massey University, Department of Information Systems
Private Bag 11222, Palmerston North, New Zealand
E-mail: K.D.Schewe@massey.ac.nz

Bernhard Thalheim
Brandenburg University of Technology at Cottbus
Institute of Computer Science
P.O. Box 10 13 44, 03013 Cottbus, Germany
E-mail: thalheim@informatik.tu-cottbus.de

Cataloging-in-Publication Data applied for

A catalog record for this book is available from the Library of Congress.

Bibliographic information published by Die Deutsche Bibliothek.
Die Deutsche Bibliothek lists this publication in the Deutsche Nationalbibliografie;
detailed bibliographic data is available in the Internet at <http://dnb.ddb.de>.

CR Subject Classification (1998): H.2, H.3, H.5, I.2.4, F.3.2, G.2

ISSN 0302-9743
ISBN 3-540-00957-4 Springer-Verlag Berlin Heidelberg New York

Springer-Verlag Berlin Heidelberg New York
a member of BertelsmannSpringer Science+Business Media GmbH

http://www.springer.de

© Springer-Verlag Berlin Heidelberg 2003
Printed in Germany

Typesetting: Camera-ready by author, data conversion by PTP-Berlin GmbH
Printed on acid-free paper   SPIN: 10872394   06/3142   5 4 3 2 1 0

Preface

As an attempt to see where the research on database semantics was and where it was heading, the first workshop *Semantics in Databases* was held in Řež near Prague in January 1995 as a collocated event with ICDT'95. The workshop was informal. It featured more than a dozen one-hour talks, and plenty of time was left for informal discussions. At the conclusion of the workshop, the participants decided to prepare a volume containing full versions of selected papers presented at the workshop. All papers were reviewed, revised, reviewed in a second round, revised and partially reviewed again. Finally the volume appeared in the Springer-Verlag LNCS series as volume 1358 in 1998. The workshop led to the development of the FoIKS symposium (Foundations of Information and Knowledge Systems; FoIKS 2000, LNCS 1762; FoIKS 2002, LNCS 2284). Due to the success and the impact the volume had on database research, the community decided to organize another workshop Semantics in Databases in Dagstuhl from January 7 till January 12, 2001. The schedule of the talks was similar to the schedule of the first workshop. Each talk was one hour long and had one responder introducing the discussion of the approach and results presented during the talk. The workshop had more than 40 participants. We decided to prepare this second volume on "Semantics in Databases" using the same approach as that adopted for the first. We invited 16 papers. During the first two rounds of reviewing, 10 papers were selected for the final collection in this volume.

We are thankful to the Dagstuhl organizers. The stimulating environment and the service provided led to an intensive and fruitful seminar in January 2001.

This volume would never have been completed in the current quality if the reviewers of the papers had not put a lot of effort into it. We are very thankful to the large number of reviewers who devoted a lot of their time reviewing papers at various stages of the volume development.

13 December 2002

L. Bertossi
G.O.H. Katona
K.-D. Schewe
B. Thalheim

List of Reviewers

Mira Balaban
Catriel Beeri
Leopoldo Bertossi
Joachim Biskup
Harold Boley
Egon Börger
Francois Bry
Chris Calude
Hendrik Decker
Alexander Dikowsky
Jürgen Dix
Gill Dobbie
Thomas Eiter
Thomas Feyer
Burkhard Freitag
Ulrich Geske
Georg Gottlob
Erich Grädel
Marc Gyssens
Sven Hartmann
Stephen J. Hegner
Roland Kaschek
Gyula O.H. Katona
Markus Kirchberg
Hans-Joachim Klein
Meike Klettke
Bart Kuipers
Gabriel Kuper

Dietrich Kuske
Neil Leslie
Mark Levene
Leonid Libkin
Sebastian Link
Rainer Manthey
Enric Mayol
Heinrich C. Mayr
Dezsõ Miklós
Jan Paredaens
Oscar Pastor
Andreas Prinz
Peter Revesz
Domenico Sacca
Vladimir Sazonov
Klaus-Dieter Schewe
Thomas Schwentick
Yuri Serdyuk
Nicolas Spyratos
Srinath Srinivasa
Leticia Tanca
Ernest Teniente
Bernhard Thalheim
Alexei Tretiakov
José María Turull Torres
Jan Van den Bussche
Millist Vincent
Gerd Wagner

Table of Contents

Semantics in Databases

Leopoldo Bertossi[1], Gyula O. H. Katona[2], Klaus-Dieter Schewe[3], and
Bernhard Thalheim[4]

[1] Carleton University, School of Computer Science, Herzberg Building,
1125 Colonel By Drive, Ottawa, Canada K1S 5B6
bertossi@scs.carleton.ca
[2] Alfréd Rényi Institute of Mathematics, Hungarian Academy of Sciences,
PostBox 127, H-1364 Budapest, Hungary
ohkatona@renyi.hu
[3] Massey University, Department of Information Systems, Private Bag 11 222,
Palmerston North, New Zealand
k.d.schewe.@massey.ac.nz
[4] Computer Science Institute, Brandenburg University of Technology at Cottbus,
PostBox 101344, D-03013 Cottbus, Germany
thalheim@informatik.tu-cottbus.de

The term "Semantics" is one of the overloaded in computer science and used
in various meaning. This variety can also be observed in database literature. In
computer linguistics or web research, semantics is a component of the language
which associates words or components of a grammar with their meaning (linguistic content). In modeling and specification, semantics assigns set-theoretic
denotations to formulas in order to characterize truth. At the same time, semantics is used as the basis for certain methods of proof (truth and proof semantics
in semiotics). In programming language technology, semantics is often used in
the sense of operational semantics, i.e. consists in an interpretation of commands
of a programming language by machine operations. This widespread usage of the
term "semantics " has led to very different goals, methods, and applications. Semantics includes at the same time the interpretation of utterances, temporal,
contextual, subjective and other aspects. Semantics is either considered opera-
tionally on the basis of applications or implementations, or logically associating
a database state or a collection of database states to a truth value or prag-
matically by relating utterances to the understanding of the user. These three
understandings may be mapped to each other.

We require, however, that a formal mathematical theory is provided beside
these substantial differences of the usage of "semantics". The same statement
must have a stable interpretation and cannot be differently interpreted by different computer systems. The formal theory provides a way to prove correctness of specifications. A specification has the liveness property if a number of
positive properties can be proven. A specification has the safety property if a
number of negative properties cannot happen. Semantics is specified by means
of formulas in a logic. Formulas reflect three different applications of semantics
in databases. First, formulas are used for communicating the meaning between
parties involved. Second, formulas support to verify whether the implementation
carries the same meaning. Finally, formulas enable in the validation of the spec-

L. Bertossi et al. (Eds.): Semantics in Databases, LNCS 2582, pp. 1–6, 2003.
© Springer-Verlag Berlin Heidelberg 2003

ification by proving liveness and safety properties. This expression of semantics of databases by formulas allows to consider semantics by means of mathematical logics, i.e. by theories in a certain logical language. Therefore, a database application may be considered as a collection of models of the given theory.

In the database research, semantics is used in a variety of meanings:

- Stable semantics is declared by static integrity constraints. A static integrity constraint is valid in each database state. Feasibility considerations have led to the introduction of a number of specific constraints called dependencies.
- Static integrity constraints are mapped to transition constraints. Transition constraints are Hoare triples (precondition, operation, postcondition) specifying that the postcondition must be valid after the operation has been applied to a database state that satisfies the precondition. The transaction approach is based on Hoare triples with equal pre- and postcondition.
- Dynamic semantics of applications is often left unspecified. Typical dynamic constraints are temporal formulas describing the life cycle of database objects.
- Database instances can be either elementary instances or deductively generated instances. Deductive database are used for generation of instances by an abstract generator. Different generators led to a variety of semantics such as stable semantics.
- Semantics can be state-dependent. In this case, phases can be distinguished. The database or parts of it satisfy a theory describing the phase.
- Semantics can be partially represented by structural properties. The research on semantic data models has led to a number of main structures such as types, subtypes, supertypes, and special types like attribute, entity and relationship types. Further, constructors are used for construction of complex types.
- Structural semantics concentrates on the meaning of singleton word fields. Research on database components and research on terminological ontologies follow these approaches.
- Conceptual modeling follows approaches developed by generative semantics. The meaning of a schema is constructed by the meaning of its constituents.
- Approaches used in the semantic web and ontology research follow those of practical semantics. The meaning of constructs is based on rules for practical application in interaction scenarios.
- The semantic aspect of utterances is expressed by the content. Therefore, content management targets too on semantics of data.

The state of practice does not reflect achievements of research in our area:

- A small set of the large variety of integrity constraint classes can be declaratively specified in database programming languages. These languages only have fragmental facilities for the expression of database semantics. Moreover, these facilities are redundant, lack orthogonality, are partially non-declarative and have side effects.
- DBMS have fragmental facilities for handling of database semantics. These facilities are incomplete and partially unsound.

- Specification languages used in database development tools vary in semantics of similar constructs. The same diagram may have another meaning in another tool.
- Integrity constraints can be specified in some database development tools. A translation of the integrity constraints to the programming language of the DBMS is, however, not provided.
- In most existing relational DBMS, integrity handling and checking is far from being efficient.
- The facilities for integrity constraint maintenance are slowly and reluctantly integrated into implementations of the SQL2 or SQL3 standard.
- Treatment of integrity constraints is not well founded in the SQL3 standard. Meaning and semantics is not well understood. There exists a large variety of interpretations for each group of integrity constraints.
- Enforcement policies used in SQL3 are rather confusing. Integrity constraints can be enforced in deferred or intermediate mode, may be enforced at row level or at statement level.

Semantics research has been neglected over the last years. In the early days of database research, issues related to database research played a prominent role, and papers discussing database models, conceptual design, integrity constraints and normalization often dominated major database conferences. This began to change at the end of the 80ies. Nowadays those issues do not appear to be part of the mainstream research. Dependencies and other constraints have been considered to be the "dying swan" in the database research. We observe a number of explanations for this development:

- The research has not found its way to major implementations of DBMS. The support for integrity constraint maintenance is still rudimentary in DBMS.
- The simple questions got solved. The remaining problems are far more difficult to tackle.
- The research has been conducted within a large variety of database models without a reference to research within other models. For instance, there is no systematic summarization of research on constraints in other models beyond those for the relational model and the entity-relationship model.
- The results on semantics are spread over a large number of conferences in the last decade. It is often difficult to find the proceedings or the references.
- The large variety of models has led to a large number of incompatible approaches and solutions. Their differences are not understood. A lot of research has been conducted for models which are not used or are not known to young researchers.

Semantics research has left a large number of open problems. We may distinguish a number of major research areas which are still open:

Satisfiability of specification should be provable for a collection of integrity constraints. If a schema is unsatisfiable the part of the specification causing this property must be identifiable.

Integration of static and operational constraints allows to consistently switch between different equivalent constraint sets. A coherent theory of integrity enforcement for different DBMS enables in switching applications between DBMS.

Strong and soft interpretation of constraints is not yet solved in a satisfiable form. Most constraints can be given in a strong logical form. Some constraints may use fuzzy or deontic approaches.

Global normalization is still not investigated in detail. Normalization is currently using only a local variant in a type-wise form.

Continuous engineering requires to cope with consistent extensions of structures, functionality and integrity support of a running database system.

Integration of quality requirements into specification will lead to provide a motivation for requirements such as normal forms and acyclicity.

Complexity of constraint sets is currently only known for simple classes of constraints such as sets of keys and functional dependencies. Average complexity results would lead to a better understanding of semantics in 'normal' applications.

Enforcement of integrity constraint sets is currently mainly supported by DBMS on the level of rule triggering systems. The alternative approaches to integrity enforcement should be integrated into these approaches.

Implication problems are tackled on the level of classes of constraints. A database application uses however a set of constraints from different classes. The interaction of constraints needs more research.

Treatment of semantics by views has not yet been satisfactorily solved. At the same time, most large database systems are heavily based on views.

Distributed integrity management allows to avoid the centralized integrity management in distributed applications. Distributed systems are still developed on the allocation of data sets to nodes without consideration of semantics.

Integration of vertical, horizontal and deductive normalization allow a more flexible optimization of database behavior. Therefore, a common framework for vertical, horizontal and deductive normalization is sought.

Treatment of incomplete specifications is necessary for practical reasons. Usually, specifications are incomplete. An incomplete specification of constraints leads to completely different normalized schemata.

Implementation of integrity constraints in database systems is still based on the waterfall model:

1. The initial aim in conceptual database design is the development of structural or semantical integrity constraints.
2. The operational optimization of integrity constraints is based on normalization approaches which restructure the database by decomposition or fragmentation and lead to equivalent sets of integrity constraints which integrity enforcement is simpler in the given DBMS.
3. During the mapping phase check constraints are derived which allow to control consistency of databases on the row level, by single sub-queries or aggregation or inter-table sub-queries.

4. Constraints which cannot be mapped to check constraints are mapped to assertions on the basis of independent and inter-table constraints if the DBMS intended to be used for the database implementation supports assertions.
5. Static integrity constraints can be mapped to triggers if the DBMS supports triggers. Trigger support may vary and uses additional semantics in the DBMS.
6. Stored procedures enable in encapsulation of operations and integrity constraints. Integrity maintenance is wrapped into the stored procedure code.
7. Integrity constraint checking at the application level is mainly performed by the transaction management or at the level of application programs.

At the same time we observe a number of pitfalls in the existing research literature:

- *Normalization* is still considered as the ultimate weapon for structural optimization. The approaches well developed so far cover vertical normalization at the local type level. Horizontal or deductive normalization is still not integrated into vertical normalization approaches. Global normalization is moreover in an embryonic state.
- *Tuning and operational optimization* lead to physical schemata which are structurally and semantically different from the conceptual schema. The later is used for programming and querying.
- The *class-centered approach to integrity specification* concentrates on the class-wise treatment of integrity constraints. Instead of that, we should concentrate on the treatment of a set of integrity constraints valid in a application.

The papers in this volume reflect a variety of approaches to semantics in databases:

P. Barceló, L. Bertossi and L. Bravo develop a theory of *consistent answers* from database that violate constraints, i.e. answers to queries that are generated in every minimal repair of the database.
J. Biskup and B. Sprick design a *unifying framework* for consistent and integrated reasoning on *semantic constraints and security constraints* in highly distributed systems. The formalization is based on a propositional multi-model logic for multi-agent systems with temporal and epistemic operators.
H. Decker discusses the potential of paraconsistent reasoning for datalog, computational logics and for foundations of *partially inconsistent databases*.
S. Hartmann introduces *soft cardinality constraints*. If cardinality constraints are conflicting then a resolution strategy may use a weakening approach and may consider only those which are not in conflict and which conflict is not causing other inconsistencies. A number of conflict resolution strategies is introduced.
S. Hegner develops a characterization of *type hierarchies under open specifications*. The algebraic characterization is based on weak partial lattices, the logical one on products of models of propositional-based specifications.

H.-J. Klein focuses on sure answers in the case of *null values* which are representing existing but unknown values or are indicating that there is no information. He uses the introduced theory to generate transformations of SQL queries that approximate sure information answers.

S. Link generalizes the integrity enforcement on the basis of greatest consistent specializations to *maximal consistent effect preservers*. He introduces a general definition and framework that formalizes effect preservation of operations while maintaining database consistency.

F. Neven and T. Schwentick survey results on *pattern languages for tree-structured data* such as XML data. It is shown that an intermediate logic between first-order logic and monadic sencond-order logic captures the expressive power of languages with regular path expressions.

D. Seipel and U. Geske investigate the semantics of *cardinality constraints* on the basis of *disjunctive deductive database programs* and consider various deductive calculi for cardinality constraints. There is no sound and complete calculus using only definite deductive rules. Instead, a mix of disjunctive rules and rules from the definite calculus can be used.

J.M. Turull-Torres considers the *expressive power of relational query languages* on the basis of an abstract notion of *similarity* or indistinguishability of databases. He develops a general framework for definition of semantic classifications of queries and shows that the hierarchy is orthogonal with the time-space hierarchy for Turing machines.

Characterizing and Computing Semantically Correct Answers from Databases with Annotated Logic and Answer Sets

Pablo Barceló[1], Leopoldo Bertossi[2], and Loreto Bravo[3]

[1] University of Toronto, Department of Computer Science, Toronto, Canada.
pablo@cs.toronto.edu
[2] Carleton University, School of Computer Science, Ottawa, Canada.
bertossi@scs.carleton.ca
[3] Pontificia Universidad Catolica de Chile, Departamento de Ciencia de
Computación, Santiago, Chile. lbravo@ing.puc.cl

Abstract. A relational database may not satisfy certain integrity constraints (ICs) for several reasons. However most likely most of the information in it is still consistent with the ICs. The answers to queries that are consistent with the ICs can be considered sematically correct answers, and are characterized [2] as ordinary answers that can be obtained from *every* minimally repaired version of the database. In this paper we address the problem of specifying those repaired versions as the minimal models of a theory written in *Annotated Predicate Logic* [27]. It is also shown how to specify database repairs using disjunctive logic program with annotation arguments and a classical stable model semantics. Those programs are then used to compute consistent answers to general first order queries. Both the annotated logic and the logic programming approaches work for any set of universal and referential integrity constraints. Optimizations of the logic programs are also analyzed.

1 Introduction

In databases, integrity constraints (ICs) capture the semantics of the application domain and help maintain the correspondence between that domain and its model provided by the database when updates on the database are performed. However, there are several reasons why a database may be or become inconsistent wrt a given set of integrity constraints (ICs) [2]. This could happen due to the materialized integration of several, possibly consistent data sources. We can also reach such a situation when we need to impose certain, new semantic constraints on legacy data. Another natural scenario is provided by a user who does not have control on the database maintenance mechanisms and wants to query the database through his/her own semantics of the database. Actually such a user could be querying several data sources and needs to impose some semantics on the combined information.

More generally speaking, we could think ICs on a database as constraints on the answers to queries rather than on the information stored in the database

L. Bertossi et al. (Eds.): Semantics in Databases, LNCS 2582, pp. 7–33, 2003.

[32]. In this case, retrieving answers to queries that are consistent wrt the ICs becomes a central issue in the development of DBMSs.

In consequence, in any of the scenarios above and others, we are in the presence of an inconsistent database, where maybe a small portion of the information is incorrect wrt the intended semantics of the database; and as a an important and natural problem we have to characterize and retrieve data that is still correct wrt the ICs when queries are posed.

The notion of consistent answer to a first order (FO) query was defined in [2], where also a computational mechanism for obtaining consistent answers was presented. Intuitively speaking, a ground tuple $\bar{t}$ to a first order query $Q(\bar{x})$ is *consistent* in a, possibly inconsistent, relational database instance DB, if it is an (ordinary) answer to $Q(\bar{x})$ in every minimal repair of DB, that is in every database instance over the same schema and domain that differs from DB by a minimal (under set inclusion) set of inserted or deleted tuples. In other words, the consistent data in an inconsistent database is invariant under sensible restorations of the consistency of the database.

The mechanism presented in [2] has some limitations in terms of the ICs and queries that can handle. Although most of the ICs found in database praxis are covered by the positive cases in [2], the queries are restricted to conjunctions of literals. In [4,6], a more general methodology based on logic programs with a stable model semantics was introduced. There is a one to one correspondence between the stable models of the logic programs and the database repairs. More general queries could be considered, but ICs were restricted to be "binary", i.e. universal with at most two database literals (plus built-in formulas). A similar, independent approach to database repair based on logic programs was also presented in [26].

The basic idea behind the logic programming based approach to consistent query answering is that since we need to deal with *all* the repairs of a database, we had better specify the class of the repairs. From a manageable logical specification of this class different reasoning tasks could be performed, in particular, computation of consistent answers to queries.

Notice that a specification of the class of database repairs must include information about (from) the database and the information contained in the ICs. Since these two pieces of information may be mutually inconsistent, we need a logic that does not collapse in the presence of contradictions. A non classical logic, like *Annotated Predicate Calculus (APC)* [27], for which a classically inconsistent set of premises can still have a model, is a natural candidate. In [3], a new declarative semantic framework was presented for studying the problem of query answering in databases that are inconsistent with respect to universal integrity constraints. This was done by embedding both the database instance and the integrity constraints into a single theory written in APC, with an appropriate, non classical truth-values lattice $Latt$.

In [3] it was shown that there is a one to one correspondence between some minimal models of the annotated theory and the repairs of the inconsistent database for universal ICs. In this way, a non monotonic logical specification of the database repairs was achieved. The annotated theory was used to derived some algorithms for obtaining consistent answers to some simple first order queries.

The results presented here extend those presented in [3] in different ways. First, we show how to annotate other important classes of ICs found in database praxis, e.g. referential integrity constraints [1], and the correspondence results are extended. Next, the problem of consistent query answering is characterized as a problem of non monotonic entailment.

We also show how the the APC theory that specifies the database repairs motivates the generation of new logic programs to specify the database repairs. Those programs have a classical stable model semantics and contain the annotations as constants that appear as new arguments of the database predicates. We establish a one to one correspondence between the stable models of the program and the repairs of the original database. The programs obtained in this way are simpler than those presented in in [4,6,26] in the sense that only one rule per IC is needed, whereas the latter may lead to an exponential number of rules.

The logic programs obtained can be used to retrieve consistent answers to arbitrary FO queries. Some computational experiments with DLV [21] are shown. The methodology for consistent query answering based on logic programs presented here works for arbitrary FO queries and universal ICs, what considerable extends the cases that could be handled in [2,4,3].

This paper improves, combines and extends results presented in [8,9]. The main extensions have to do with the analysis and optimizations of the logic programs for consistent query answering introduced here.

This paper is structured as follows. In Section 2 we give some basic background. In section 3, we show how to annotate referential ICs, taking them, in addition to universal ICs, into a theory written in annotated predicate calculus. The correspondence between minimal models of the theory and database repairs is also established. Next, in Section 4, we show how to annotate queries and formulate the problem of consistent query answering as a problem of non-monotonic (minimal) entailment from the annotated theory. Then, in Section 5, on the basis of the generated annotated theory, disjunctive logic programs with annotation arguments to specify the database repairs are presented. It is also shown how to use them for consistent query answering. Some computational examples are presented in Section 6. Section 7 gives the first full treatment of logic program for computing repairs wrt referential integrity constraints. In Section 8 we introduce some optimizations of the logic programs. Finally, in Section 9 we draw some conclusions and consider related work. Proofs and intermediate results can be found in http://www.scs.carleton.ca/~bertossi/papers/proofsChap.ps.

2 Preliminaries

2.1 Database Repairs and Consistent Answers

In the context of relational databases, we will consider a fixed relational schema $\Sigma = (D, \mathcal{P} \cup \mathcal{B})$ that determines a first order language. It consists of a fixed, possibly infinite, database domain $D = \{c_1, c_2, ...\}$, a fixed set of database predicates $\mathcal{P} = \{p_1, \ldots, p_n\}$, and a fixed set of built-in predicates $\mathcal{B} = \{e_1, \ldots, e_m\}$.

A database instance over Σ is a finite collection DB of facts of the form $p(c_1, ..., c_n)$, where p is a predicate in $\mathcal{P}$ and $c_1, ..., c_n$ are constants in D. Built-in predicates have a fixed and same extension in every database instance, not subject to changes.

A *universal integrity constraint* (IC) is an implicitly universally quantified clause of the form

$$q_1(\bar{t}_1) \vee \cdots \vee q_n(\bar{t}_n) \vee \neg p_1(\bar{s}_1) \vee \cdots \vee \neg p_m(\bar{s}_m) \tag{1}$$

in the *FO* language $\mathcal{L}(\Sigma)$ based on Σ, where each p_i, q_j is a predicate in $\mathcal{P} \cup \mathcal{B}$ and the $\bar{t}_i, \bar{s}_j$ are tuples containing constants and variables. We assume we have a fixed set IC of ICs that is consistent as a FO theory. The database DB is always logically consistent if considered in isolation from the ICs.

It may be the case that $DB \cup IC$ is inconsistent. Equivalently, if we associate to DB a first order structure, also denoted with DB, in the natural way, i.e. by applying the domain closure and unique names assumptions and the closed world assumption [33] that makes false any ground atom not explicitly appearing in the set of atoms DB, it may happen that DB, as a structure, does not satisfy the IC. We denote with $DB \models_\Sigma IC$ the fact that the database satisfies IC. In this case we say that DB is consistent wrt IC; otherwise we say DB is inconsistent.

The *distance* [2] between two database instances DB_1 and DB_2 is their symmetric difference $\Delta(DB_1, DB_2) = (DB_1 - DB_2) \cup (DB_2 - DB_1)$. Now, given a database instance DB, possibly inconsistent wrt IC, we say that the instance DB' is a *repair* [2] of DB wrt IC iff $DB' \models_\Sigma IC$ and $\Delta(DB, DB')$ is minimal under set inclusion in the class of instances that satisfy IC and are compatible with the schema Σ.

Example 1. Consider the relational schema $Book(author, name, publYear)$, a database instance $DB = \{Book(kafka, metamorph, 1915), Book(kafka, meta-morph, 1919)\}$; and the functional dependency $FD : author, name \rightarrow publYear$, that can be expressed by $IC : \neg Book(x, y, z) \vee \neg Book(x, y, w) \vee z = w$. Instance DB is inconsistent with respect to IC. The original instance has two possible repairs: $DB_1 = \{Book(kafka, metamorph, 1915)\}$, and $DB_2 = \{Book(kafka, meta-morph, 1919)\}$.   □

Let DB be a database instance, possibly not satisfying a set IC of integrity constraints. Given a query $Q(\bar{x}) \in \mathcal{L}(\Sigma)$, we say that a tuple of constants $\bar{t}$ is a *consistent answer* to $Q(\bar{x})$ in DB wrt IC, denoted $DB \models_c Q(\bar{t})$, if for every repair DB' of DB, $DB' \models_\Sigma Q(\bar{t})$ [2].[1] If Q is a closed formula, *i.e.* a sentence, then *true* is a *consistent answer* to Q, denoted $DB \models_c Q$, if for every repair DB' of DB, $DB' \models_\Sigma Q$.

Example 2. (example 1 continued) The query $Q_1 : Book(kafka, metamorph, 1915)$ does not have *true* as a consistent answer, because it is not true in every repair. Query $Q_2(y) : \exists x \exists z Book(x, y, z)$ has $y = metamorph$ as a consistent answer. Query $Q_3(x) : \exists z Book(x, metamorph, z)$ has $x = kafka$ as a consistent answer.   □

---

[1] $DB' \models_\Sigma Q(\bar{t})$ means that when the variables in $\bar{x}$ are replaced in Q by the constants in $\bar{t}$ we obtain a sentence that is true in DB'.

2.2   Annotating DBs and ICs

Annotated Predicate Calculus (*APC*) was introduced in [27] and also studied in [12] and [28]. It constitutes a non classical logic, where classically inconsistent information does not trivializes logical inference, reasoning about causes of inconsistency becomes possible, making one of its goals to study the differences in the contribution to the inconsistency made by the different literals in a theory, what is related to the problem of consistent query answers.

The syntax of *APC* is similar to that of classical logic, except for the fact that the atoms (and only the atoms) are annotated with values drawn from a *truth-values lattice*. The lattice *Latt* we will use throughout this paper is shown in Figure 1, first introduced in [3].

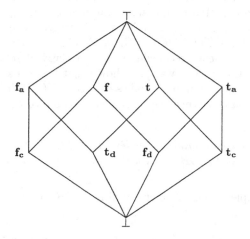

Fig. 1. *Latt* with *constraints values*, *database values* and *advisory values*

The lattice contain the usual truth values $\mathbf{t}, \mathbf{f}, \top, \bot$, for true, false, inconsistent and unknown, resp., but also six new truth values. Intuitively, we can think of values $\mathbf{t_c}$ and $\mathbf{f_c}$ as specifying what is needed for constraint satisfaction and will be used to annotate atoms appearing in ICs. The values $\mathbf{t_d}$ and $\mathbf{f_d}$ represent the truth values according to the original database and will be used to annotate atoms inside, resp. outside, the database. Finally, $\mathbf{t_a}$ and $\mathbf{f_a}$ are considered *advisory* truth values. These are intended to solve conflicts between the original database and the integrity constraints. Notice that $lub(\mathbf{t_d}, \mathbf{f_c}) = \mathbf{f_a}$ and $lub(\mathbf{f_d}, \mathbf{t_c}) = \mathbf{t_a}$. This means that whenever we have an atom, e.g. annotated with both $\mathbf{t_d}$ and $\mathbf{f_c}$, i.e. it is true according to the DB, but false according to the ICs, then it becomes automatically annotated with $\mathbf{f_a}$, meaning that the advise is to make it false. This will be made precise through the notion of formula satisfaction in *APC* below.

The intuition behind is that, in case of a conflict between the constraints and the database, we should obey the constraints, because the database instance only can be changed to restore consistency. This lack of symmetry between data and ICs is precisely captured by the lattice. Advisory value $\mathbf{t_a}$ is an indication that the atom annotated with it must be inserted into the DB; and deleted from the DB when annotated with $\mathbf{f_a}$.

Herbrand interpretations are now sets of annotated ground atoms. The notion of formula satisfaction in an *Herbrand interpretation* I is defined classically, except for atomic formulas p: we say that $I \models p\!:\!\mathbf{s}$, with $\mathbf{s} \in Latt$, iff for some $\mathbf{s}'$ such that $\mathbf{s} \le \mathbf{s}'$ we have that $p\!:\!\mathbf{s}' \in I$ [27].

Given an APC theory $\mathcal{T}$, we say that an Herbrand interpretation I is a Δ-minimal model of $\mathcal{T}$, with $\Delta = \{\mathbf{t_a}, \mathbf{f_a}\}$, if I is a model of $\mathcal{T}$ and no other model of $\mathcal{T}$ has a proper subset of atoms annotated with elements in Δ, *i.e.* the set of atoms annotated with $\mathbf{t_a}$ or $\mathbf{f_a}$ in I is minimal under set inclusion. Considering Δ-minimal models is natural, because they minimize the set of changes, which in their turn are represented by the atoms annotated with $\mathbf{t_a}$ or $\mathbf{f_a}$.[2]

Given a database instance DB and a set of integrity constraints IC of the form (1), an embedding $\mathcal{T}(DB, IC)$ of DB and IC into a new APC theory was defined [3]. The new theory reconciles in a non classical setting the conflicts between data and ICs. In [3] it was also shown that there is a one-to-one correspondence between the Δ-minimal models of theory $\mathcal{T}(DB, IC)$ and the repairs of the original database instance. Actually, repairs can be obtained from minimal models as follows:

Definition 1. [3] *Given a minimal model* $\mathcal{M}$ *of* $\mathcal{T}(DB, IC)$, *the corresponding DB instance is defined by* $DB_{\mathcal{M}} = \{p(\bar{a}) \mid \mathcal{M} \models p(\bar{a})\!:\!\mathbf{t} \ \vee \ p(\bar{a})\!:\!\mathbf{t_a}\}$.   □

Example 3. (example 1 cont.) The embedding $\mathcal{T}(DB)$ of DB into APC is given by the following formulas:

1. $Book(kafka, metamorph, 1915)\!:\!\mathbf{t_d}, \quad Book(kafka, metamorph, 1919)\!:\!\mathbf{t_d}$.
   Every ground atom that is not in DB is (possibly implicitly) annotated with $\mathbf{f_d}$.
2. Predicate closure axioms:
   $((x = kafka)\!:\!\mathbf{t_d} \ \wedge \ (y = metamorph)\!:\!\mathbf{t_d} \ \wedge \ (z = 1915)\!:\!\mathbf{t_d}) \ \vee$
   $((x = kafka)\!:\!\mathbf{t_d} \ \wedge \ (y = metamorph)\!:\!\mathbf{t_d} \ \wedge \ (z = 1919)\!:\!\mathbf{t_d}) \ \vee \ Book(x, y, z)\!:\!\mathbf{f_d}$.

The embedding $\mathcal{T}(IC)$ of IC into APC is given by:

3. $Book(x, y, z)\!:\!\mathbf{f_c} \vee Book(x, y, w)\!:\!\mathbf{f_c} \vee (z = w)\!:\!\mathbf{t_c}$.
4. $Book(x, y, z)\!:\!\mathbf{f_c} \vee Book(x, y, z)\!:\!\mathbf{t_c}, \quad \neg Book(x, y, z)\!:\!\mathbf{f_c} \vee \neg Book(x, y, z)\!:\!\mathbf{t_c}$.[3]
   These formulas specify that every fact must have one and just one constraint value.

Furthermore

---

[2] Most of the time we will simply say "minimal" instead of Δ-minimal. In this case there should be no confusion with the other notion of minimality in this paper, namely the one that applies to repairs.

[3] Since only atomic formulas are annotated, the non atomic formula $\neg p(\bar{x})\!:\!\mathbf{s}$ is to be read as $\neg(p(\bar{x})\!:\!\mathbf{s})$. We will omit the parenthesis though.

5. For every true *built-in* atom φ we include φ:t in $\mathcal{T}(\mathcal{B})$, and φ:f for every false built-in atom, e.g. $(1915 = 1915)$:t, but $(1915 = 1919)$:f.

The Δ-minimal models of $\mathcal{T}(DB, IC) = \mathcal{T}(DB) \cup \mathcal{T}(IC) \cup \mathcal{T}(\mathcal{B})$ are:

$$\mathcal{M}_1 = \{Book(kafka, metamorph, 1915)\text{:}t, \ Book(kafka, metamorph, 1919)\text{:}f_a\},$$
$$\mathcal{M}_2 = \{Book(kafka, metamorph, 1915)\text{:}f_a, \ Book(kafka, metamorph, 1919)\text{:}t\}.$$

They also contain annotated false DB atoms and built-ins, but we will show only the most relevant data in them. The corresponding database instances, $DB_{\mathcal{M}_1}, DB_{\mathcal{M}_2}$ are the repairs of DB shown in Example 1. □

From the definition of the lattice and the fact that no atom from the database is annotated with both t_d and f_d, it is possible to show that, in the minimal models of the annotated theory, a DB atom may get as annotation either t or f_a if the atom was annotated with t_d; similarly either f or t_a if the atom was annotated with f_d. In the transition from the annotated theory to its minimal models, the annotations t_d, f_d "disappear", as we want the atoms to be annotated at the highest possible layer in the lattice; except for $\top$, that can always we avoided in the minimal models.

3 Annotating Referential ICs

Referential integrity constraints (RICs) like

$$\forall \bar{x}(p(\bar{x}) \rightarrow \exists y q(\bar{x}', y)), \tag{2}$$

where the variables in $\bar{x}'$ are a subset of the variables in $\bar{x}$, cannot be expressed as an equivalent clause of the form (1). RICs are important and common in databases. For that reason, we need to extend our embedding methodology. Actually, we embed (2) into APC by means of

$$p(\bar{x})\text{:}f_c \vee \exists y(q(\bar{x}', y)\text{:}t_c). \tag{3}$$

In the rest of this section we allow the given set of ICs to contain, in addition to universal ICs of the form (1), also RICs like (2). The one-to-one correspondence between minimal models of the new theory $\mathcal{T}(DB, IC)$ and the repairs of DB still holds. Most important for us is to obtain repairs from minimal models.

Given a pair of database instances DB_1 and DB_2 over the same schema (and domain), we construct the Herbrand structure $\mathcal{M}(DB_1, DB_2) = \langle D, I_{\mathcal{P}}, I_{\mathcal{B}} \rangle$, where D is the domain of the database and $I_{\mathcal{P}}, I_{\mathcal{B}}$ are the interpretations for the predicates and the built-ins, respectively. $I_{\mathcal{P}}$ is defined as follows:

$$I_{\mathcal{P}}(p(\bar{a})) = \begin{cases} t & p(\bar{a}) \in DB_1, \ p(\bar{a}) \in DB_2 \\ f & p(\bar{a}) \notin DB_1, \ p(\bar{a}) \notin DB_2 \\ f_a & p(\bar{a}) \in DB_1, \ p(\bar{a}) \notin DB_2 \\ t_a & p(\bar{a}) \notin DB_1, \ p(\bar{a}) \in DB_2 \end{cases}$$

The interpretation I_B is defined as expected: if q is a built-in, then $I_P(q(\bar{a})) = \mathbf{t}$ iff $q(\bar{a})$ is true in classical logic, and $I_P(q(\bar{a})) = \mathbf{f}$ iff $q(\bar{a})$ is false.

Lemma 1. *Given two database instances DB and DB', if $DB' \models_{\Sigma} IC$, then $\mathcal{M}(DB, DB') \models \mathcal{T}(DB, IC)$.* □

Lemma 2. *If $\mathcal{M}$ is a model of $\mathcal{T}(DB, IC)$ such that $DB_{\mathcal{M}}$ is finite[4], then $DB_{\mathcal{M}} \models_{\Sigma} IC$.* □

The following results shows the one-to one correspondence between the minimal models of $\mathcal{T}(DB, IC)$ and the repairs of DB.

Proposition 1. *If DB' is a repair of DB with respect to the set of integrity constraints IC, then $\mathcal{M}(DB, DB')$ is minimal among the models of $\mathcal{T}(DB, IC)$.* □

Proposition 2. *Let $\mathcal{M}$ be a model of $\mathcal{T}(DB, IC)$. If $\mathcal{M}$ is minimal and $DB_{\mathcal{M}}$ is finite, then $DB_{\mathcal{M}}$ is a repair of DB with respect to IC.* □

Example 4. Consider the relational schema of Example 1 extended with the table *Author(name, citizenship)*. Now, *IC* also contains the RIC: $Book(x, y, z) \rightarrow \exists w\, Author(x, w)$, expressing that every writer of a book in the database instance must be registered as an author. The theory $\mathcal{T}(IC)$ now also contains:

$Book(x, y, z)$:$\mathbf{f_c} \vee \exists w(Author(x, w)$:$\mathbf{t_c})$, $Author(x, w)$:$\mathbf{f_c} \vee Author(x, w)$:$\mathbf{t_c}$, $\neg Author(x, w)$:$\mathbf{f_c} \vee \neg Author(x, w)$:$\mathbf{t_c}$.

We might also have the functional dependency FD : *name* $\rightarrow$ *citizenship*, that in conjunction with the RIC, produces a foreign key constraint. The database instance $\{Book(neruda, 20\,lovepoems, 1924)\}$ is inconsistent wrt the given RIC. If we have the following subdomain $D(Author.citizenship) = \{chilean, canadian\}$ for the attribute "citizenship", we obtain the following database theory:

$\mathcal{T}(DB) = \{Book(neruda, 20\,lovepoems, 1924)$: $\mathbf{t_d}$, $Author(neruda, chilean)$: $\mathbf{f_d}$, $Author(neruda, canadian)$:$\mathbf{f_d}$, ...$\}$.

The minimal models of $\mathcal{T}(DB, IC)$ are:

$\mathcal{M}_1 = \{Book(neruda, 20\,lovepoems, 1924)$:$\mathbf{f_a}$, $Author(neruda, chilean)$:$\mathbf{f}$, $\quad\quad Author(neruda, canadian)$:$\mathbf{f}$, ...$\}$
$\mathcal{M}_2 = \{Book(neruda, 20\,lovepoems, 1924)$:$\mathbf{t}$, $Author(neruda, chilean)$:$\mathbf{t_a}$, $\quad\quad Author(neruda, canadian)$:$\mathbf{f}$, ...$\}$

---

[4] That is, the extensions of the database predicates are finite. These are the models that may lead to database instances, because the latter have finite database relations.

$\mathcal{M}_3 = \{Book(neruda, 20\,lovepoems, 1924)\text{:}\mathbf{t},\ Author(neruda, chilean)\text{:}\mathbf{f},$
     $Author(neruda, canadian)\text{:}\mathbf{t_a}, \dots\}.$

We obtain $DB_{\mathcal{M}_1} = \emptyset$, $DB_{\mathcal{M}_2} = \{Book(neruda, 20\,lovepoems, 1924), Author($ $neruda, chilean)\}$ and $DB_{\mathcal{M}_3}$ similar to $DB_{\mathcal{M}_2}$, but with a Canadian Neruda. According to Proposition 2, these are repairs of the original database instance, actually the only ones.   □

As in [3], it can be proved that when the original instance is consistent, then it is its only repair and it corresponds to a unique minimal model of the APC theory.

3.1   Annotating General ICs

The class of ICs found in database praxis is contained in the class of FO formulas of the form:

$$\forall \bar{x}\ (\varphi(\bar{x}) \rightarrow \exists \bar{z}\psi(\bar{y})) \tag{4}$$

where φ and ψ are (possibly empty) conjunctions of literals, and $\bar{z} = \bar{y} - \bar{x}$. This class [1, chapter 10] includes the ICs of the form (1), in particular, range constraints (e.g. $\forall x\ (p(x) \rightarrow x > 30)$), join dependencies, functional dependencies, full inclusion dependencies; and also referential integrity constraints, and in consequence, also foreign key constraints.

The annotation methodology introduced so far can be extended to the whole class (4). We only sketch this extension here.

If in (4) $\varphi(\bar{x})$ is $\bigwedge_{i=1}^{k} p_i(\bar{x}_i) \wedge \bigwedge_{i=k+1}^{m} \neg p_i(\bar{x}_i)$ and $\psi(\bar{y})$ is $\bigwedge_{j=1}^{l} q_j(\bar{y}_j) \wedge \bigwedge_{j=l+1}^{r} \neg q_j(\bar{y}_j)$, we embed the constraint into APC as follows:

$$\bigvee_{i=1}^{k} p_i(\bar{x}_i)\text{:}\mathbf{f_c} \vee \bigvee_{i=k+1}^{m} p_i(\bar{x}_i)\text{:}\mathbf{t_c} \vee \exists \bar{z}(\bigwedge_{j=1}^{l} q_j(\bar{y}_j)\text{:}\mathbf{t_c} \wedge \bigwedge_{j=l+1}^{r} q_j(\bar{y}_j)\text{:}\mathbf{f_c}).$$

If we allow now that IC contains ICs of the form (4), it is still possible to establish the one-to-one correspondence between minimal models of $\mathcal{T}(DB, IC)$ and the repairs of DB.

4   Annotation of Queries

According to Proposition 2, a ground tuple $\bar{t}$ is a consistent answer to a FO query $Q(\bar{x})$ iff $Q(\bar{t})$ is true in every minimal model of $\mathcal{T}(DB, IC)$. However, if we want to pose the query directly to the theory, it is necessary to reformulate it as an annotated formula.

Definition 2. *Given a FO query $Q(\bar{x})$ in language $\mathcal{L}(\Sigma)$, we denote by $Q^{an}(\bar{x})$ the APC formula obtained from Q by simultaneously replacing, for $p \in \mathcal{P}$, the negative literal $\neg p(\bar{s})$ by the APC formula $p(\bar{s})\text{:}\mathbf{f} \vee p(\bar{s})\text{:}\mathbf{f_a}$, and the positive literal $p(\bar{s})$ by the APC formula $p(\bar{s})\text{:}\mathbf{t} \vee p(\bar{s})\text{:}\mathbf{t_a}$. For $p \in \mathcal{B}$, the atom $p(\bar{s})$ is replaced by the APC formula $p(\bar{s})\text{:}\mathbf{t}$.*   □

According to this definition, logically equivalent versions of a query could have different annotated versions, but it can be shown (Proposition 3), that they retrieve the same consistent answers.

Example 5. (example 1 cont.) If we want the consistent answers to the query $Q(x) : \neg\exists y\exists z\exists w\exists t(Book(x, y, z) \wedge Book(x, w, t) \wedge y \neq w)$, asking for those authors that have at most one book, we generate the annotated query $Q^{an}(\bar{x})$: $\neg\exists y\exists z\exists w\exists t((Book(x, y, z){:}\mathbf{t} \vee Book(x, y, z){:}\mathbf{t_a}) \wedge (Book(x, w, t){:}\mathbf{t} \vee Book(x, w, t){:}$ $\mathbf{t_a}) \wedge (y \neq w){:}\mathbf{t})$, to be posed to the annotated theory with its minimal model semantics. □

Definition 3. *If φ is an APC sentence in the language of $\mathcal{T}(DB, IC)$, we say that $\mathcal{T}(DB, IC)$ Δ-minimally entails φ, written $\mathcal{T}(DB, IC) \models_{\Delta} \varphi$, iff every Δ-minimal model $\mathcal{M}$ of $\mathcal{T}(DB, IC)$, such that $DB_{\mathcal{M}}$ is finite, satisfies φ, i.e. $\mathcal{M} \models_{APC} \varphi$.* □

Now we characterize consistent query answers wrt the annotated theory.

Proposition 3. *Let DB be a database instance, IC a set of integrity constraints and $Q(\bar{x})$ a query in FO language $\mathcal{L}(\Sigma)$. It holds:*

$$DB \models_c Q(\bar{t}) \quad iff \quad \mathcal{T}(DB, IC) \models_{\Delta} Q^{an}(\bar{t}).$$ □

Example 6. (example 5 continued) For consistently answering the query $Q(x)$, we pose the query $Q^{an}(x)$ to the minimal models of $\mathcal{T}(DB, IC)$. The answer we obtain from *every* minimal model is $x = kafka$. □

According to this proposition, in order to consistently answer queries, we are left with the problem of evaluating minimal entailment wrt the annotated theory. In [3] some limited FO queries were evaluated without passing to their annotated versions. The algorithms for consistent query answering were rather ad hoc and were extracted from the theory $\mathcal{T}(DB, IC)$. However, no advantage was taken from a characterization of consistent answers in terms of minimal entailment from $\mathcal{T}(DB, IC)$. In the next section we will address this issue by taking the original DB instance with the ICs into a logic program that is inspired by the annotated theory $\mathcal{T}(DB, IC)$. Furthermore, the query to be posed to the logic program will be built from Q^{an}.

5   Logic Programming Specification of Repairs

In this section we will consider ICs of the form (1). Our aim is to specify database repairs using classical first order logic programs. However, those programs will be suggested by the non classical annotated theory.

In order to accommodate annotations in this classical framework, we will first consider the annotations in the lattice *Latt* as new constants in the language. Next, we will replace each predicate $p(\bar{x}) \in \mathcal{P}$ by a new predicate $p(\bar{x}, \cdot)$, with an extra argument to be occupied by annotation constants. In this way we can simulate the annotations we had before, but in a classical setting. With all this, we have a new FO language, $\mathcal{L}(\Sigma)^{an}$, for annotated $\mathcal{L}(\Sigma)$.

Definition 4. *The repair logic program, $\Pi(DB, IC)$, for DB and IC, is written with predicates from $\mathcal{L}(\Sigma)^{an}$ and contains the following clauses:*

1. *For every atom $p(\bar{a}) \in DB$, $\Pi(DB, IC)$ contains the fact $p(\bar{a}, \mathbf{t_d})$.*
2. *For every predicate $p \in P$, $\Pi(DB, IC)$ contains the clauses:*

$$p(\bar{x}, \mathbf{t^\star}) \leftarrow p(\bar{x}, \mathbf{t_d}). \qquad p(\bar{x}, \mathbf{t^\star}) \leftarrow p(\bar{x}, \mathbf{t_a}).$$
$$p(\bar{x}, \mathbf{f^\star}) \leftarrow p(\bar{x}, \mathbf{f_a}). \qquad p(\bar{x}, \mathbf{f^\star}) \leftarrow \; not \; p(\bar{x}, \mathbf{t_d}).,$$

   where $\mathbf{t^\star}, \mathbf{f^\star}$ are new, auxiliary elements in the domain of annotations.
3. *For every constraint of the form (1), $\Pi(DB, IC)$ contains the clause:*

$$\bigvee_{i=1}^{n} p_i(\bar{t}_i, \mathbf{f_a}) \vee \bigvee_{j=1}^{m} q_j(\bar{s}_j, \mathbf{t_a}) \;\longleftarrow\; \bigwedge_{i=1}^{n} p_i(\bar{t}_i, \mathbf{t^\star}) \wedge \bigwedge_{j=1}^{m} q_j(\bar{s}_j, \mathbf{f^\star}) \wedge \bar{\varphi},$$

   where $\bar{\varphi}$ represents the negation of φ. □

Intuitively, the clauses in 3. say that when the IC is violated (the body), then *DB* has to be repaired according to one of the alternatives shown in the head. Since there may be interactions between constraints, these single repairing steps may not be enough to restore the consistency of *DB*. We have to make sure that the repairing process continues and stabilizes in a state where all the ICs hold. This is the role of the clauses in 2. containing the new annotations $\mathbf{t^\star}$, that groups together those atoms annotated with $\mathbf{t_d}$ and $\mathbf{t_a}$, and $\mathbf{f^\star}$, that does the same with $\mathbf{f_d}$ and $\mathbf{f_a}$. Notice that the annotations $\mathbf{t^\star}$, $\mathbf{f^\star}$, obtained through the combined effect of rules 2. and 3., can be fed back into rules 3. until consistency is restored. This possibility is what allows us to have just one program rule for each IC.

Example 7 shows the interaction of a functional dependency and a full inclusion dependency. When atoms are deleted in order to satisfy the functional dependency, the inclusion dependency could be violated, and in a second step it should be repaired. At that second step, the annotations $\mathbf{t^\star}$ and $\mathbf{f^\star}$, computed at the first step where the functional dependency was repaired, will detect the violation of the inclusion dependency and trigger the corresponding repairing process.

Example 7. (example 1 continued) We extend the schema with the table *Eurbook(author, name, publYear)*, for European books. Now, *DB* also contains the literal *Eurbook(kafka, metamorph, 1919)}*. If in addition to the ICs we had before, we consider the full inclusion dependency $\forall xyz \; (Eurbook(x, y, z) \rightarrow Book(x, y, z))$, we obtain the following program $\Pi(DB, IC)$:

1. $EurBook(kafka, metamorph, 1919, \mathbf{t_d})$. $Book(kafka, metamorph, 1919, \mathbf{t_d})$. $Book(kafka, metamorph, 1915, \mathbf{t_d})$.

2. $Book(x, y, z, \mathbf{t^\star}) \leftarrow Book(x, y, z, \mathbf{t_d})$. $Book(x, y, z, \mathbf{t^\star}) \leftarrow Book(x, y, z, \mathbf{t_a})$. $Book(x, y, z, \mathbf{f^\star}) \leftarrow Book(x, y, z, \mathbf{f_a})$. $Book(x, y, z, \mathbf{f^\star}) \leftarrow \; not \; Book(x, y, z, \mathbf{t_d})$. $Eurbook(x, y, z, \mathbf{t^\star}) \leftarrow Eurbook(x, y, z, \mathbf{t_d})$. $Eurbook(x, y, z, \mathbf{t^\star}) \leftarrow Eurbook(x, y, z, \mathbf{t_a})$. $Eurbook(x, y, z, \mathbf{f^\star}) \leftarrow Eurbook(x, y, z, \mathbf{f_a})$. $Eurbook(x, y, z, \mathbf{f^\star}) \leftarrow \; not \; Eurbook(x, y, z, \mathbf{t_d})$.

3. $Book(x, y, z, \mathbf{f_a}) \vee Book(x, y, w, \mathbf{f_a}) \leftarrow Book(x, y, z, \mathbf{t^*}), Book(x, y, w, \mathbf{t^*}),$
$$z \neq w.$$
$Eurbook(x, y, z, \mathbf{f_a}) \vee Book(x, y, z, \mathbf{t_a}) \leftarrow Eurbook(x, y, z, \mathbf{t^*}), Book(x, y, z, \mathbf{f^*}).$

□

Our programs are standard logic programs (as opposed to annotated logic programs [28]) and, finding in them negation as failure, we will give them an also standard stable model semantics.

Let Π be the ground logic program obtained by instantiating the disjunctive program $\Pi(DB, IC)$ in its Herbrand universe. A set of ground atoms $\mathcal{M}$ is a *stable model* of $\Pi(DB, IC)$ iff it is a minimal model of $\Pi^{\mathcal{M}}$, where $\Pi^{\mathcal{M}} = \{A_1 \vee \cdots \vee A_n \leftarrow B_1, \cdots, B_m \mid A_1 \vee \cdots \vee A_n \leftarrow B_1, \cdots, B_m, not\ C_1, \cdots, not\ C_k \in \Pi$ and $C_i \notin \mathcal{M}$ for $1 \leq i \leq k\}$ [23,24].

Definition 5. *A Herbrand model $\mathcal{M}$ is* coherent *if it does not contain a pair of literals of the form* $\{p(\bar{a}, \mathbf{t_a}), p(\bar{a}, \mathbf{f_a})\}$. □

Example 8. (example 7 continued) The coherent stable models of the program presented in Example 7 are:

$\mathcal{M}_1 = \{Book(kafka, metamorph, 1919, \mathbf{t_d}),\ Book(kafka, metamorph, 1919, \mathbf{t^*}),$
$Book(kafka, metamorph, 1915, \mathbf{t_d}),\ Book(kafka, metamorph, 1915, \mathbf{t^*}),$
$Book(kafka, metamorph, 1915, \mathbf{f_a}),\ Book(kafka, metamorph, 1915, \mathbf{f^*}),$
$Eurbook(kafka, metamorph, 1919, \mathbf{t_d}),\ Eurbook(kafka, metamorph, 1919, \mathbf{t^*})\};$

$\mathcal{M}_2 = \{Book(kafka, metamorph, 1919, \mathbf{t_d}),\ Book(kafka, metamorph, 1919, \mathbf{t^*}),$
$Book(kafka, metamorph, 1919, \mathbf{f_a}),\ Book(kafka, metamorph, 1919, \mathbf{f^*}),$
$Book(kafka, metamorph, 1915, \mathbf{t_d}),\ Book(kafka, metamorph, 1915, \mathbf{t^*}),$
$Eurbook(kafka, metamorph, 1919, \mathbf{t_d}),\ Eurbook(kafka, metamorph, 1919, \mathbf{t^*}),$
$Eurbook(kafka, metamorph, 1919, \mathbf{f_a}),\ Eurbook(kafka, metamorph, 1919, \mathbf{f^*})\}.$ □

The stable models of the program will include the database contents with its original annotations ($\mathbf{t_d}$). Every time there is an atom in a model annotated with $\mathbf{t_d}$ or $\mathbf{t_a}$, it will appear annotated with $\mathbf{t^*}$. From these models we should be able to "read" database repairs. Every stable model of the logic program has to be interpreted. In order to do this, we introduce two new annotations, $\mathbf{t^{**}}, \mathbf{f^{**}}$, in the last arguments. The first one groups together those atoms annotated with $\mathbf{t_a}$ and those annotated with $\mathbf{t_d}$, but not $\mathbf{f_a}$. Intuitively, they correspond to those annotated with $\mathbf{t}$ in the models of $\mathcal{T}(DB, IC)$. A similar role plays the other new annotation wrt the "false" annotations. These new annotations will simplify the expression of the queries to be posed to the program. Without them, instead of simply asking $p(\bar{x}, \mathbf{t^{**}})$ (for the tuples in p in a repair), we would have to ask for $p(\bar{x}, \mathbf{t_a}) \vee (p(\bar{x}, \mathbf{t_d}) \wedge \neg p(\bar{x}, \mathbf{f_a}))$. The interpreted models can be easily obtained by adding new rules.

Definition 6. *The interpretation program $\Pi^*(DB, IC)$ extends $\Pi(DB, IC)$ with the following rules:*

$p(\bar{a}, \mathbf{f^{**}}) \leftarrow p(\bar{a}, \mathbf{f_a}). \qquad p(\bar{a}, \mathbf{f^{**}}) \leftarrow not\ p(\bar{a}, \mathbf{t_d}),\ not\ p(\bar{a}, \mathbf{t_a}).$
$p(\bar{a}, \mathbf{t^{**}}) \leftarrow p(\bar{a}, \mathbf{t_a}). \qquad p(\bar{a}, \mathbf{t^{**}}) \leftarrow p(\bar{a}, \mathbf{t_d}),\ not\ p(\bar{a}, \mathbf{f_a}).$

□

Example 9. (example 8 continued) The coherent stable models of the interpretation program extend

$\mathcal{M}_1$ with $\{Eurbook(kafka, metamorph, 1919, \mathbf{t}^{**}),$
    $Book(kafka, metamorph, 1919, \mathbf{t}^{**}), Book(kafka, metamorph, 1915, \mathbf{f}^{**})\}$;

$\mathcal{M}_2$ with $\{Eurbook(kafka, metamorph, 1919, \mathbf{f}^{**}),$
    $Book(kafka, metamorph, 1919, \mathbf{f}^{**}), Book(kafka, metamorph, 1915, \mathbf{t}^{**})\}$. $\square$

From an interpretation model we can obtain a database instance.

Definition 7. *Let $\mathcal{M}$ be a coherent stable model of program $\Pi^*(DB, IC)$. The database associated to $\mathcal{M}$ is $DB_{\mathcal{M}} = \{p(\bar{a}) \mid p(\bar{a}, \mathbf{t}^{**}) \in \mathcal{M}\}$.* $\square$

The following theorem establishes the one-to-one correspondence between coherent stable models of the program and the repairs of the original instance.

Theorem 1. *If $\mathcal{M}$ is a coherent stable model of $\Pi^*(DB, IC)$, and $DB_{\mathcal{M}}$ is finite, then $DB_{\mathcal{M}}$ is a repair of DB with respect to IC. Furthermore, the repairs obtained in this way are all the repairs of DB.* $\square$

Example 10. (example 9 continued) The following database instances obtained from Definition 7 are the repairs of DB:

$DB_{\mathcal{M}_1} = \{Eurbook(kafka, metamorph, 1919),\ Book(kafka, metamorph, 1919)\}$,

$DB_{\mathcal{M}_2} = \{Book(kafka, metamorph, 1915)\}$. $\square$

5.1   The Query Program

Given a first order query Q, we want the consistent answers from DB. In consequence, we need those atoms that are simultaneously true of Q in every stable model of the program $\Pi(DB, IC)$. They are obtained through the query Q^{**}, obtained from Q by replacing, for $p \in \mathcal{P}$, every positive literal $p(\bar{s})$ by $p(\bar{s}, \mathbf{t}^{**})$ and every negative literal $\neg p(\bar{s})$ by $p(\bar{s}, \mathbf{f}^{**})$. Now Q^{**} can be transformed into a query program $\Pi(Q^{**})$ by a standard transformation [30,1]. This query program will be run in combination with $\Pi^*(DB, IC)$.

Example 11. For the query $Q(y) : \exists z Book(kafka, y, z)$, we generate $Q^{**}(y) :$ $\exists z Book(kafka, y, z, \mathbf{t}^{**})$, that is transformed into the query program clause $Answer(y) \leftarrow Book(kafka, y, z, \mathbf{t}^{**})$. $\square$

6   Computing from the Program

The database repairs could be computed using an implementation of the disjunctive stable models semantics like DLV [21], that also supports denial constraints as studied in [13]. In this way we are able to prune out the models that are not coherent, imposing for every predicate p the constraint $\leftarrow p(\bar{x}, \mathbf{t_a}), p(\bar{x}, \mathbf{f_a})$.

Example 12. Consider the database instance $\{p(a)\}$ that is inconsistent wrt the full inclusion dependency $\forall x(p(x) \rightarrow q(x))$. The program $\Pi^*(DB, IC)$ contains the following clauses:

1. Database contents: $p(a, \mathbf{t_d})$.
2. Rules for the closed world assumption:

   $p(x, \mathbf{f^*}) \leftarrow not\ p(x, \mathbf{t_d})$. $q(x, \mathbf{f^*}) \leftarrow not\ q(x, \mathbf{t_d})$.
3. Annotation rules:

   $p(x, \mathbf{f^*}) \leftarrow p(x, \mathbf{f_a})$. $p(x, \mathbf{t^*}) \leftarrow p(x, \mathbf{t_a})$. $p(x, \mathbf{t^*}) \leftarrow p(x, \mathbf{t_d})$.
   $q(x, \mathbf{f^*}) \leftarrow q(x, \mathbf{f_a})$. $q(x, \mathbf{t^*}) \leftarrow q(x, \mathbf{t_a})$. $q(x, \mathbf{t^*}) \leftarrow q(x, \mathbf{t_d})$.
4. Rule for the IC: $p(x, \mathbf{f_a}) \vee q(x, \mathbf{t_a}) \leftarrow p(x, \mathbf{t^*}), q(x, \mathbf{f^*})$.
5. Denial constraints for coherence

   $\leftarrow p(\bar{x}, \mathbf{t_a}), p(\bar{x}, \mathbf{f_a})$. $\leftarrow q(\bar{x}, \mathbf{t_a}), q(\bar{x}, \mathbf{f_a})$.
6. Interpretation rules:

   $p(x, \mathbf{t^{**}}) \leftarrow p(x, \mathbf{t_a})$. $p(x, \mathbf{t^{**}}) \leftarrow p(x, \mathbf{t_d}),\ not\ p(x, \mathbf{f_a})$.
   $p(x, \mathbf{f^{**}}) \leftarrow p(x, \mathbf{f_a})$. $p(x, \mathbf{f^{**}}) \leftarrow\ not\ p(x, \mathbf{t_d}),\ not\ p(x, \mathbf{t_a})$.
   $q(x, \mathbf{t^{**}}) \leftarrow q(x, \mathbf{t_a})$. $q(x, \mathbf{t^{**}}) \leftarrow q(x, \mathbf{t_d}),\ not\ q(x, \mathbf{f_a})$.
   $q(x, \mathbf{f^{**}}) \leftarrow q(x, \mathbf{f_a})$. $q(x, \mathbf{f^{**}}) \leftarrow\ not\ q(x, \mathbf{t_d}),\ not\ q(x, \mathbf{t_a})$.

Running program $\Pi^*(DB, IC)$ with *DLV* we obtain two stable models:

$\mathcal{M}_1 = \{p(a, \mathbf{t_d}), p(a, \mathbf{t^*}), q(a, \mathbf{f^*}), q(a, \mathbf{t_a}), p(a, \mathbf{t^{**}}), q(a, \mathbf{t^*}), q(a, \mathbf{t^{**}})\}$,

$\mathcal{M}_2 = \{p(a, \mathbf{t_d}), p(a, \mathbf{t^*}), p(a, \mathbf{f^*})), q(a, \mathbf{f^*}), p(a, \mathbf{f^{**}}), q(a, \mathbf{f^{**}}), p(a, \mathbf{f_a})\}$.

The first model says, through its atom $q(a, \mathbf{t^{**}})$, that $q(a)$ has to be inserted in the database. The second one, through its atom $p(a, \mathbf{f^{**}})$, that $p(a)$ has to be deleted.   □

The coherence denial constraints did not play any role in the previous example, we obtain exactly the same model with or without them. The reason is that we have only one IC; in consequence, only one step is needed to obtain a repair of the database. There is no way to obtain an incoherent stable model due to the application of the rules 1. and 2. in Example 12 in a second repair step.

Example 13. (example 12 continued) Let us now add an extra full inclusion dependency, $\forall x(q(x) \rightarrow r(x))$, keeping the same instance. One repair is obtained by inserting $q(a)$, what causes the insertion of $r(a)$. The program is as before, but with the additional rules

   $r(x, \mathbf{f^*}) \leftarrow not\ r(x, \mathbf{t_d})$. $r(x, \mathbf{f^*}) \leftarrow r(x, \mathbf{f_a})$. $r(x, \mathbf{t^*}) \leftarrow r(x, \mathbf{t_a})$.
   $r(X, \mathbf{t^*}) \quad \leftarrow \quad r(X, \mathbf{t_d})$. $r(x, \mathbf{t^{**}}) \quad \leftarrow \quad r(x, \mathbf{t_a})$. $r(x, \mathbf{t^{**}}) \quad \leftarrow$
   $r(x, \mathbf{t_d}),\ not\ r(x, \mathbf{f_a})$.
   $r(x, \mathbf{f^{**}}) \leftarrow r(x, \mathbf{f_a})$. $r(x, \mathbf{f^{**}}) \leftarrow not\ r(x, \mathbf{t_d}), not\ r(x, \mathbf{t_a})$.
   $q(x, \mathbf{f_a}) \vee r(x, \mathbf{t_a}) \leftarrow q(x, \mathbf{t^*}), r(x, \mathbf{f^*})$. $\leftarrow r(x, \mathbf{t_a}), r(x, \mathbf{f_a})$.

If we run the program we obtain the expected models, one that deletes $p(a)$, and a second one that inserts both $q(a)$ and $r(a)$. However, if we omit the coherence denial constraints, more precisely the one for table q, we obtain a third model, namely $\{p(a, \mathbf{t_d}), p(a, \mathbf{t^*}), q(a, \mathbf{f^*}), r(a, \mathbf{f^*}), q(a, \mathbf{f_a}), q(a, \mathbf{t_a}), p(a, \mathbf{t^{**}}), q(a, \mathbf{t^*}),$ $q(a, \mathbf{t^{**}}), q(a, \mathbf{f^{**}}), r(a, \mathbf{f^{**}})\}$, that is not coherent, because it contains both $q(a, \mathbf{f_a})$ and $q(a, \mathbf{t_a})$, and cannot be interpreted as a repair of the original database. $\qquad\square$

Notice that the programs with annotations obtained are very simple in terms of their dependency on the ICs. As mentioned before, consistent answers can be obtained "running" a query program together with the repair program $\Pi^*(DB, IC)$, under the skeptical stable model semantics, that sanctions as true what is true of all stable models.

Example 14. (example 12 continued) Assume now that the original database is $\{p(a), p(b), q(b)\}$, and we want the consistent answers to the query $p(x)$. In this case we need to add the facts $p(b, \mathbf{t_d}), q(b, \mathbf{t_d})$, and the query rule $ans(x) \leftarrow p(x, \mathbf{t^{**}})$ to the program.

Now the stable models we had before are extended with ground query atoms. In $\mathcal{M}_1$ we find $ans(a), ans(b)$. In $\mathcal{M}_2$ we find $ans(b)$ only. In consequence, the tuple b is the only consistent answer to the query. $\qquad\square$

7   Repair Programs for Referential ICs

So far we have presented repair programs for universal ICs. Now we also want to consider referential ICs (RICs) of the form (2). We assume that the variables range over the underlying database domain D that now may may contain the *null* value (a new constant). A RIC can be repaired by cascaded deletion, but also by insertion of this null value, i.e. through insertion of the atom $q(\bar{a}, null)$. If this second case, it is expected that this change will not propagate through other ICs like a full inclusion dependency of the form $\forall \bar{x}(q(\bar{x}, y) \rightarrow r(\bar{x}, y))$. The program should not detect such inconsistency wrt this IC. This can be easily avoided at the program level by appropriately qualifying the values of variables in the disjunctive repair clause for the other ICs, like the full inclusion dependency above.

The program $\Pi^*(DB, IC)$ we presented in previous sections is, therefore, extended with the following formulas:

$$p(\bar{x}, \mathbf{f_a}) \vee q(\bar{x}', null, \mathbf{t_a}) \leftarrow p(\bar{x}, \mathbf{t^*}), \ not \ aux(\bar{x}'), \ not \ q(\bar{x}', null, \mathbf{t_d}). \quad (5)$$
$$aux(\bar{x}') \leftarrow q(\bar{x}', y, \mathbf{t_d}), \ not \ q(\bar{x}', y, \mathbf{f_a}). \quad (6)$$
$$aux(\bar{x}') \leftarrow q(\bar{x}', y, \mathbf{t_a}). \quad (7)$$

Intuitively, clauses (6) and (7) detect if the formula $\exists y(q(\bar{a}', y){:}\mathbf{t} \vee q(\bar{a}', y){:}\mathbf{t_a}))$ is satisfied by the model. If this is not the case and $p(\bar{a}, \mathbf{t^*})$ belongs to the model (in which case (2) is violated by $\bar{a}$), and $q(\bar{a}', null)$ is not in the database, then, according to rule (5), the repair is done either by deleting $p(\bar{a})$ or inserting $q(\bar{a}', null)$.

Notice that in this section we have been departing from the definition of repair given in Section 2, in the sense that repairs are obtained now by deletion of tuples or insertion of null values only, the usual ways in which RICs are maintained. In particular, if the instance is $\{p(\bar{a})\}$ and IC contains only $p(\bar{x}) \to \exists y q(\bar{x}, y)$, then $\{p(\bar{a}), q(\bar{a}, b)\}$, with $b \in D$, will not be obtained as a repair (although it is according to the initial definition), because it will not be captured by the program. This makes sense, because allowing such repairs would produce infinitely many of them, all of which are not natural from the perspective of usual database praxis.

If we want to establish a correspondence between stable models of the new repair program and the database repairs, we need first a precise definition of a repair in the new sense, according to which repairs can be achieved by insertion of null values that do not propagate through other ICs. We proceed by first redefining when a database instance, possibly containing null values, satisfies a set of ICs.

Definition 8. *For a database instance DB, whose domain D may contain the constant null and a set of integrity constraints $IC = IC_U \cup IC_R$, where IC_U is a set of universal integrity constraints of the form $\forall \bar{x} \varphi$, with φ quantifier free, and IC_R is a set of referential integrity constraints of the form $\forall \bar{x}(p(\bar{x}) \to \exists y q(\bar{x}', y))$, with $\bar{x}' \subseteq \bar{x}$, we say that r satisfies IC, written $DB \models_\Sigma IC$ iff:*

1. *For each $\forall \bar{x} \varphi \in IC_U$, $DB \models_\Sigma \varphi[\bar{a}]$ for all $\bar{a} \in D - \{null\}$, and*
2. *For each $\forall \bar{x}(p(\bar{x}) \to \exists y q(\bar{x}', y)) \in IC_R$, if $DB \models_\Sigma p(\bar{a})$, with $\bar{a} \in D - \{null\}$, then $DB \models_\Sigma \exists y q(\bar{a}, y)$.* □

Definition 9. *Let DB, DB_1, DB_2 be database instances over the same schema and domain D (that may now contain null). It holds $DB_1 \leq_{DB} DB_2$ iff:*

1. *for every atom $p(\bar{a}) \in \Delta(DB, DB_1)$, with $\bar{a} \in D - \{null\}$, it holds $p(\bar{a}) \in \Delta(DB, DB_2)$, and*
2. *for every atom $p(\bar{a}, null) \in \Delta(DB, DB_1)$, it holds $p(\bar{a}, null) \in \Delta(DB, DB_2)$ or $p(\bar{a}, \bar{b}) \in \Delta(DB, DB_2)$ with $\bar{b} \in D - \{null\}$.* □

Definition 10. *Given a database instance DB and a set of universal and referential integrity constraints IC, a repair of DB wrt IC is a database instance DB' over the same schema and domain (plus possibly null if it was not in the domain of DB), such that $DB' \models_\Sigma IC$ (in the sense of Definition 8) and DB' is $\leq_{DB}$-minimal in the class of database instances that satisfy IC.* □

Example 15. Consider the universal integrity constraint $\forall xy(q(x, y) \to r(x, y))$ together with the referential integrity constraints $\forall x(p(x) \to \exists y q(x, y))$ and $\forall x(s(x) \to \exists y r(x, y))$ and the inconsistent database instance $DB = \{q(a, b), p(c), s(a)\}$. The repairs for the latter are:

i	DB_i	$\Delta(DB, DB_i)$
1	$\{q(a, b), r(a, b), p(c), q(c, null), s(a)\}$	$\{r(a, b), q(c, null)\}$
2	$\{q(a, b), r(a, b), s(a)\}$	$\{p(c), r(a, b)\}$
3	$\{p(c), q(c, null), s(a), r(a, null)\}$	$\{q(a, b), q(c, null), r(a, null)\}$
4	$\{p(c), q(c, null)\}$	$\{q(a, b), q(c, null), s(a)\}$
5	$\{s(a), r(a, null)\}$	$\{q(a, b), p(c), r(a, null)\}$
6	$\emptyset$	$\{q(a, b), p(c), s(a)\}$

In the first repair it can be seen that the atom $q(c, null)$ does not propagate through the universal constraint to $r(c, null)$. For example, the instance $DB_7 = \{q(a, b), r(a, b), p(c), q(c, a), r(c, a), s(a))\}$, where we have introduced $r(c, a)$ in order to satisfy the second RIC, does satisfy IC, but is not a repair because $\Delta(DB, DB_1) \leq_{DB} \Delta(DB, DB_7) = \{r(a, b), q(c, a), r(c, a)\}$.

If $r(a, b)$ was inserted due to the universal constraint, we do want $r(a, null)$ to be inserted in order to satisfy the second referential constraint. This fact is captured by both the definition of repair and the repair program. Actually, the instance $DB_8 = \{q(a, b), r(a, b), s(a), r(a, null))\}$ is not a repair, because $\Delta(DB, DB_2) \subseteq \Delta(DB, DB_8) = \{p(c), r(a, b), r(a, null)\}$ and, in consequence, $\Delta(DB, DB_2) \leq_{DB} \Delta(DB, DB_8)$. The program also does not consider DB_8 as a repair, because the clauses (6) and (7) detect that $r(a, b)$ is already in the repair. □

If the set of IC contains both universal ICs and referential ICs, then the repair program $\Pi^*(DB, IC)$ contains now the extra rules we introduced at the beginning of this section. As before, for a stable model $\mathcal{M}$ of the program, $DB_{\mathcal{M}}$ denotes the corresponding database as in Definition 7. With the class of repairs introduced in Definition 10 it holds as before

Theorem 2. *If $\mathcal{M}$ is a coherent stable model of $\Pi^*(DB, IC)$, and $DB_{\mathcal{M}}$ is finite, then $DB_{\mathcal{M}}$ is a repair of DB with respect to IC. Furthermore, the repairs obtained in this way are all the repairs of DB.* □

Example 16. Consider the database instance $\{p(\bar{a})\}$ and the following set of ICs: $p(x) \rightarrow \exists y q(x, y)$, $q(x, y) \rightarrow r(x, y)$. The program $\Pi^*(DB, IC)$ is written in DLV as follows (ts, tss, ta, etc. stand for $\mathbf{t}^\star, \mathbf{t}^{\star\star}, \mathbf{t_a}$, etc.):[5]
Database contents

```
domd(a).
p(a,td).
```

Rules for CWA

```
p(X,fs)    :- domd(X), not p(X,td).
q(X,Y,fs) :- domd(X), domd(Y), not q(X,Y,td).
r(X,Y,fs) :-  not r(X,Y,td), domd(X), domd(Y).
```

Annotation rules

```
p(X,fs)    :- p(X,fa), domd(X).
p(X,ts)    :- p(X,ta),domd(X).
p(X,ts)    :- p(X,td), domd(X).
q(X,Y,fs) :- q(X,Y,fa),domd(X),domd(Y).
q(X,Y,ts) :- q(X,Y,ta), domd(X), domd(Y).
q(X,Y,ts) :- q(X,Y,td), domd(X), domd(Y).
r(X,Y,fs) :- r(X,Y,fa), domd(X), domd(Y).
r(X,Y,ts) :- r(X,Y,ta), domd(X), domd(Y).
r(X,Y,ts) :- r(X,Y,td), domd(X), domd(Y).
```

---

[5] The domain predicate used in the program should contain all the constants different from *null* that appear in the active domain of the database.

Rules for the ICs

```
    p(X,fa) v q(X,null,ta) :- p(X,ts), not aux(x), not q(X,null,td),
                                                              domd(X).
                   aux(X) :- q(X,Y,td), not q(X,Y,fa), domd(X), domd(Y).
                   aux(X) :- q(X,Y,ta), domd(X), domd(Y).
  q(X,Y,fa) v r(X,Y,ta)  :- q(X,Y,ts), r(X,Y,fs), domd(X), domd(Y).
```

Interpretation rules

```
    p(X,tss)    :- p(X,ta).
    p(X,tss)    :- p(X,td), not p(X,fa).
    p(X,fss)    :- p(X,fa).
    p(X,fss)    :- domd(X), not p(X,td), not p(X,ta).
    q(X,Y,tss) :- q(X,Y,ta).
    q(X,Y,tss) :- q(X,Y,td), not q(X,Y,fa).
    q(X,Y,fss) :- q(X,Y,fa).
    q(X,Y,fss) :- domd(X), domd(Y),not q(X,Y,td), not q(X,Y,ta).
    r(X,Y,tss) :- r(X,Y,ta).
    r(X,Y,tss) :- r(X,Y,td), not q(X,Y,fa).
    r(X,Y,fss) :- r(X,Y,fa).
    r(X,Y,fss) :- domd(X), domd(Y), not r(X,Y,td), not r(X,Y,ta).
```

Denial constraints

```
    :- p(X,ta), p(X,fa).
    :- q(X,Y,ta), q(X,Y,fa).
    :- r(X,Y,ta),r(X,Y,fa).
```

The stable models of the program are:

{domd(a), p(a,td), p(a,ts), p(a,fs), p(a,fss), p(a,fa), q(a,a,fs),
r(a,a,fs), q(a,a,fss), r(a,a,fss)}

{domd(a), p(a,td), p(a,ts), p(a,tss), q(a,null,ta), q(a,a,fs), r(a,a,fs),
q(a,a,fss), r(a,a,fss), q(a,null,tss)},

corresponding to the database instances $\emptyset$ and $\{p(a), q(a,null)\}$.

If the fact $q(a,null)$ is added to the original instance, the fact q(a,null,td)
becomes a part of the program. In this case, the program considers that the
new instance $\{p(a), q(a,null)\}$ satisfies the RIC. It also considers that the full
inclusion dependency $q(x,y) \rightarrow r(x,y)$ is satisfied, because we do not want null
values to be propagated. All this is reflected in the only model of the program,
namely

{domd(a), p(a,td), p(a,ts), q(a,null,td), p(a,tss), q(a,a,fs), r(a,a,fs),
q(a,a,fss), r(a,a,fss), q(a,null,tss)}.   □

If we want to impose the policy of repairing the violation of a RIC just by
deleting tuples, then, rule (5) should be changed by

$$p(\bar{x}, \mathbf{f_a}) \leftarrow p(\bar{x}, \mathbf{t}^*),\ not\ aux(\bar{x}'),\ not\ q(\bar{x}', null, \mathbf{t_d}),$$

that says that if the RIC is violated, then the fact $p(\bar{a})$ that produces such
violation must be deleted.

If we insist in keeping the original definition of repair (Section 2), i.e. allowing $\{p(\bar{a}), q(\bar{a}, b)\}$ to be a repair for every element $b \in D$, clause (5) could be replaced by:

$$p(\bar{x}, \mathbf{f_a}) \vee q(\bar{x}', y, \mathbf{t_a}) \leftarrow p(\bar{x}, \mathbf{t^*}), \; not \; aux(\bar{x}'), \; not \; q(\bar{x}', null, \mathbf{t_d}), choice(\bar{x}', y). \tag{8}$$

where $choice(\bar{X}, \bar{Y})$ is the static non-deterministic choice operator [25] that selects one value for attribute tuple $\bar{Y}$ for each value of the attribute tuple $\bar{X}$. In equation (8), $choice(\bar{x}', y)$ would select one value for y from the domain for each combination of values $\bar{x}'$. Then, this rule forces the one to one correspondence between stable models of the program and the repairs as introduced in Section 2.

8   Optimization of Repair Programs

The logic programs used to specify database repairs can be optimized in several ways. In Section 8.1 we examine certain program transformations that can lead to programs with a lower computational complexity. In Section 8.2, we address the issue of avoiding the explicit computation of negative information or of materialization of absent data, what in database applications can be a serious problem from the point of view of space and time complexity.

Other possible optimizations, that are not further discussed here, have to do with avoiding the complete computation of all stable models (the repairs) whenever a query is to be answered. The query rewriting methodology introduced in [2] had this advantage: inconsistencies were solved locally, without having to restore the consistency of the complete database. In contrast, the logic programming base methodology, at least if implemented in a straightforward manner, computes all stable models. This issue is related to finding methodologies for minimizing the number of rules to be instantiated, the way ground instantiations are done, avoiding evaluation of irrelevant subgoals, etc. Further implementation issues are discussed in Section 9.

8.1   Head Cycle Free Programs

In some cases, the repair programs we have introduced may be transformed into equivalent non disjunctive programs. This is the case of head-cycle-free programs [10] introduced below. These programs have better computational complexity than general disjunctive programs in the sense that the complexity is reduced from Π_2^P-complete to coNP-complete [18,29].

The *dependency graph* of a ground disjunctive program Π is defined as a directed graph where each literal is a node and there is an arch from L to L' iff there is a rule in which L appears positive in the body and L' appears in the head. Π is *headcycle free* (HCF) iff its dependency graph does not contain directed cycles that go through two literals that belong to the head of the same rule.

A disjunctive program Π is HCF if its ground version is HCF. If this is the case, Π can be transformed into a non disjunctive normal program $sh(\Pi)$ with the same stable models that is obtained by replacing every disjunctive rule of the form: $\bigvee_{i=1}^{n} p_i(\bar{x}_i) \leftarrow \bigwedge_{j=1}^{m} q_j(\bar{y}_j)$ by the n following rules $p_i(\bar{x}_i) \leftarrow \bigwedge_{j=1}^{m} q_j(\bar{y}_j) \wedge \bigwedge_{k \neq i} not\ p_k(\bar{x}_k)$, $i = 1, ..., n$. Such transformations can be justified or discarded on the basis of a careful analysis of the intrinsic complexity of consistent query answering [15]. If the original program can be transformed into a normal program, then also other efficient implementations could be used for query evaluation, e.g. *XSB* [34], that has been already successfully applied in the context of consistent query answering via query transformation, with non-existentially quantified conjunctive queries [14].

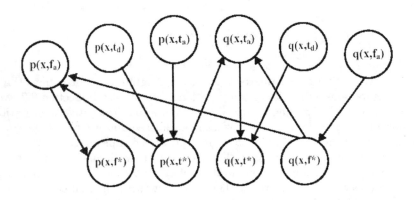

The repair program is HCF because, as it can be seen from the (relevant part of the) dependency graph, there is no cycle involving both $p(x, \mathbf{f_a})$ and $q(x, \mathbf{t_a})$), the atoms that appear in the only disjunctive head.

The non disjunctive version of the program has the disjunctive clause replaced by the two definite clauses $p(x, \mathbf{f_a}) \leftarrow p(x, \mathbf{t^\star}), q(x, \mathbf{f^\star})$, $not\ q(x, \mathbf{t_a})$, and $q(x, \mathbf{t_a}) \leftarrow p(x, \mathbf{t^\star}), q(x, \mathbf{f^\star})$, $not\ p(x, \mathbf{f_a})$. The two programs have the same stable models. $\qquad\square$

In the rest of this section we will consider a set IC of universal ICs of the form

$$q_1(\bar{t}_1) \vee \cdots \vee q_n(\bar{t}_n) \leftarrow p_1(\bar{s}_1) \wedge \cdots \wedge p_m(\bar{s}_m). \qquad (9)$$

(the rule version of (1)). We denote with $ground(IC)$ the set of ground instantiations of the clauses in IC in D. A ground literal l is *bilateral* with respect to $ground(IC)$ if appears in the head of a rule in $ground(IC)$ and in the body of a possibly different rule in $ground(IC)$.

Example 17. In $ground(IC) = \{s(a, b) \rightarrow s(a, b) \vee r(a, b),\ r(a, b) \rightarrow r(b, a)\}$, the literals $s(a, b)$ and $r(a, b)$ are bilateral, because they appear in a head of a rule and in a body of a rule. Instead, $r(b, a)$ is not bilateral. $\qquad\square$

The following theorem tells us how to check if the repair program is HCF by analyzing just the set of ICs.

Theorem 3. *The program $\Pi^*(DB, IC)$ is HCF iff ground(IC) does not have a pair of bilateral literals in the same rule.* □

Example 18. If $IC = \{s(x,y) \rightarrow r(x),\ r(x) \rightarrow p(x)\}$ and the domain is $D = \{a, b\}$, we have $ground(IC) = \{s(a, a) \rightarrow r(a),\ s(a, b) \rightarrow r(a),\ s(b, a) \rightarrow r(b),\ s(b, b) \rightarrow r(b),\ r(a) \rightarrow p(a),\ r(b) \rightarrow p(b)\}$. The bilateral literals are $r(a)$ and $r(b)$. The program $\Pi^*(DB, IC)$ is HCF because $r(a)$ and $r(b)$ do not appear in a same rule in $ground(IC)$. As a consequence, the clause $s(x, y, \mathbf{f_a}) \vee r(x, \mathbf{t_a}) \leftarrow s(x, y, \mathbf{f}^*), r(x, \mathbf{t}^*)$ of $\Pi^*(DB, IC)$, for example, can be replaced in $sh(\Pi^*(DB, IC))$ by the two clauses $s(x, y, \mathbf{f_a}) \leftarrow s(x, y, \mathbf{f}^*), r(x, \mathbf{t}^*), not\ r(x, \mathbf{t_a})$ and $r(x, \mathbf{t_a}) \leftarrow s(x, y, \mathbf{f}^*), r(x, \mathbf{t}^*), not\ s(x, y, \mathbf{f_a})$. □

Example 19. If $IC = \{s(x) \rightarrow r(x),\ p(x) \rightarrow s(x),\ u(x, y) \rightarrow p(x)\}$ and the domain is $D = \{a, b\}$, we have $ground(IC) = \{s(a) \rightarrow r(a),\ p(a) \rightarrow s(a),\ u(a, a) \rightarrow p(a),\ s(b) \rightarrow r(b),\ p(b) \rightarrow s(b),\ u(b, b) \rightarrow p(b),\ u(a, b) \rightarrow p(a),\ u(b, a) \rightarrow p(b)\}$. The bilateral literals in $ground(IC)$ are $s(a), s(b), p(a), p(b)$. The program $\Pi^*(DB, IC)$ is not HCF, because there are pairs of bilateral literals appearing in the same rule in $ground(IC)$, e.g. $\{s(a), p(a)\}$ and $\{s(b), p(b)\}$. □

Corollary 1. *If IC contains only denial constraints, i.e. formulas of the form $\bigvee_{i=1}^{n} p_i(\bar{t}_i) \rightarrow \varphi$, where $p_i(\bar{t}_i)$ is an atom and φ is a formula containing built-in predicates only, then $\Pi^*(DB, IC)$ is HCF.* □

Example 20. For $IC = \{\forall xyzuv(p(x, y, z) \wedge p(x, u, v) \rightarrow y = u),\ \forall xyzuv(p(x, y, z) \wedge p(x, u, v) \rightarrow z = v),\ \forall xyzv(q(x, y, z) \wedge p(x, y, v) \rightarrow z = v)\}$, and any ground instantiation, there are no bilateral literals. In consequence, the program $\Pi^*(DB, IC)$ will be always HCF. □

This corollary includes important classes of *ICs* as key constraints, functional dependencies and range constraints. In [15] it was shown that, for this kind of ICs, the intrinsic lower bound complexity of consistent query answering is coNP-complete. The corollary shows that by means of the transformed program this lower bound is achieved.

8.2   Avoiding Materialization of the CWA

The repair programs introduced in Section 5 contain clauses of the form $p(\bar{x}, f^*) \leftarrow not\ p(\bar{x}, t_d)$, that have the consequence of materializing negative information in the stable models of the programs. The repairs programs can be optimized, making them compute only that negative data that is needed to obtain the database repairs.

First, by unfolding, atoms of the form $p(\bar{x}, \mathbf{f}^*)$ that appear as subgoals in the bodies are replaced by their definitions. More precisely, replace every rule that contains an atom of the form $p(\bar{x}, \mathbf{f}^*)$ in the body, by two rules, one replacing the atom by $p(\bar{x}, \mathbf{f_a})$, and another replacing the atom by $not\ p(\bar{x}, \mathbf{t_d})$. Next, eliminate from the repair program those rules that have atoms annotated with f** or f* in their heads, because they compute data that should not be explicitly contained in the repairs. If $\Pi^{*opt}(DB, IC)$ denotes the program obtained after applying these two transformations, it is easy to see that the following holds

Proposition 4. $\Pi^{*opt}(DB, IC)$ *and* $\Pi^{*}(DB, IC)$ *produce the same database repairs, more precisely, they compute exactly the same database instances in the sense of Definition 7.*   □

Example 21. (example 16 continued) The optimized program $\Pi^{*opt}(DB, IC)$ is as below and determines the same repairs as the original program. Notice that the second disjunctive rule in the original program was replaced by two new rules in the new program.

```
domd(a).
p(a,td).

p(X,ts)    :- p(X,ta),domd(X).
p(X,ts)    :- p(X,td), domd(X).

q(X,Y,ts) :- q(X,Y,ta), domd(X), domd(Y).
q(X,Y,ts) :- q(X,Y,td), domd(X), domd(Y).

r(X,Y,ts) :- r(X,Y,ta), domd(X), domd(Y).
r(X,Y,ts) :- r(X,Y,td), domd(X), domd(Y).

p(X,fa) v q(X,null,ta) :- p(X,ts), not aux(x), not q(X,null,td),
                                                       domd(X).
               aux(X) :- q(X,Y,td), not q(X,Y,fa), domd(X), domd(Y).
               aux(X) :- q(X,Y,ta), domd(X), domd(Y).
q(X,Y,fa) v r(X,Y,ta)  :- q(X,Y,ts), r(X,Y,fa), domd(X), domd(Y).
q(X,Y,fa) v r(X,Y,ta)  :- q(X,Y,ts), not r(X,Y,td), domd(X), domd(Y).

p(X,tss)   :- p(X,ta).
p(X,tss)   :- p(X,td), not p(X,fa).

q(X,Y,tss) :- q(X,Y,ta).
q(X,Y,tss) :- q(X,Y,td), not q(X,Y,fa).

r(X,Y,tss) :- r(X,Y,ta).
r(X,Y,tss) :- r(X,Y,td), not q(X,Y,fa).

:- p(X,ta), p(X,fa).
:- q(X,Y,ta), q(X,Y,fa).
:- r(X,Y,ta),r(X,Y,fa).                                              □
```

The optimization for HCF programs of Section 8.1 and the one that avoids the materialization of unnecessary negative data can be combined.

Theorem 4. *If* $\Pi^{*}(DB, IC)$ *is HCF, then* $sh(\Pi^{*}(DB, IC))^{opt}$ *and* $\Pi^{*}(DB, IC)$ *compute the same database repairs in the sense of Definition 7.*   □

9   Conclusions

We have presented a general treatment of consistent query answering for first order queries and universal and referential ICs that is based on annotated predicate calculus (APC). Integrity constraints and database information are translated into a theory written in APC in such a way that there is a correspondence between the minimal models of the new theory and the repairs of the original database.

We have also shown how to specify database repairs by means of classical disjunctive logic programs with stable model semantics. Those programs have annotations as new arguments, and are inspired by the APC theory mentioned above. In consequence, consistent query answers can be obtained by "running" a query program together with the specification program. We illustrated their use by means of the DLV system. Finally, some optimizations of the repair programs were introduced.

The problem of consistent query answering was explicitly presented in [2], where also the notions of repair and consistent answer were formally defined. In addition, a methodology for consistent query answering based on a rewriting of the original query was developed (and further investigated and implemented in [14]). Basically, if we want the consistent answers to a FO query expressed in, say SQL2, a new query in SQL2 can be computed, such that its usual answers from the database are the consistent answers to the original query. That methodology has a polynomial data complexity, and that is the reason why it works for some restricted classes of FO ICs and queries, basically for non existentially quantified conjunctive queries [1]. Actually, in [15] it is shown that the problem of CQA is coNP-complete for simple functional dependencies and existential queries.

In this paper, we have formulated the problem of CQA as a problem of nonmonotonic reasoning, more precisely of minimal entailment, whose complexity, even in the propositional case, can be at least Π_2^P-complete [19]. Having a problem of nonmonotonic reasoning with such complexity, it is not strange to try to use disjunctive logic programs with negation with a stable or answer set semantics to solve the problem of CQA, because such programs have nonmonotonic consequences and a Π_2^P-complete complexity [18]. Answer set programming has been successfully used in formalizing and implementing complex nommonotonic reasoning tasks [7].

Under those circumstances, the problem then is to come up with the best logic programming specifications and the best way to use them, so that the computational complexity involved does not go beyond the intrinsic, theoretical lower bound complexity of consistent query answering.

Implementation and applications are important directions of research. The logic programming environment will interact with a DBMS, where the inconsistent DB will be stored. As much of the computation as possible should be pushed into the DBMS instead of doing it at the logic programming level.

The problem of developing query evaluation mechanisms from disjunctive logic programs that are guided by the query, most likely containing free variables and then expecting a set of answers, like magic sets [1], deserves more attention from the logic programming and database communities. The current alternative relies on finding those ground query atoms that belong to all the stable models once they have been computed via a ground instantiation of the original program (see Example 11). In [20] intelligent grounding strategies for pruning in advance the instantiated program have been explored and incorporated into *DLV*. It would be interesting to explore to what extent the program can be further pruned from irrelevant rules and subgoals using information obtained by querying the original database.

As shown in [6], there are classes of ICs for which the intersection of the stable models of the repair program coincides with the well-founded semantics, which can be computed more efficiently than the stable model semantics. It could be possible to take advantage of this efficient "core" computation for consistent query answering if ways of modularizing or splitting the whole computation into a core part and a query specific part are found. Such cases were identified in [5] for FDs and aggregation queries.

In [26], a general methodology based on disjunctive logic programs with stable model semantics is used for specifying database repairs wrt universal ICs. In their approach, preferences between repairs can be specified. The program is given through a schema for rule generation.

Independently, in [4] a specification of database repairs for binary universal ICs by means of disjunctive logic programs with a stable model semantics was presented. Those programs contained both "triggering" rules and "stabilizing" rules. The former trigger local, one-step changes, and the latter stabilize the chain of local changes in a state where all the ICs hold. The same rules, among others, are generated by the rule generation schema introduced in [26].

The programs presented here also work for the whole class of universal ICs, but they are much simpler and shorter than those presented in [26,4]. Actually, the schema presented in [26] and the extended methodology sketched in [4], both generate an exponential number of rules in terms of the number of ICs and literals in them. Instead, in the present work, due to the simplicity of the program, that takes full advantage of the relationship between the annotations, a linear number of rules is generated. Our treatment of referential ICs considerably extends what has been sketched in [4,26].

There are several similarities between our approach to consistency handling and those followed by the belief revision/update community. Database repairs coincide with revised models defined by Winslett in [35]. The treatment in [35] is mainly propositional, but a preliminary extension to first order knowledge bases can be found in [16]. Those papers concentrate on the computation of the models of the revised theory, i.e., the repairs in our case, but not on query answering. Comparing our framework with that of belief revision, we have an empty domain theory, one model: the database instance, and a revision by a set of ICs. The revision of a database instance by the ICs produces new database instances, the repairs of the original database.

Nevertheless, our motivation and starting point are quite different from those of belief revision. We are not interested in computing the repairs *per se*, but in answering queries, hopefully using the original database as much as possible, possibly posing a modified query. If this is not possible, we look for methodologies for representing and querying simultaneously and implicitly all the repairs of the database. Furthermore, we work in a fully first-order framework.

The semantics of database updates is treated in [22], a treatment that is close to belief revision. That paper represents databases as collections of theories, in such a way that under updates a new collection of theories is generated that minimally differ from the original ones. So, there is some similarity to our database repairs. However, that paper does not consider inconsistencies, nor query answering in any sense.

Another approach to database repairs based on a logic programming semantics consists of the *revision programs* [31]. The rules in those programs explicitly declare how to enforce the satisfaction of an integrity constraint, rather than explicitly stating the ICs, e.g. $in(a) \leftarrow in(a_1), \dots, in(a_k), out(b_1), \dots, out(b_m)$ has the intended procedural meaning of inserting the database atom a whenever $a_1, \dots, a_k$ are in the database, but not $b_1, \dots, b_m$. Also a declarative, stable model semantics is given to revision programs. Preferences for certain kinds of repair actions can be captured by declaring the corresponding rules in program and omitting rules that could lead to other forms of repairs.

In [12,28] paraconsistent and annotated logic programs, with non classical semantics, are introduced. However, in [17] some transformation methodologies for paraconsistent logic programs [12] are shown that allow assigning to them extensions of classical semantics. Our programs have a fully standard stable model semantics.

Acknowledgments. Work funded by DIPUC, MECESUP, FONDECYT Grant 1000593, Carleton University Start-Up Grant 9364-01, NSERC Grant 250279-02. We are grateful to Marcelo Arenas, Alvaro Campos, Alberto Mendelzon, and Nicola Leone for useful conversations.

References

1. Abiteboul, S., Hull, R. and Vianu, V. *Foundations of Databases.* Addison-Wesley, 1995.
2. Arenas, M., Bertossi, L. and Chomicki, J. Consistent Query Answers in Inconsistent Databases. In *Proc. ACM Symposium on Principles of Database Systems (ACM PODS'99)*, 1999, pp. 68–79.
3. Arenas, M., Bertossi, L. and Kifer, M. Applications of Annotated Predicate Calculus to Querying Inconsistent Databases. In *'Computational Logic - CL2000' Stream: 6th International Conference on Rules and Objects in Databases (DOOD'2000)*. Springer Lecture Notes in Artificial Intelligence 1861, 2000, pp. 926–941.
4. Arenas, M., Bertossi, L. and Chomicki, J. Specifying and Querying Database Repairs using Logic Programs with Exceptions. In *Flexible Query Answering Systems. Recent Developments*, H.L. Larsen, J. Kacprzyk, S. Zadrozny, H. Christiansen (eds.), Springer, 2000, pp. 27–41.

5. Arenas, M., Bertossi, L. and Chomicki, J. Scalar Aggregation in FD-Inconsistent Databases. In *Database Theory - ICDT 2001*, Springer, LNCS 1973, 2001, pp. 39–53.
6. Arenas, M., Bertossi, L. and Chomicki, J. Answer Sets for Consistent Query Answers. To appear in Theory and Practice of Logic Programming.
7. Baral, C. *Knowledge Representation, Reasoning and Declarative Problem Solving.* Cambridge University Press, 2003.
8. Barcelo, P. and Bertossi, L. Repairing Databases with Annotated Predicate Logic. In *Proc. Nineth International Workshop on Non-Monotonic Reasoning (NMR'2002), Special session: Changing and Integrating Information: From Theory to Practice*, S. Benferhat and E. Giunchiglia (eds.), 2002, pp. 160–170.
9. Barcelo, P. and Bertossi, L. Logic Programs for Querying Inconsistent Databases. Proc. Practical Aspects of Declarative Languages (PADL03), Springer LNCS 2562, 2003, pp. 208–222.
10. Ben-Eliyahu, R. and Dechter, R. Propositional Semantics for Disjunctive Logic Programs. *Annals of Mathematics in Artificial Intelligence*, 1994, 12:53-87.
11. Bertossi, L., Chomicki, J., Cortes, A. and Gutierrez, C. Consistent Answers from Integrated Data Sources. In 'Flexible Query Answering Systems', Proc. of the 5th International Conference, FQAS 2002. T. Andreasen, A. Motro, H. Christiansen, H. L. Larsen (eds.). Springer LNAI 2522, 2002, pp. 71–85.
12. Blair, H.A. and Subrahmanian, V.S. Paraconsistent Logic Programming. *Theoretical Computer Science,* 1989, 68:135–154.
13. Buccafurri, F., Leone, N. and Rullo, P. Enhancing Disjunctive Datalog by Constraints. *IEEE Transactions on Knowledge and Data Engineering*, 2000, 12(5):845–860.
14. Celle, A. and Bertossi, L. Querying Inconsistent Databases: Algorithms and Implementation. In 'Computational Logic - CL 2000', J. Lloyd et al. (eds.). Stream: 6th International Conference on Rules and Objects in Databases (DOOD'2000). Springer Lecture Notes in Artificial Intelligence 1861, 2000, pp. 942–956.
15. Chomicki, J. and Marcinkowski, J. On the Computational Complexity of Consistent Query Answers. Submitted in 2002 (CoRR paper cs.DB/0204010).
16. Chou, T. and Winslett, M. A Model-Based Belief Revision System. *Journal of Automated Reasoning*, 1994, 12:157–208.
17. Damasio, C. V. and Pereira, L.M. A Survey on Paraconsistent Semantics for Extended Logic Programas. In *Handbook of Defeasible Reasoning and Uncertainty Management Systems*, Vol. 2, D.M. Gabbay and Ph. Smets (eds.), Kluwer Academic Publishers, 1998, pp. 241–320.
18. Dantsin, E., Eiter, T., Gottlob, G. and Voronkov, A. Complexity and Expressive Power of Logic Programming. *ACM Computing Surveys*, 2001, 33(3): 374–425.
19. Eiter, T. and Gottlob, G. Propositional Circumscription and Extended Closed World Assumption are Π_2^p-complete. Theoretical Computer Science, 1993, 114, pp. 231–245.
20. Eiter, T., Leone, N., Mateis, C., Pfeifer, G. and Scarcello, F. A Deductive System for Non-Monotonic Reasoning. Proc. LPNMR'97, Springer LNAI 1265, 1997, pp. 364–375.
21. Eiter, T., Faber, W.; Leone, N. and Pfeifer, G. Declarative Problem-Solving in DLV. In *Logic-Based Artificial Intelligence*, J. Minker (ed.), Kluwer, 2000, pp. 79–103.
22. Fagin, R., Kuper, G., Ullman, J. and Vardi, M. Updating Logical Databases. In *Advances in Computing Research*, JAI Press, 1986, Vol. 3, pp. 1–18.

23. Gelfond, M. and Lifschitz, V. The Stable Model Semantics for Logic Programming. In *Logic Programming, Proceedings of the Fifth International Conference and Symposium*, R. A. Kowalski and K. A. Bowen (eds.), MIT Press, 1988, pp. 1070–1080.

24. Gelfond, M. and Lifschitz, V. Classical Negation in Logic Programs and Disjunctive Databases. *New Generation Computing*, 1991, 9:365–385.

25. Giannotti, F., Greco, S.; Sacca, D. and Zaniolo, C. Programming with Non-determinism in Deductive Databases. *Annals of Mathematics and Artificial Intelligence*, 1997, 19(3–4).

26. Greco, G., Greco, S. and Zumpano, E. A Logic Programming Approach to the Integration, Repairing and Querying of Inconsistent Databases. In *Proc. 17th International Conference on Logic Programming, ICLP'01*, Ph. Codognet (ed.), LNCS 2237, Springer, 2001, pp. 348–364.

27. Kifer, M. and Lozinskii, E.L. A Logic for Reasoning with Inconsistency. *Journal of Automated reasoning*, 1992, 9(2):179–215.

28. Kifer, M. and Subrahmanian, V.S. Theory of Generalized Annotated Logic Programming and its Applications. *Journal of Logic Programming*, 1992, 12(4):335–368.

29. Leone, N., Rullo, P. and Scarcello, F. Disjunctive Stable Models: Unfounded Sets, Fixpoint Semantics, and Computation. *Information and Computation*, 1997, 135(2):69–112.

30. Lloyd, J.W. *Foundations of Logic Programming*. Springer Verlag, 1987.

31. Marek, V.W. and Truszczynski, M. Revision Programming. *Theoretical Computer Science*, 1998, 190(2):241–277.

32. Pradhan, S. Reasoning with Conflicting Information in Artificial Intelligence and Database Theory. PhD thesis, Department of Computer Science, University of Maryland, 2001.

33. Reiter, R. Towards a Logical Reconstruction of Relational Database Theory. In *On Conceptual Modelling*, M.L. Brodie, J. Mylopoulos, J.W. Schmidt (eds.), Springer, 1984.

34. Sagonas, K.F., Swift, T. and Warren, D.S. XSB as an Efficient Deductive Database Engine. In *Proc. of the 1994 ACM SIGMOD International Conference on Management of Data*, ACM Press, 1994, pp. 442–453.

35. Winslett, M. Reasoning about Action using a Possible Models Approach. In *Proc. Seventh National Conference on Artificial Intelligence (AAAI'88)*, 1988, pp. 89–93.

Towards Unifying Semantic Constraints and Security Constraints

Joachim Biskup and Barbara Sprick

Fachbereich Informatik, Lehrstuhl 6
Universität Dortmund, Germany
{biskup;sprick}@ls6.cs.uni-dortmund.de

Abstract. Modern information systems must respect certain restrictions in order to guarantee the proper and desired functionality. Semantic constraints help to prevent inconsistencies in the stored data resulting from faulty updates. Security constraints are to maintain integrity, secrecy and availability over updates and over queries. In this paper we design an unifying framework for both kinds of constraints in order to study interactions between them. We view a distributed information system as a multi-agent system in which all components of the system are seen as agents. We present a temporal and epistemic logic for the defined framework and show in an example how security constraints and semantic constraints can be expressed in this framework.

1 Introduction

A modern information system must respect certain restrictions in order to guarantee the proper and desired functionality. These restrictions can be declared on the stored data as well as on the behaviour of the information system and its users. The "real world" implies certain restrictions on the stored data, such as *An employee of 25 years age cannot have a work history of 30 years*. *Semantic constraints* help to prevent such inconsistencies in the stored data, resulting from faulty updates.

Often the application context imposes certain restrictions and obligations on the behaviour of the system and its users. For example *An employee may not increase her own salary* or *Secret information may not be disclosed to an unclassified user*. *Security constraints* help to maintain integrity, secrecy and availability over updates as well as over queries.

We consider an *information system* as a *highly distributed system* with heterogeneous distributed components. It seems unrealistic to assume anything like a global schema or a centralized mechanism for updates with enforcement of semantic constraints and for query evaluation. Instead, we have to assume that any individual component has only a restricted view on the whole. Accordingly, security is considered as being concerned with the different interests and views of the various components. Here, it seems to be rather undesirable to assume anything like a global security policy or centralized supervising mechanisms.

L. Bertossi et al. (Eds.): Semantics in Databases, LNCS 2582, pp. 34–62, 2003.

Instead, whenever possible, security interests should be enforced by the components autonomously. In the remainder of this paper, whenever we consider semantic constraints and security constraints, we will have to keep in mind, that these constraints are employed in a distributed environment. In particular, even if an ideal observer thinks in terms of a global state, which, however, is questionable because of the problem of synchronized time, in practice any component will have only partial knowledge about that fictitious state.

Our aim is to provide a unifying framework for both kinds of constraints, semantic constraints as well as security constraints declared for distributed information systems without global situations in order to study interactions between them on the conceptual level at the design time in a manageable way.

1.1   Various Types of Constraints for Information Systems

Let us first investigate the various types of constraints in information systems in order to justify the required features of our framework.

Semantic constraints reflect properties of an outside "mini-world" that is modelled by the information system. These properties can either be found in the real world (like *An employee of 25 years age cannot have a work history of 30 years*) or they can be imposed by the application requirements (like *Every employee must have a social security number*).

Semantic constraints are invariants that need to be maintained by the information system whenever the current instance of the system is changed by some update operation. Further, semantic constrains are intended to support the users of an information system when they interpret instances of a system in terms of a real world.

Maintaining *secrecy* means preventing the improper disclosure of data. Users may access information directly by querying the information system or indirectly through logical conclusions of their knowledge. Thus, we can distinguish between two types of confidentiality. The first type is "authorization confidentiality", roughly meaning queries should be successfully invoked only by authorized users. The second type of confidentially roughly means that the users' knowledge should comply with the task specific restrictions, they should not be able to infer information they are not entitled to know.

Maintaining *integrity* means preventing the improper modification of data. In the context of security requirements this can be seen as "authorization integrity": updates should be successfully invoked only by authorized users, e.g. in a company, employees may not increase their own salary.

Maintaining *availability* means avoidance of denial of service. Availability has two aspects: Firstly, whenever information is requested by a user, the system should eventually provide this information or, in other words, user knowledge should eventually comply with the task specific richness. Secondly, whenever a user submits an (update) operation, this operation should eventually be executed.

Security constraints are invariants that are to be maintained by the system whenever some operation sequences are executed on behalf of an actor. Here, update operations (e.g. for maintaining integrity) as well as read operations (e.g. for maintaining secrecy) are important. Further, security constraints are intended to support system users when they employ the system as a communication tool, e.g. by controlling the flow of information.

In the logical approach to information systems, an administrator expresses the semantic constraints in a declarative style. He specifies formulae in an appropriately restricted formal language of a logic calculus. The logic calculus is assumed to provide, besides a syntax for formulae, a model-theoretic semantics, i.e. a definition of *a formula is satisfied by a structure (or an interpretation)*, and based on that, a definition of *a set of formulae implies another formula*. In the context of information systems, instances of an information system are considered as structures in the sense of the logic calculus.

In a policy oriented approach to security, a security administrator expresses the security constraints in a declarative style. He specifies some formulae in an appropriate formal language. It is assumed that besides a syntax for formulae, a semantics is provided too, i.e. a definition of *a formula is satisfied by a history* that comprises, possibly among others, a sequence of actual executions of operations, and based on that, of *a set of formulae implies another formula*.

1.2   Our Aim: A Unifying Framework for the Specification of Semantic Constraints and Security Constraints

There are examples of subtle interactions of semantic constraints and security constraints. We will discuss one such example in the last section of this paper. *Interaction of semantic constraints and security constraints can be investigated only if all constraints are expressed in the same framework.*

Semantic constraints restrict the possible *states* of the information system, *updates* and *sequences of updates. Thus, our framework has to be suited to talk about states and (sequences of) updates.*

Security constraints restrict the possible flow of information by restricting read and write access from/to objects and by keeping track of users' knowledge. *Thus, we need a notion of execute rights for read operations and update operations as well as a notion of users' knowledge.* Authorization integrity restricts possible (sequences) of operations. To model these constraints, *we need a notion of execute rights for update operations.* Availability constraints impose obligations on the system. In a weak sense, these obligations can be modelled using temporal notions: If a user submits an operation, the system should *eventually* execute it.

In both contexts, semantic constraints as well as security constraints, we sometimes have to talk about *sequences of operations.* Such sequences can be modelled using temporal notions.

Observations made above justify the need for a framework in which we can formalize *epistemic* constraints (to observe the knowledge of actors) and *temporal* constraints (to restrict the behaviour of the objects and actors over updates and read operations and operation sequences).

Further, the framework must be capable of formalizing *actors* operating in a distributed environment and having only a partial view on the whole system. These actors perform *operations* concurrently, thus *we need a notion of (concurrent) operations*. Over the time, these actors gain *knowledge* by performing operations and by inferring new knowledge. Thus *we need to capture the actors' knowledge over the time*.

Our framework must also be capable of formalizing *stored data*, e.g. by formalizing *objects* whose current state represents the stored data. On these objects, *operations* are executable. Also, the current state of the objects may change over the time by the execution of operations, which means that *the state of objects over the time has to be formalized*.

As discussed in the next section, modal logics used so far for the formalization of database constraints cover the required constraints only partially. But to avoid flaws already at the conceptual level, a comprehensive specification is needed. Such a specification could then also be used as a starting point for implementations or as reference of verifications.

2   Related Work

In the past a lot of work concerning formalization of information system constraints has been done. While semantical constraints have been widely investigated, still a lot of work is required with regard to security constraints and in particular with regard to the combination of both types. In most papers only few aspects of security constraints were investigated, other aspects were faded out. Before we choose one appropriate modal logic in the next sections, we try to get an overview of related logics.

2.1   Temporal Aspects and Distribution

Information systems can be seen as open, reactive and usually distributed systems that administrate persistent data. A lot of work has been done in the field of temporal logics for reactive systems and for multi-agent systems (see for example [MP92], [Rei91] [Lam94], [Woo00], [Thi95], [Nie97]).

A very common approach in the field of multi-agent systems are BDI-models (*belief-desire-intention*-models), [Wei99], [HS98], [Woo00]. Agents are seen as entities that are capable of observing their environment and reasoning about it and of independently and autonomously performing actions based upon decisions made by these agents dependent on their observations of the environment. In this perspective, each agent is assumed to have its own belief, desires and intentions. In [Woo00], Wooldridge introduces a very powerful first-order BDI-logic called *LORA* (Logic for Reasoning Agents). *LORA* consists of several components: a first-order component, a belief/desire/intention component and a temporal component. The temporal component is based on CTL^*, a well known temporal logic with an interleaving semantics.

Reactive systems can be seen a bit wider than multi-agent-systems. In reactive systems, not all components necessarily need to be agents in the sense described above. A very common model for reactive systems are so-called Mazurkiewicz traces, refer [Maz95], [MOP89]. Mazurkiewicz traces are a semantic concept for modelling concurrency. Unlike other semantic concepts, e.g. Kripke structures, the main characteristic of such trace systems is to explicitly differentiate between concurrency and nondeterministic choice. Another main feature of trace systems is the abstraction of interleaving; modelling of true concurrency is possible. Although this model is not more expressive than semantic concepts with interleaving, it can avoid the known (notorious) state explosion problem.

Temporal logics for distributed systems based on Mazurkiewicz-trace systems have been developed by Penczek, Thiagarajan, Niebert and others [Thi95,Pen93, Nie97]. Thiagarajan and Niebert develop trace based propositional temporal logics to express nonsequential behaviour directly without using interleaving.

Their logics base upon traces seen as partially ordered runs of a distributed system. Such a system consists of an arbitrary but fix number k of sequential agents which synchronize by performing actions jointly. Each agent i is assigned a non-empty local alphabet Σ_i of actions , $\tilde{\Sigma} = (\Sigma_1, \Sigma_2, \dots, \Sigma_k)$, ($k$ = number of agents of the system) is called a *distributed alphabet*. Agent i must take part in each action $a \in \Sigma_i$. Thus synchronization between individual agents is modelled by the joint execution of actions. If two actions a and b are not contained in the same alphabet, they can be executed independently. Infinite traces can be seen as Σ labelled partial orders (fulfilling certain characteristics), where $\Sigma = \bigcup_{i \in \{1,\dots,k\}} \Sigma_i$.

Inspired by this development of logics for distributed systems, Ehrich, Caleiro, Sernadas and Denker presented in [ECSD98] two object-oriented logics capable of expressing communication among objects. In these logics, all objects of a database system are seen as sequential agents and are thus components of the distributed system.

2.2   Epistemic Aspects in Distributed Systems

There are several approaches in which a database is seen as a collection of knowledge (see e.g. [Rei90] or [CD96,CD97]). If we follow this view, we can uniformly see database objects, users, administrators etc. as reasoning agents in a distributed environment. Over the last decade, modelling of knowledge has been a field of great research interest, especially for multi-agent systems. [FHMV96] gives an excellent overview of the state of the art in this topic.

Traditionally, modal logics of knowledge are interpreted over global states of the distributed system. As motivated in the introduction, we cannot assume such a global state: Every agent only has a local view on the system and when the system changes due to an action by a group of agents, only agents of the acting group typically know the effect of that action. The knowledge of the non-participating agents remains unchanged, and this fact is known by all agents.

Ramanujam analyses in [Ram96a,KR94] the meaning of knowledge in distributed systems. A similar discussion carried out by Van der Hoek and Meyer in [vdHM92] addresses the following questions: What exactly is a knowledge state? Is the knowledge that is logically implied by the available information actually computable? Given a knowledge state, can we decide, which other states are reachable? Can the actions of one agent have influence on the knowledge of another?

Ramanujam develops a decidable action based temporal and epistemic logic ([Ram96a]) for a distributed environment, in which knowledge changes caused by actions of agents can be expressed. His logic has two layers: formulae local to one agent can be combined to formulae about the whole system.

2.3   Deontic Aspects

We considered different types of constraints in the introduction. Security constraints in a database system always deal with obligations and authorizations. The tendency to formalize such constraints through deontic logic has been mainly followed by Cuppens, Demolombe, Carmo and Jones. In this method, a database is considered as a normative system and the corresponding security and semantic constraints are seen as deontic constraints: *It ought to be, that ... , It is necessary, that ... , It is permitted, that ...* . The meaning of a normative system in this context is, that a set of agents (software systems, humans etc.) interact according to some rules. It is not explicitly mentioned, that the misconduct of the agents is impossible. Normative systems rather suggest, how to handle the misconduct of an agent.

In our view, authorizations for performing operations can be modelled by assigning particular rights to the components of the information system. Then we can formalize a temporal constraint, that roughly expresses the following: All operations that are not explicitely allowed for a component, cannot happen. We could also formalize a dual temporal constraint: All operations that are not explicitly forbidden, may eventually happen. These rights are not static in the system but may be changed by other components.

The distinction between soft deontic constraints (those, that must eventually be fulfilled) and hard deontic constraints (those, that must always be fulfilled) can also be done in a temporal way: Hard deontic constraints *always* have to be fulfilled, whereas soft deontic constraints may be violated at some states but must *eventually* be fulfilled. As a consequence, in our approach deontic aspects are completely reduced to temporal aspects and the added feature of explicite rights, or explicate forbiddances, respectively.

2.4   Combination of Deontic and Epistemic Aspects

We look back to the epistemic aspects of security constraints. There are a lot of works, mainly by Reiter [Rei90], Cuppens and Demolombe [CD96,CD97], in which a database is seen as a collection of knowledge: the database *knows* or *believes in* a set of facts about the real world, users of the database get to know the parts of these facts by querying the database. In association with inference control, the question arises, which knowledge is allowed to be accumulated by a user? What is he unauthorized to know? In [CD97], Cuppens and Demolombe define a modal logic with epistemic and deontic operators and based on Kripke structures. They show, that this logic is axiomatisable. It can be expressed in this logic, which contents of a database a user knows (KB_i), or may know (PKB_i) as well as which contents a user in a particular role may or may not know (PKB_r, FKB_r).

The knowledge, the prohibitions or permissions to know are always related to the facts in the database. Knowledge about the knowledge of other users or the knowledge about the actions performed in the database is however not formulatable. Changes in the database, which lead to changes of knowledge of the users, were also disregarded. It is assumed, that the contents of the database is fixed. In our context we must assume, that users can gain knowledge not only about the data stored in the information system but also about the behaviour of the system itself as well as of the knowledge (or belief) of other users.

3   Our View of an Information System

In this section, we capture our view of an information system in a computational model for the temporal and epistemic logic defined in the next section. The model consists of three aspects: static aspects, dynamic aspects concerning control and dynamic aspects concerning knowledge and belief. In the first two aspects we roughly follow the definitions in [Nie97] of a partial order model for a reactive system. We then extend the model by an epistemic component which is related to that of [Ram96a].

3.1   Static Aspects

We view a distributed information system as consisting of several components, e.g. data objects, users, administrators, and security managers. If the information system is viewed as a piece of software, then data objects lie inside the system whereas users, administrators, etc. appear outside the system. If we provide a representative user-agent inside the system for each outside actor, and a repository-agent for each data object, then we can model all the components of an information system uniformly as agents. We call the set of all k agents of an information system Ag. That is,

$$Ag := \{1, \dots, k\}.$$

The current state of each object, i.e. its data content, is represented by a set of propositions local to the corresponding repository-agents. Similarly, the users' current state, i.e. what data she has read or which rights have been granted to her, is represented by a set of propositions local to the corresponding user-agent. Each agent i of our system is thus associated with a set of local propositions $\mathcal{P}_i$. Let

$$\tilde{\mathcal{P}} := (\mathcal{P}_1, \dots, \mathcal{P}_k)$$

be the distributed set of propositions, such that $\mathcal{P}_i \cap \mathcal{P}_j = \emptyset$ for $i \neq j$. Further, we denote with $\mathcal{P} = \bigcup_{i \in Ag} \mathcal{P}_i$ the set of all propositions.

Seen as a whole system, all these components work concurrently: E.g. while actor a_1 inserts some value x in object o, another actor a_2 could at the same time (concurrently) grant some execute right r to a third actor a_3. However, each single component performs its operations sequentially. Thus, we equip each agent i with her own finite, non-empty set of atomic operations $\mathcal{O}_i$. Let

$$\tilde{\mathcal{O}} = (\mathcal{O}_1, \dots, \mathcal{O}_k)$$

be a distributed set of operations, and $\mathcal{O} = \bigcup_{i \in Ag} \mathcal{O}_i$.

With $ag(op) = \{i \in Ag | op \in \mathcal{O}_i\}$ we refer to the set of agents which are involved in the execution of a joint operation op. An operation $op \in \mathcal{O}_i \cap \mathcal{O}_j$ is called a *synchronization* operation between agents i and j. Two operations op_1 and op_2 are called *independent* iff $ag(op_1) \cap ag(op_2) = \emptyset$.

The informal meaning of these operations will be represented by the changes of the interpretation (see definition 8 below) of local propositions.

Summarizing the discussion above, we see the static declarations of a distributed information system as a concurrent system of sequential agents, where each agent is equipped with a set of operations and a set of local propositions.

Definition 1 (static declarations of an information system). *Let the static declarations of an information system be defined as a tuple $(Ag, \tilde{\mathcal{O}}, \tilde{\mathcal{P}})$, consisting of a set of agents, their distributed set of operations and their distributed set of local propositions.*

3.2   Dynamic Aspects Concerning Control

Given the static declarations of an information system, we now describe the control flow of its possible dynamic behaviours. Below, this control flow is formalized as runs. Each occurrence of an operation within a behaviour is denoted by an event. Thus a possible behaviour is captured by a set of events which should satisfy some requirements in order to represent a reasonable control flow.

- It should be finite or at most denumerably infinite.
- It should be partially ordered according to the relative occurrence of events in time.
- It should distinguish some specific events as the explicit beginning of a behaviour. More precisely, each single event should have only a finite set of predecessors according to the partial order.
- At each event, exactly one operation occurs.
- For each agent i, the set of events this agent is involved in is even totally ordered according to the relative sequence in time.

In order to formalize these requirements we need the following auxiliary definition.

Definition 2 (downward closure). *Let E be a set and $\leq \subseteq E \times E$ be a partial order on E. For $M \subseteq E$,*

$$\downarrow M := \{e \in E \mid \exists e' \in M : e \leq e'\}$$

denotes the downward closure *of M. For $e' \in E$ we write $\downarrow e'$ instead of $\downarrow \{e'\}$.*

Now we can formally define the control flow of a possible behaviour of an information system as a run.

Definition 3 (run). *A run $F_{(Ag,\tilde{O},\tilde{P})} = (E, \leq, \lambda)$ of an information system with the static declarations $(Ag, \tilde{O}, \tilde{P})$ is a partially ordered, labelled set of events E, s.t. the following holds:*

- *E is finite or denumerably infinite.*
- *$\leq$ is a partial order on E.*
- *For all $e \in E$ the downward closure $\downarrow e$ is finite.*
- *$\lambda : E \longrightarrow \mathcal{O}$ is a labelling function yielding the operation $\lambda(e)$ occurred at event e.*
- *For all $i \in Ag$, the reduction of $\leq$ on $E_i := \{e \in E \mid \lambda(e) \in \mathcal{O}_i\}$, i.e. $\leq \cap (E_i \times E_i)$, is a total order.*

We define $\mathcal{F}_{(Ag,\tilde{O},\tilde{P})}$ as the set of all possible runs $F_{(Ag,\tilde{O},\tilde{P})}$ over the same static declarations $(Ag, \tilde{O}, \tilde{P})$. We write F instead of $F_{(Ag,\tilde{O},\tilde{P})}$ and $\mathcal{F}$ instead of $\mathcal{F}_{(Ag,\tilde{O},\tilde{P})}$ where this does not lead to misunderstandings.

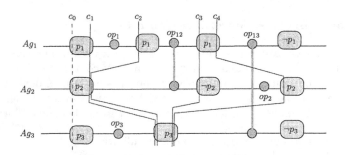

Fig. 1. Example of a run and local views

See figure 1 for an illustration of a possible run: The example run is executed by three agents Ag_1, Ag_2 and Ag_3. Each agent is represented by her own (horizontal) time line. The agents perform operations sequentially: Ag_1 performs the sequence of operations op_1, op_{12}, op_{13}. Agent Ag_2 performs the sequence of operations op_{12}, op_2. And agent Ag_3 performs the sequence of operations op_3, op_{13}. Each event, i.e. occurrence of an operation, is represented by a vertical bar with circles for the participating agents. Though the events for each agent are totally ordered, the set of all events of the system are only partially ordered: for example, operations op_1 and op_3 are performed concurrently, no order is given between them.

Each agent only has a local view on the system and a local configuration. The local configuration of a run F for an agent i consists of the downward closure of all events which agent i has already performed up to a particular local time point. Formally, we define a configuration as follows:

Definition 4 ((local) configuration; i-view). *Let $F = (E, \leq, \lambda)$ be a run. A configuration c of F is a finite, downward closed set of events (i.e. $c \subseteq E$ and $\downarrow c = c$). Let $\mathcal{C}_F$ denote the set of all configurations of run F.*

The localization of a configuration $\downarrow^i c := \downarrow (c \cap E_i)$ is called the i-view of configuration c, i.e. the i-view is the least configuration, which coincides with c on all i-events.

Two configurations c, c' of F are said to be i-equivalent ($c \equiv_i c'$), iff $\downarrow^i c = \downarrow^i c'$.

On configurations of F we define a successor relation so that $c \xrightarrow{op} c'$ iff $c' = c \cup \{e\}$ for some $e \in E$ with $\lambda(e) = op$.

Two configurations of the same run that differ from each other only through events in which agent i does not participate are equivalent from the point of view of agent i. We call such configurations i-equivalent.

Consider again figure 1. A *configuration* is indicated by a (more or less vertical) line that crosses each agent's time line at a box, and these boxes contain the agents' *current state* (where, by default, local propositions not shown are assumed to be interpreted by false). The configurations c_1, c_2, c_3 and c_4 are Ag_3-equivalent, which is illustrated by the solid lines through agent Ag_3's second box. These configurations differ only through operations performed by Ag_1 or Ag_2 but not Ag_3.

In the rest of this paper we will often talk about configurations of different runs. In these cases we have to name both the configuration and the run to which it belongs.

Definition 5 (situation). *We call (F, c) a situation, when $F = (E, \leq, \lambda)$ is a run and $c \in \mathcal{C}_F$ is a configuration of this run.*

Note, that our definition of situation mainly describes progess in time and leaves propositions still uninterpreted.

3.3   Dynamic Aspects Concerning Knowledge and Belief

In order to capture an agent's knowledge and belief, we need a more general notion of "indistinguishability" than the one introduced above. At all times, an agent only has a partial view on the system, e.g. she can notice the behaviour of other agents only via synchronization operations, other operations are independent of and invisible for agent i. Thus, the agent cannot distinguish the actual configuration from other configurations, that differ only in operations local to other agents. The agent is not even aware of the actual run, she for example cannot distinguish between the actual configuration of the actual run and a configuration of a different run, in which she has acquired the same knowledge. Thus, for any agent some configurations in various runs may be indistinguishable of the actual current situation. Later, in chapter 4.2 we define, that *an agent knows a fact* if this fact is true in all situations indistinguishable for this agent. (In terms of standard modal logics such situations are called *accessible*, the corresponding relation is called *accessibility relation*.) We require that the accessibility relation for knowledge is an equivalence relation[1]:

- reflexive: the agent considers the actual world to be possible,
- transitive: the agent knows, what she knows (positive introspection),
- Euclidean: the agent knows, what she does not know (negative introspection).

We also require, that in a run two different configurations must be distinguishable for at least one agent.

Further, an agent i considers all configurations to be possible, that are i-equivalent to the actual configuration. The difference between two i-equivalent configurations can only be a set of operations executed by an agent other than i and thus invisible to agent i.

The fourth requirement ensures, that the "enabling" of an operation as well as its effect, that is, the changes in the truth values of local propositions (see definition 8 below), are only dependent on the knowledge of the acting agents and not on any outside factor. Additionally, the execution of an operation by a group of agents has no effect on the knowledge of other agents not participating in this operation.

Definition 6 (accessibility relations (for knowledge)). *Let $\mathcal{A} \subseteq \mathcal{F}$ be a set of runs. For each agent $i \in Ag$ we define her accessibility relation for knowledge*

$$R_i^K \subseteq \{((F,c),(F',c')) \mid F, F' \in \mathcal{A}, c \in \mathcal{C}_F, c' \in \mathcal{C}_{F'}\}$$

such that the following properties hold:

1. *R_i^K is an equivalence relation.*
2. *If $((F,c),(F,c')) \in R_i^K$ forall $i \in Ag$ then $c = c'$.*
3. *If for $c, c' \in \mathcal{C}_F$ we have that $c \equiv_i c'$ then $((F,c),(F,c')) \in R_i^K$.*
4. *If for $c_1, c_2 \in \mathcal{C}_F$ with $F \in \mathcal{A}$ holds, that $c_1 \xrightarrow{op} c_2$ and it exists $c_1' \in \mathcal{C}_{F'}$ with $F' \in \mathcal{A}$ such that for every $i \in ag(op)$ it holds that $((F,c_1),(F',c_1')) \in R_i^K$ then there exists $c_2' \in \mathcal{C}_{F'}$ such that $c_1' \xrightarrow{op} c_2'$ and $((F,c_2),(F',c_2')) \in R_i^K$ for every $i \in ag(op)$.*

---

[1] The standard definition of an equivalence relation requires reflexivity, transitivity and symmetry. However, it is easy to prove, that a relation is reflexive, transitive and Euclidian iff it is reflexive, symmetric and transitive. See e.g. [FHMV96].

Sometimes it might be more appropriate not to talk about *knowledge* but about *belief* of an agent. Concerning belief, an agent might refuse to consider the actual situation as possible. Thus we want to weaken the requirements for her accessibility relation for belief such that it does not need to be reflexive:

- serial: the belief of the agent may not be contradictory,
- transitive: the agent believes, what she believes (positive introspection),
- euclidean: the agent believes, what she does not believe (negative introspection).

Further we require, that each agent at least believes, what she knows, such that the accessibility relation for belief must be a subset of the accessibility relation for knowledge.

These observations lead to the following definition:

Definition 7 (accessibility relations (for belief)). *Let $\mathcal{A} \subseteq \mathcal{F}$ be a set of runs.*

For each agent $i \in Ag$ we define her accessibility relation for belief

$$R_i^B \subseteq \{((F,c),(F',c')) \mid F,F' \in \mathcal{A}, c \in \mathcal{C}_F, c' \in \mathcal{C}_{F'}\}$$

such that the following properties hold:

1. $-$ R_i^B *is serial, i.e. for each situation (F,c) it exists a situation (F',c') such that $((F,c),(F',c')) \in R_i^B$.*
   - R_i^B *is transitive, i.e. if $((F,c),(F',c')) \in R_i^B$ and $((F',c'),(F'',c'')) \in R_i^B$ then $((F,c),(F'',c'')) \in R_i^B$.*
   - R_i^B *is euclidean, i.e. if $((F,c),(F',c')) \in R_i^B$ and $((F,c),(F'',c'')) \in R_i^B$ then $((F',c'),(F'',c'')) \in R_i^B$.*
2. *Each relation R_i^B is a subset of the corresponding accessibility relation for knowledge R_i^K: $R_i^B \subseteq R_i^K$.*

All agents are equipped with a set of local propositions. Agents, that represent data objects in an information system, can be seen as repository-agents: They are equipped with a set of local propositions the interpretation of which reflects their current content. User-agents are equipped with a set of local propositions the interpretation of which is something like "the information this user currently holds", e.g. as a printout on her desk.

More formally, treating the notion of *current* by situations, we consider an interpretation as mapping each pair (situation, local proposition) to a truth-value.

The content (or state) of repository-agents and the information user-agents have, can be changed via operations, e.g. if a user inserts some fact into an object, then after this operation the local proposition that represents this fact for her must be true. If a user-agent reads some fact A from the information system, then after the read operation, the local proposition that represents this fact for her must be true. The informal meaning of operations will be represented by the changes of the interpretation of local propositions:

Definition 8 (interpretation). *An* interpretation *of a set of runs* $\mathcal{A} \subseteq \mathcal{F}$ *is a mapping* $\mathcal{I} : \{(F, c) \mid F \in \mathcal{A}, c \in \mathcal{C}_F\} \times \mathcal{P} \longrightarrow \{\top, \bot\}$, *such that if* $p \in \mathcal{P}_i$ *and* $((F, c), (F', c')) \in R_i^K$ *then* $\mathcal{I}((F, c), p) = \mathcal{I}((F', c'), p)$.

In definition 8 we have the requirement, that in two situations, which are indistinguishable for an agent i, the interpretation of all propositions local to this agent must be the same. In terms of our information system framework this means, that each user-agent is aware of all information she holds (on her desk) and all repository-agents are aware of their content.

3.4   Summarized View

In section 3.1 we have defined the static part of an information system to consist of a set of agents, their distributed set of propositions and their distributed set of operations. In section 3.2 we defined partially ordered runs as an underlying structure for dynamic aspects. Finally, in section 3.3 we introduced a notion of indistinguishable situations. These indistinguishable situations together with interpretations are the basis for internal states of agents, their knowledge and their belief.

This summarized view can be represented like in the example of figure 2, which extends figure 1.

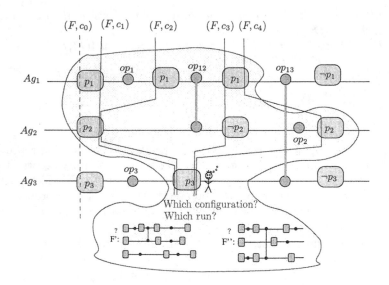

Fig. 2. Illustration of an example model

Assume the same temporal model as in figure 1. The truth-value of a proposition local to an agent can change when this agent performs an operation: for example, before agent Ag_2 performs operation op_2, the value of proposition p_2 is $false$. After operation op_2 has happened, the value of proposition p_2 is $true$. An agent i knows a fact if the fact is true in all situations agent i considers to be possible. Let for example the actual situation be (F, c_2). Agent Ag_3 considers all Ag_3-equivalent configurations as possible. Further, she considers some configurations of runs F', F'' as possible as well. We now combine all these aspects and define a $model$ $\mathcal{M} = (\mathcal{A}, \mathcal{R}^K, \mathcal{R}^B, \mathcal{I})$ of an information system as a tuple consisting of

- a set $\mathcal{A}$ of runs over a fixed static part of an information system $(Ag, \tilde{\mathcal{O}}, \tilde{\mathcal{P}})$,
- a family $\mathcal{R}^K$ containing an accessibility relation for knowledge R_i^K for each agent $i \in Ag$,
- a family $\mathcal{R}^B$ containing an accessibility relation for belief R_i^B for each agent $i \in Ag$, and
- an interpretation $\mathcal{I}$ of $\mathcal{A}$.

4 A Temporal and Epistemic Logic

4.1 Syntax

Following [Nie97], the key idea of our language is that the formulae "look" at the configurations from a local point of view. Some formulae look at the configuration of a run from the point of view of a single agent (e.g. in the case of local propositions or epistemic operators), others may involve a joint look from several agents (e.g. after a joint action of these agents). This idea is reflected in the syntax by a $family$ of sets of formulae Φ_A where $A \subseteq Ag$ is a set of agents. Formulae $\phi \in \Phi_A$ are called of $type$ A: $type(\phi) = A$. Note, that every formula has exactly one type, i.e. for two sets of agents $A, B \subseteq Ag$ with $A \neq B$ we have $\Phi_A \cap \Phi_B = \emptyset$.

As we will see in the following definition, these types are used for the syntactic restriction of the construction of temporal formulae.

Definition 9 (syntax of typed formulae). *Given a fixed static part of an information system $(Ag, \tilde{\mathcal{O}}, \tilde{\mathcal{P}})$ as defined in section 3.*
We define a family of formulae, such that for each set $A \subseteq Ag$ the set Φ_A is the least set satisfying the following properties:

propositions

$$\bot, \top \in \Phi_\emptyset$$

for $p \in \mathcal{P}_i$ we have $p \in \Phi_{\{i\}}$

logical connectives

$$\phi \in \Phi_A \ implies \ \neg\phi \in \Phi_A$$

$$\phi \in \Phi_A, \psi \in \Phi_B \ implies \ \phi \vee \psi \in \Phi_{A \cup B}$$

temporal operators

$$i \in ag(op), A \subseteq ag(op), \phi \in \Phi_A \ implies \ \langle op \rangle_i \phi \in \Phi_{\{i\}}$$

$$\phi, \psi \in \Phi_A, A \subseteq \{i\} \ implies \ \phi \, \mathcal{U}_i \psi \in \Phi_{\{i\}}$$

epistemic operators

$$\mathcal{K}_i \phi \in \Phi_{\{i\}}$$

$$\mathcal{B}_i \phi \in \Phi_{\{i\}}$$

Note, that in the following we will write Φ_i instead of $\Phi_{\{i\}}$ where appropriate. The intuitive meaning of a temporal formula $\langle op \rangle_i \phi$ is that from agent i's point of view the next operation is op and after the execution of op, ϕ holds. There is a syntactic restriction on the construction of such temporal formulae: within a formula of form $\langle op \rangle_i \phi$ (next-modality) the subformula ϕ may only refer to agents that participate in operation op. Also, agent i has to participate in operation op, i.e. changing of the point of view of a temporal next-formula is only possible via a common action of the "old" (before the execution of the operation) and the "new" point of view (after execution of the operation). Consider for example a set of agents $Ag = \{ag_1, ag_2\}$, with $\mathcal{P}_1 = \{p_1\}, \mathcal{P}_2 = \{p_2\}$ and with $\mathcal{O}_1 = \{o_1, o_{12}\}, \mathcal{O}_2 = \{o_{12}\}$. Then the formula $\langle o_{12} \rangle_1 p_2$ is well-formed whereas the formula $\langle o_1 \rangle_1 p_2$ is not. In the first formula the view is changed from agent ag_1 to agent ag_2 via a common operation o_{12}, whereas in the latter formula the view is changed from agent ag_1 to agent ag_2 via operation o_1. However, o_1 is only local to agent ag_1 and after the execution of o_1 the local configuration of ag_2 could result from several configurations differing for agent ag_1. Without this restriction, these temporal formulae could lead to "uncontrolled jumps into the past". (Further explanations are given in [Nie97]).

A temporal formula of kind $\phi \mathcal{U}_i \psi$ intuitively means that either ψ holds already in the *current* configuration or the subformula ϕ holds in the *current* configuration and in all *following* configurations *until* at some point of time ψ holds. (In this context, *until* can be seen as a *strong* until: At some point in the future ψ will hold.) The intuitive meaning of this until operator is quite standard, however note, that in our logic we have imposed the syntactic restriction, that both the formulae ϕ and ψ have to be local to agent i.

The epistemic operators are quite standard again. The intuitive meaning of $\mathcal{K}_i \phi$ is that *agent i knows ϕ*, i.e. in all situations, that agent i considers to be possible ϕ holds, in particular we have that ϕ holds in the current configuration. The intuitive meaning of $\mathcal{B}_i \phi$ is similar, i.e. again ϕ holds in all situations agent i considers to be possible with respect to belief. However, in this case, i does not need to consider the actual configuration to be possible.

4.2   Semantics

We will now give the formal semantics of the logic introduced above.

Definition 10 (semantics). *Given a model* $\mathcal{M} = (\mathcal{A}, \mathcal{R}^K, \mathcal{R}^B, \mathcal{I})$, *a run* $F \in \mathcal{A}$ *and a configuration* $c \in \mathcal{C}_F$ *of run* F. *The semantics of a formula* ϕ *is then inductively defined by:*

$$(F, c) \models_{\mathcal{M}} \top$$

$$(F, c) \not\models_{\mathcal{M}} \bot$$

$$(F, c) \models_{\mathcal{M}} p \;\; :iff\; \mathcal{I}((F, c), p) = true$$

$$(F, c) \models_{\mathcal{M}} \neg\phi \;\; :iff\; not\; (F, c) \models_{\mathcal{M}} \phi$$

$$(F, c) \models_{\mathcal{M}} \phi \vee \psi \;\; :iff\; (F, c) \models_{\mathcal{M}} \phi \; or \; (F, c) \models_{\mathcal{M}} \psi$$

$$(F, c) \models_{\mathcal{M}} \langle op \rangle_i \phi \;\; :iff\; there\; exists\; c', r \in \mathcal{C}_F, \; such\; that$$
$$c \equiv_i c' \; and \; c' \xrightarrow{op} r \; and \; (F, r) \models_{\mathcal{M}} \phi$$

$$(F, c) \models_{\mathcal{M}} \phi \, \mathcal{U}_i \psi \;\; :iff\; there\; exists\; c' \in \mathcal{C}_F \; such\; that$$
$$\downarrow^i c \subseteq \downarrow^i c' \; and \; (F, \downarrow^i c') \models_{\mathcal{M}} \psi \; and \; forall$$
$$c'' \in \mathcal{C}_F \; with \; \downarrow^i c \subseteq \downarrow^i c'' \subset \downarrow^i c'$$
$$it\; holds\; that\; (F, \downarrow^i c'') \models_{\mathcal{M}} \phi$$

$$(F, c) \models_{\mathcal{M}} \mathcal{K}_i \phi \;\; :iff\; forall\; (F', c') \; with \; ((F, c), (F', c')) \in R_i^K$$
$$it\; holds\; that\; (F', c') \models_{\mathcal{M}} \phi$$

$$(F, c) \models_{\mathcal{M}} \mathcal{B}_i \phi \;\; :iff\; forall\; (F', c') \; with \; ((F, c), (F', c')) \in R_i^B$$
$$it\; holds\; that\; (F', c') \models_{\mathcal{M}} \phi$$

Note, that the semantics of logical connectives and of local propositions is dependent on a single configuration of a single run.

We have already stated the intuitive semantics of the temporal formulae. The semantics of temporal formulae involves several configurations, but is, however, dependent only on a single run.

Further note, that formulae of form $\langle a \rangle_i \phi$ are local to a single agent. $\langle op \rangle_i \phi$ does *not* mean, that the next operation of the run is op, but the next operation of agent i. It might well be, that some other agent which is involved in operation op performs operations other than op first.

The semantics of the knowledge/belief operators is similar to the one given in [Ram96b]. An agent i *believes* that a formula ϕ holds, $\mathcal{B}_i \phi$, if ϕ is true in all situations which i cannot distinguish from the actual situation. Since R_i^K are equivalence relations, we have that each situation that satisfies $\mathcal{K}_i \phi$ also satisfies ϕ. This is, however, not the case for the belief operator: The relations R_i^B do not necessarily need to be reflexive. Thus, from $\mathcal{B}_i \phi$ we cannot conclude ϕ. The semantics of knowledge or belief operators is dependent on several configurations and several runs.

In the example in section 5 we will use some derived operators as short cuts to make the formulae more easily understandable. For better readability we will use the following abbreviations:

- $[a]_i \phi \equiv \neg \langle a \rangle_i \neg \phi$

  The operator $[\cdot]_i$ is the dual operator to $\langle \cdot \rangle_i$. Intuitively, $[a]_i \phi$ means that *if* a is the next operation from agent i's point of view, then after the execution of a the formula ϕ holds.
- $\phi \wedge \psi \equiv \neg(\neg \phi \vee \neg \psi)$
- $\Diamond_i \phi \equiv \top \, \mathcal{U}_i \phi$

  Intuitively, $\Diamond_i \phi$ means that eventually from agent i's point of view ϕ holds.
- $\Box_i \phi \equiv \neg \Diamond_i \neg \phi$

  The box operator $\Box_i \phi$ is the dual to the diamond operator above. It means, that from agent i's point of view, ϕ holds always in the future.
- $\phi \Rightarrow \psi \equiv \neg \phi \vee \psi$

Remark: Note, that $\langle a \rangle_i \phi \Rightarrow [a]_i \phi$, i.e.,for all situations (F, c) of a model $\mathcal{M} = (\mathcal{A}, \mathcal{R}^K, \mathcal{R}^B, \mathcal{I})$, for all agents $i \in Ag$, for all operations $a \in \mathcal{O}_i$ and for all formulae $\phi \in \Phi_A$ with $A \subseteq ag(a)$ the following holds:
If $(F, c) \models_{\mathcal{M}} \langle a \rangle_i \phi$ then $(F, c) \models_{\mathcal{M}} [a]_i \phi$.

Proofsketch:

$\quad (F, c) \models_{\mathcal{M}} \langle a \rangle_i \phi$

$\Longrightarrow$ (* by definition 10 *)

$\quad$ there exists $c', r \in \mathcal{C}_F$, such that $c \equiv_i c'$ and $c' \xrightarrow{op} r$ and $(F, r) \models_{\mathcal{M}} \phi$

$\Longrightarrow$ (*Definition 3 ensures that the events of each agent are totally ordered*)

$\quad$ for all $c', r \in \mathcal{C}_F$, such that $c \equiv_i c'$ we have $c' \xrightarrow{op} r$ and $(F, r) \models_{\mathcal{M}} \phi$

$\Longrightarrow$ not exists $c', r \in \mathcal{C}_F$, such that $c \equiv_i c'$ and $c' \xrightarrow{op} r$ and $(F, r) \models_{\mathcal{M}} \neg \phi$

$\Longrightarrow (F, c) \models_{\mathcal{M}} \neg \langle a \rangle_i \neg \phi$

$\Longrightarrow (F, c) \models_{\mathcal{M}} [a]_i \phi$ $\hfill \Box$

5   An Example

By giving an a little complex example we now show how information systems constraints can be specified in our framework. First, we specify the computational model of our example and define a key constraint as semantic constraint on this model. Next, we define security constraints on this model and show how they collide with the semantic constraint.

5.1   Basic Model

Assume an information system with three agents: Two user-agents S and U and one repository-agent O. We then have $Ag = \{S, O, U\}$.

Next, we define the sets of local propositions for each agent. We use propositions with a syntactic structure which hopefully helps to understand the example more easily. The distributed set of propositions $\tilde{\mathcal{P}} = (\mathcal{P}_S, \mathcal{P}_O, \mathcal{P}_U)$ is a tuple of sets of propositions with the following structure:

proposition ::= **fact @ agent**
agent   ::= $S|O|U$
fact   ::= **key_value**   A proposition consists of a fact plus the name
key   ::= $k_1|\ldots|k_n$
value   ::= $v_1|\ldots|v_m$

of the agent to whose set of propositions it belongs. This supplement ensures, that all sets of propositions are disjunct. Intuitively, it tells us, where (at which location) the fact is stored. A fact consists of a key and a value. The set of propositions for agent i is then defined as follows: $\mathcal{P}_i := \{\text{fact@agent} \mid \text{agent} = i\}$

The distributed set of operations $\tilde{\mathcal{O}} = (\mathcal{O}_S, \mathcal{O}_O, \mathcal{O}_U)$ is a tuple of sets of operations with the following structure:

operation ::= **actor_object::opcode(parameter)**
actor   ::= $S|U$
object   ::= O
opcode   ::= read | insert | delete
parameter ::= **fact**
fact   ::= **key_value**
key   ::= $k_1|\ldots|k_n$
value   ::= $v_1|\ldots|v_m$

The set of operations for agent i then consists of exactly those operations, in which agent i takes part:

$$\mathcal{O}_S := \{actor\_object :: opcode(parameter)|actor = S\}$$
$$\mathcal{O}_O := \{actor\_object :: opcode(parameter)|object = O\}$$
$$\mathcal{O}_U := \{actor\_object :: opcode(parameter)|actor = U\}$$

Now we fix the static part of the information system as $(Ag, \tilde{\mathcal{O}}, \tilde{\mathcal{P}})$.

Since operations are only syntactic entities without any semantics, we define their meaning over the interpretation of relevant propositions. For every operation we have to specify a set of preconditions and postconditions. And for every proposition we have to specify the conditions under which it may or may not change its value. The latter part is known as the "frame problem". As already explained before, local propositions can only be changed through operations of the agent they are affiliated with. Thus we have to specify the interactions between operations and propositions only locally for each agent, which reduces the number of formulae needed significantly.

We look at a local proposition $A@i$ as *Agent i "possesses" fact A* (if agent i is a user-agent, she could e.g. have a printout of A on her desk) or *Fact A is stored at agent i* (if agent i is a repository-agent, A could be literally stored in the data object). The intuitive meaning of an operation $S\_O :: insert(A)$ is that agent S inserts a fact A into the repository-agent O. Thus, no matter which propositions (local to repository-agent O) hold before the execution of this operation, afterwards we want the proposition $A@O$ to hold. This holds analogously for agent U as well and leads to the following requirements for each fact A:

$$\Box_O([S\_O :: insert(A)]_O \ A@O) \quad (1) \qquad \Box_O([U\_O :: insert(A)]_O \ A@O) \quad (2)$$

The requirements for delete-operations $S\_O :: delete(A)$ and $U\_O :: delete(A)$ are very similar: No matter, which propositions hold before the execution of the delete-operations, afterwards we want the proposition $A@O$ not to be satisfied. This leads to the following requirements:

$$\Box_O([S\_O :: delete(A)]_O \ \neg A@O) \quad (3) \qquad \Box_O([U\_O :: delete(A)]_O \ \neg A@O) \quad (4)$$

The intuitive meaning of the read-operations is somewhat more complicated. Read-operations change the knowledge and the belief of the reading user. What exactly happens, when user S reads (queries) the fact A which is stored in repository-agent O?

First of all, she (user S) gets an answer, which will be represented by a proposition $A@S$ local to agent S. Secondly, afterwards she *knows*, that she has got this answer, which will be expressed by the formula $K_S \ A@S$. However, though she has got this answer and she knows, that she has got this answer, she does not know, whether this fact is actually stored in the repository-agent. It could for example already be modified by another agent or the system could have made a mistake, or it could have even lied. However, in our context we assume, that the system answers correctly, it does not lie, nor does it make a mistake. Under this assumption, the agent *believes*, that this fact is stored in the repository-agent.

$$\Box_O \ (A@O \Rightarrow [S\_O :: read(A)]_O \ (A@S \ \wedge \ K_S A@S \ \wedge \ B_S A@O)) \quad (5)$$

If the fact A is not stored in repository-agent O, then the answer will be $\neg A@S$ and the requirements will be corresponding:

$$\Box_O \ (\neg A@O \Rightarrow [S\_O :: read(A)]_O \ (\neg A@S \ \wedge \ K_S \neg A@S \ \wedge \ B_S \neg A@O)) \quad (6)$$

For agent U the constraints hold accordingly.

It might not be obvious, why we consider the read-operations to be the next operation from the point of view of the repository-agent. One might expect the user-agent's view here instead. However, the precondition is that the proposition $A@O$, which is local to the repository-agent O, is satisfied. To prevent that the proposition $A@O$ changes its value before the read-operation is executed we require, that the operation $S\_O :: read(A)$ is the next operation from agent O's point of view. This problem is illustrated in figure 3.

Suppose, we have a run F as sketched in figure 3. Note that in this figure we left out the accessibility relations for knowledge and belief.

Suppose now, $(F, c_0) \models A@O$. We then have $(F, c_0) \models \langle S\_O :: read(A) \rangle_S \top$ but $(F, c_0) \not\models \langle S\_O :: read(A) \rangle_O \top$. Before the read operation $S\_O :: read(A)$ actually

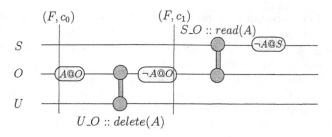

Fig. 3. Local view of formulae

happens, the repository-agent O performs a delete-operation $U\_O :: delete(A)$ together with agent U. This leads to $(F, c_1) \models \neg A@O$. So, after the read operation we want agent S to have the information $\neg A@S$ "on her desk" and to *know*, that she has this information "on her desk". Only if the read-operation $S\_O :: read(A)$ was the next operation from c_0 from the point of view of the repository-agent, we could be sure, that agent S reads A and thus after the read-operation knows $A@S$ and believes $A@O$.

We must further ensure, that no other operations but the corresponding read, write or insert operations change the value of the propositions:

- If $A@O$ does not hold, only an $insert(A)$ operation may change the value of $A@O$:

$$\Box_O \, (\neg A@O \Rightarrow (\neg A@O \; \mathcal{U}_O \; (\neg A@O \wedge \langle S\_O :: insert(A) \rangle_O \top)$$
$$\vee \, (\neg A@O \wedge \langle U\_O :: insert(A) \rangle_O \top)) \qquad (7)$$
$$\vee \, (\Box_O \neg A@O))$$

- If $A@O$ holds, then only deletion of $A@O$ may change the value of $A@O$:

$$\Box_O \, (A@O \Rightarrow (A@O \; \mathcal{U}_O \; (A@O \wedge \langle S\_O :: delete(A) \rangle_O \top)$$
$$\vee \, (A@O \wedge \langle U\_O :: delete(A) \rangle_O \top)) \qquad (8)$$
$$\vee \, (\Box_O A@O))$$

5.2 Semantic Constraints

A very common *semantic constraint* is that *keys* are unique in a relation. Our propositions have a syntactic structure which helps us to define the key constraint in our framework.

At the first sight, the key constraint only needs to be defined for the repository-agent O: Data stored in the repository-agent must comply with the key constraint.

What about the user-agents? Suppose, agent U first reads k_v_1 which means that the proposition $k\_v_1@U$ holds after the read operation. Later she reads k_v_2, so that $k\_v_2@U$ holds. Is it conceivable, that $k\_v_1@U$ and $k\_v_2@U$ both hold in the same situation and thus agent U also "believes" facts $k\_v_1@O$ and $k\_v_2@O$ at the same time? If agent U is aware of the key constraint and she first reads k_v_1 and then k_v_2, she will certainly "throw her local printout of $k\_v_1@U$ away" after she has read k_v_2 (which means, $k\_v_1@U$ will be false after the second read-operation). She will then not believe $k\_v_1@O$ and $k\_v_2@O$ to hold in the same situation either. These observations are not compatible with the basic model of section 5.1.

However, there might be scenarios, in which U is not aware of the key constraint. In these scenarios there is no reason why the agent should "throw the previous answer of the system away", meaning why $k\_v_1@U$ should not hold any more (as long as there is no other constraint which contradicts against the existence of both the propositions).

In this example we assume, that all agents are aware of the key constraint, which implies that for no agent i two different propositions $k\_v_1@i$ and $k\_v_2@i$ are satisfied at the same configuration:

$$\bigwedge_{l\in\{S,O,U\}} \quad \bigwedge_{k\in\{k_1,\dots,k_n\}} \quad \bigwedge_{v_i,v_j\in\{v_1,\dots,v_m\},v_i\neq v_j} \quad \Box_l\,(k\_v_i@l \Rightarrow \neg k\_v_j@l) \tag{9}$$

Note, that the big conjunctions are on a syntactic level.

Assume now a situation in which first user S inserts a fact k_v_i into repository-agent O and then user U inserts a fact k_v_j with $v_i \neq v_j$:

$$\langle S\_O :: insert(k\_v_i)\rangle_O \,\langle U\_O :: insert(k\_v_j)\rangle_O \top \tag{10}$$

This sequence violates the above stated key constraint:

Suppose, a situation (F,c) of a model $\mathcal{M}$ satisfies statement (10), i.e.

$(F,c) \models_{\mathcal{M}} \langle S\_O :: insert(k\_v_i)\rangle_O \,\langle U\_O :: insert(k\_v_j)\rangle_O \top$

$\Longrightarrow (*$ because of equation (1) and the remark in chapter $(4.2)*)$

$(F,c) \models_{\mathcal{M}} \langle S\_O :: insert(k\_v_i)\rangle_O \,(k\_v_i@O \wedge \langle U\_O :: insert(k\_v_j)\rangle_O \top)$

$\Longrightarrow (*$ because of equation (2) and the remark in chapter $(4.2)*)$

$(F,c) \models_{\mathcal{M}} \langle S\_O :: insert(k\_v_i)\rangle_O \,(k\_v_i@O \wedge \langle U\_O :: insert(k\_v_j)\rangle_O k\_v_j@O)$

$\Longrightarrow (*$ because of equation $(7)*)$

$(F,c) \models_{\mathcal{M}} \langle S\_O :: insert(k\_v_i)\rangle_O \,\langle U\_O :: insert(k\_v_j)\rangle_O (k\_v_j@O \wedge k\_v_i@O)$

$\Longrightarrow (F,c) \not\models \Box_O(k\_v_i@O \Rightarrow \neg K\_v_i@O)$

Violation to (9)

One possibility for agent U to evade this violation is to first delete the fact k_v_i from repository-agent O and then insert k_v_j into O. More generally, the basic model could be appropriately adapted in order to ensure the key constraint, say by introducing a concept of "transaction".

5.3 Security Constraints

One area in which semantic constraints and security constraints collide is in mandatory access control. In mandatory access control, we assume a set of security labels on which a partial order is defined. Each user of an information system is associated with one of these labels.

We continue our example of section 5.1 by defining the security labels *secret* and *unclassified* and the lattice *secret* > *unclassified* on these labels. Agent U is associated with security label *unclassified* and agent S with security label *secret*. Reconsider now the situation at the end of the previous chapter. First the secret agent S inserts a fact k_v_i into the repository-agent O. Before the unclassified agent U can insert the fact k_v_j, she first has to delete the fact k_v_i from repository-agent O. In practice, in a secure information system we do not want users associated with a lower security label to modify (delete) data that was inserted by users with a higher security label. The direct implication of this would be for the information system to simply reject agent U's request to insert k_v_j (but this would reveal the information, that there already exists a tuple with this key, which this user is not able to view).

Denning et al ([LDS+88]) and Jajodia et al([JS90]) developed security models in which they attacked this problem through the method of polyinstantiation: It allows both tuples to simultaneously exist in the information system. Unfortunately the models described in [JS90] and [LDS+88] suffer from semantical ambiguity due to the lack of a clear semantics. In [WSQ94] Winslett et al proposed a solution by defining a belief based semantics for secure databases. We will now follow this approach in a slightly modified way and show how under the requirement of mandatory access control an information system can be modelled in our framework. Note, that the meaning of "belief" in the context of [WSQ94] is not the same as the meaning of belief in our framework.

We assume, that repository-agent O can be polyinstantiated, i.e. secret as well as unclassified data can be stored in the repository-agent at the same time. Accordingly, we require that the key constraint holds within each single label only. To do so we need to adjust our model defined in the previous chapter.

Whenever agent S inserts data into the repository-agent O, this data is classified as *secret*, and whenever agent U inserts data into the repository-agent O this data is classified as *unclassified*. Each agent may read data classified at her own label or a label lower than her own.

For this purpose, we slightly modify the structure of the propositions:

proposition ::= **fact**$^{\textbf{label}}$ @ **agent**

agent	::= $S\|O\|U$
label	::= $s\|u$
fact	::= **key_value**
key	::= $k_1\|\ldots\|k_n$
value	::= $v_1\|\ldots\|v_m$

The sets of propositions are then defined as:

$$\mathcal{P}'_S := \{\text{fact}^{label}@\text{agent} \mid \text{agent} = S\}$$
$$\mathcal{P}'_O := \{\text{fact}^{label}@\text{agent} \mid \text{agent} = O\}$$
$$\mathcal{P}'_U := \{\text{fact}^{label}@\text{agent} \mid \text{agent} = U \text{ and label} = u\}$$

The distributed set of propositions is defined as $\tilde{\mathcal{P}}' = (\mathcal{P}'_S, \mathcal{P}'_O, \mathcal{P}'_U)$.

Agent S may now read data stored at label s as well as data stored at label u. We have to adjust the distributed set of operations as well by adding the security label to the fact:

operation ::= **actor_object::opcode(parameter)**
actor   ::= $U|S$
object   ::= O
opcode   ::= read | insert | delete
parameter ::= **fact**^{label}
fact   ::= **key_value**
label   ::= $s|u$
key   ::= $k_1|\ldots|k_n$
value   ::= $v_1|\ldots|v_m$

$\tilde{\mathcal{O}}' = (\mathcal{O}'_S, \mathcal{O}'_U, \mathcal{O}'_O)$ then is the new distributed set of operations, with:

$$\mathcal{O}'_S := \quad \{actor\_object :: opcode(fact^{label}) | actor = S \text{ and}$$
$$(opcode \in \{insert, delete\} \text{ and } label = s)$$
$$\text{or } (opcode = read \text{ and } label \in \{s, u\}\}$$
$$\mathcal{O}'_O := \quad \{actor\_object :: opcode(fact^{label}) | object = O\}$$
$$\mathcal{O}'_U := \{actor\_object :: opcode(fact^l) | actor = U \text{ and } label = u\}$$

Thus we now fix the static part of our model as $(Ag, \tilde{\mathcal{O}}', \tilde{\mathcal{P}}')$.

The set of restrictions that we defined in section 5.1 to give a semantics to the operations must be adapted as well:

When agent S inserts some fact A into the repository-agent O then this fact should be stored in O with security label s. Analogously, when agent U inserts some fact A, then this fact should be stored with security label u:

$$\Box_O([S\_O :: insert(A^s)]_O A^s@O) \quad (11) \quad \Box_O([U\_O :: insert(A^u)]_O A^u@O) \quad (12)$$

The requirements for the delete-operations change accordingly.

The equations (7) and (8) must be adjusted as well. We now have the additional constraint, that only agent S may insert and delete secret facts and agent U may only insert and delete unclassified facts: For insert we have the requirements

$$\Box_O (\neg A^s@O \Rightarrow (\neg A^s@O \, \mathcal{U}_O \, (\neg A^s@O \wedge \langle S\_O :: insert(A^s) \rangle_O \top))$$
$$\vee (\Box_O \neg A^s@O)) \tag{13}$$

and

$$\Box_O \left(\neg A^u @O \Rightarrow (\neg A^u @O \, \mathcal{U}_O \, (\neg A^u @O \wedge \langle U\_O :: insert(A^u) \rangle_O \top)) \right. \tag{14}$$
$$\left. (\Box_O \neg A^u @O) \right)$$

Accordingly, for delete we have the requirements

$$\Box_O \left(A^s @O \Rightarrow (A^s @O \, \mathcal{U}_O \, (A^s @O \wedge \langle S\_O :: delete(A^s) \rangle_O \top)) \right. \tag{15}$$
$$\left. \vee \, (\Box_O A^s @O) \right)$$

and

$$\Box_O \left(A^u @O \Rightarrow (A^u @O \, \mathcal{U}_O \, (A^u @O \wedge \langle U\_O :: delete(A^u) \rangle_O \top)) \right. \tag{16}$$
$$\left. \vee \, (\Box_O A^u @O) \right)$$

The modification of read-operations is more essential. According to the requirements of mandatory access control, every user may read facts that are stored with her label or with a label lower than hers, i.e., agent S may read a fact A, which is stored with label s and may also read a fact A, that is stored with label u.

According to the semantics Winslett et al. defined in [WSQ94] one should distinguish between *What does a user see?* and *What does a user believe?* Their answers can be roughly outlined as follows:

- The secret agent S always *believes* facts stored with her own label to be correct.
- When she reads facts stored with label u, she *sees* this fact but she doesn't believe it to be correct. Further, she *believes*, that the unclassified agent U believes this fact.

Winslett et al use the term *correct* in relation to the *real world*. This is unambiguous, if we assume, that the real world does not change. However, under the assumption that the real world may change, it is not clear what exactly the user believes: Does she believe the facts to be correct at the time when they were inserted into the information system, by the time, when she read them from the information system, for any time in the future (under the assumption that the belief remains consistent)?

Using our approach, for the example we decided to assume, a user who has read a fact A from the information system *knows* that she has read the fact and *believes*, that the fact is stored in the database. Note, that we do not talk about the *real world* in our example.

Now, assume agent S gets answer A^s from the system. After the read-operation $(S\_O :: read(A^s))$ the proposition $A^s @S$ is satisfied. Again, as before, she knows, that she has got this answer $(K_S A^s @S)$ and she believes, that A^s is stored at the repository-agent $(B_S A^s @O)$.

$$\Box_O \left(A^s @O \Rightarrow [S\_O :: read(A^s)]_O (A^s @S \wedge K_S A^s @S \wedge B_S A^s @O) \right) \tag{17}$$

If agent S reads an unclassified fact we have the corresponding implication to the one above.

$$\Box_O \ (A^u@O \Rightarrow [S\_O :: read(A^u)]_O(A^u@S \ \wedge \ K_S A^u@S \ \wedge \ B_S A^u@O)) \quad (18)$$

Further, agent S believes, that the unclassified user U believes, that A^u is stored in the repository-agent, if U reads A^u from O before it gets deleted:

$$\Box_O \ (A^u@O \Rightarrow [S\_O :: read(A^u)]_O$$
$$(B_S((([U\_O :: read(A^u)]_O B_U A^u@O)\mathcal{U}_O(\langle U\_O :: delete(A^u)\rangle_O\top))$$
$$\vee \ (\Box_O(([U\_O :: read(A^u)]_O B_U A^u@O)\wedge\neg\langle U\_O :: delete(A^u)\rangle_O\top)))) \tag{19}$$

The unclassified agent U can only read facts stored in the repository-agent O with security label u. The implications of the query of an unclassified agent are the same as for a secret agent: She also gets an answer $A^u@U$, knows, that she got this answer ($K_U A^u@U$), and believes that $A^u@O$ is stored in the repository-agent.

$$\Box_O \ (A^u@O \Rightarrow [U\_O :: read(A^u)]_O(A^u@U \ \wedge \ K_U A^u@U \ \wedge \ B_U A^u@O)) \quad (20)$$

The read constraints change accordingly for the case that the read fact is not stored in the repository-agent O.

The main difference between a multi-level information system and a single-level information system is that in a multi-level system the data stored in a repository-agents can be polyinstantiated. We can have one instantiation of the content at each security label. This changes the key requirement as follows:

$$\bigwedge_{l\in\{S,O\}} \ \bigwedge_{k\in\{k_1,\dots,k_n\}} \ \bigwedge_{v_i,v_j\in\{v_1,\dots,v_m\},v_i\neq v_j} \ \Box_l \ (k\_v_i^s@l \Rightarrow \neg k\_v_j^s@l) \quad (21)$$

and

$$\bigwedge_{l\in\{S,O,U\}} \ \bigwedge_{k\in\{k_1,\dots,k_n\}} \ \bigwedge_{v_i,v_j\in\{v_1,\dots,v_m\},v_i\neq v_j} \ \Box_l \ (k\_v_i^u@l \Rightarrow \neg k\_v_j^u@l) \quad (22)$$

Now the sequence of inserts regarded at the end of the previous chapter,

$$\langle S\_O :: insert(k\_v_i^s)\rangle_O \ \langle U\_O :: insert(k\_v_j^u)\rangle_O\top$$

does not violate the key requirement any more. Figure 4 shows a run with its interpretation. In this run, the initial configuration satisfies the key constraint as well as the sequence of inserts given above.

In this example we assume that all propositions that are not explicitly shown in the figure are false in a configuration. Further we assume the following (minimal) accessibility relations for knowledge and belief:

– Accessibility relations for knowledge:

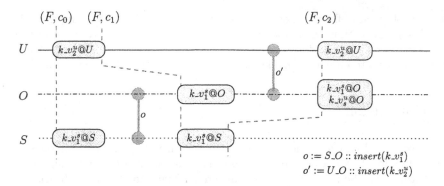

Fig. 4. A model satisfying the modified key requirements

- $\mathcal{R}_S^K = \{((F, c_0), (F, c_0)), ((F, c_1), (F, c_1)), ((F, c_2), (F, c_2)),$
  $((F, c_1), (F, c_2)), ((F, c_2), (F, c_1))\}$
- $\mathcal{R}_O^K = \{((F, c_0), (F, c_0)), ((F, c_1), (F, c_1)), ((F, c_2), (F, c_2))\}$
- $\mathcal{R}_U^K = \{((F, c_0), (F, c_0)), ((F, c_1), (F, c_1)), ((F, c_2), (F, c_2)),$
  $((F, c_0), (F, c_1)), ((F, c_1), (F, c_0))\}$
- Accessibility relations for belief:
  - $\mathcal{R}_S^B = \{((F, c_0), (F, c_0)), ((F, c_1), (F, c_1)), ((F, c_2), (F, c_1))\}$
  - $\mathcal{R}_O^B = \{((F, c_0), (F, c_0)), ((F, c_1), (F, c_1)), ((F, c_2), (F, c_2))\}$
  - $\mathcal{R}_U^B = \{((F, c_0), (F, c_0)), ((F, c_1), (F, c_0)), ((F, c_2), (F, c_2))\}$

In this example, we can easily see that the key requirements (21) and (22) hold in all configurations. Agent S only inserts the fact $k\_v_1^s$. Since no secret proposition was true before this insertion operation, the insertion of $k\_v_1^s$ obviously cannot contradict to the key requirement. Then agent U inserts the fact $k\_v_2^u$. Since agent U can only insert facts labelled with u, the insertion of $k\_v_1^s$ and $k\_v_2^u$ cannot collide.

Let us further check the requirement (11). Since the constraint must hold "always in the future" it is enough to check it for the initial configuration. The initial configuration c_0 is the only configuration in which $S\_O :: insert(k\_v_1^s)$ is the next operation from agent O's point of view. We can see in the figure, that $c_0 \xrightarrow{S\_O::insert(k\_v_1^s)} c_1$ and in c_1 the proposition $k\_v_1^s@O$ holds which means, the requirement 11 is satisfied. We can easily check the other requirements as well.

One main requirement in mandatory access control is that no user may know, which data labelled with a security label higher than her own is stored in a data object. In our example, this means, that the unclassified user-agent U may not believe, which data labelled with label s is stored in the repository-agent O:

$$\Box_U \left(\neg B_U A^s@O \right) \wedge \left(\neg B_U \neg A^s@O \right)$$

Since knowledge implies belief, it is clear, that if an agent does not believe a fact, she does not know it either.

The model given in figure 4 does not satisfy this constraint. For example in the initial configuration, in which the unclassified user only considers the actual configuration to be possible, the formula $B_U \neg A^s @ O$ holds. To ensure the constraint above, we have to enrich the model and the accessibility relation for belief.

6   Conclusion

We have introduced a framework in which semantic constraints and security constraints for databases can be uniformly formalized such that interaction between them can be studied. The introduced language for the formalization of constraints is a propositional multi-modal logic for multi-agent systems with temporal and epistemic operators. Each agent can perform operations sequentially, either autonomously or as joint operations with other agents. The language allows to talk about the dynamic behaviour of the system as well as of the knowledge and belief of the agents. We gave an example of how the defined language can be used to formalize such constraints.

In many frameworks, knowledge of agents is related about the real world. When an agent knows some fact, this fact must hold in the real world (in the actual real world as well as in all worlds, the agent considers to be possible). However, in our setting we do not model the real world. Agents' knowledge and belief is modelled only in relation to the contents of repository-agents. If we wanted to relate knowledge about the real world, we had to introduce a representing agent for the real world inside the system, that explicitly synchronizes with the system for example through a global clock (again modelled as agent of the system). As long as there is no explicit synchronisation with the "real-world", we could view agents as sitting in a black box. They can only reason about the system "inside the box", but since they do not notice changes in the real world outside the system it is in our opinion difficult to reason about "the actual state" of the real world.

As it is well-known, any attempt to formalize local effects of operations somehow has to deal with the non-effects of operations, too. This requirement is particularly important for security considerations. In our work, we approach this so-called frame problem [MH69,Sha97] by the definition 6(4) of the accessibility relation for knowledge which ensures that the knowledge of an agent cannot be affected by operations performed by other agents. This definition is related to various solutions to the frame problem: In [CGH99] Castilho et al combine a logic of actions and plans with an explicit notion of dependence/independence between propositions and actions. An action can only affect the truth value of a proposition if the proposition is dependent on the action. In [SL93] Levesque and Scherl propose successor state axioms as a solution to the frame problem for epistemic fluents. An "accessibility" predicate K for situations and a knowledge predicate $Knows$ is defined. The successor state axiom then requires, that two situations s_1 and s_1' are "accessible" (in terms of this predicate) and situation s_1 leads to a successor situation s_2 via an action a, iff action a performed in situation s_1' leads to a successor state s_2' which is "accessible" from situation s_2.

In the introduction we analyzed which kinds of constraints must be expressible in such a unifying framework for semantic constraints and security constraints. The example in section 5 does not mention all constraints. We for example left out execute-rights for operations. However, we could model a security-agent for each user-agent that performs all operations together with the corresponding user. The execute-rights can then be modelled as propositions of these security-agents. If another agent wants to grant a right to an agent or to revoke a right from an agent, she performs a grant or revoke operation with the corresponding security-agent. We can then formalize the requirement that each agent only performs operations for which she has the right, meaning for which the corresponding proposition is true at the security-agent.

Thus, to show that this framework is actually suited to unify semantic constraints and security constraints we will have to work out a more complex example in which all relevant constraints are formalized. Further, the logic given in section 4 has to be studied concerning possible tools for decision problems. Current work suggests that the logics is not compact but still allows a tableau based proof system for satisfiability.

References

[CD96]   Frederic Cuppens and Robert Demolombe. A deontic logic for reasoning about confidentiality. In J. Carmo M. Brown, editor, *Deontic Logic, Agency and Normative Systems*, Workshops in Computing. Springer, 1996.

[CD97]   Frederic Cuppens and Robert Demomlombe. Modal logical framework for security policies. In Ybigniew Ras and Andrzeij Skowron, editors, *Foundations of Intelligent Systems*, volume 1325 of *Lecture Notes in Artificial Intelligence*, pages 579–589. Springer, October 1997.

[CGH99]   Marcos A. Castilho, Olivier Gasquet, and Andreas Herzig. Formalizing action and change in modal logic I: the frame problem. *Journal of Logic and Computation*, 9(5):701–735, 1999.

[ECSD98] H-D. Ehrich, C. Caleiro, A. Sernadas, and G. Denker. *Logics for Databases and Information Systems*, chapter 6, pages 167–198. Kluwer Academic Publishers, 1998.

[FHMV96] Ronald Fagin, Joseph Y. Halpern, Yoram Moses, and Moshe Y. Vardi. *Reasoning about Knowledge*. The MIT Press, 1996.

[HS98]   Michael N. Huhns and Munindar P. Singh, editors. *Readings in Agents*. Morgan Kaufman, 1998.

[JS90]   Sushil Jajodia and Ravi Sandhu. Polyinstantiation integrity in multilevel relations. In *Proceedings of the IEEE Symposium on Security and Privacy*, pages 104–115, May 1990.

[KR94]   Paul J. Krasucki and R. Ramanujam. Knowledge and the ordering of events in distributed systems; extended abstract. In *Proceedings Theoretical Aspects of Reasoning about Knowledge*, pages 267–283. Morgan Kaufmann, 1994.

[Lam94]   Leslie Lamport. A temporal logic of actions. *ACM Transactions on Programming Languages and Systems*, 16(3):872–923, 1994.

[LDS+88]   T. Lunt, D. Denning, R. Schell, M.Heckmann, and W. Shockley. The seaview security model. *IEEE Symposium on Security and Privacy*, pages 218–233, 1988.

[Maz95]   Antoni Mazurkiewicz. Introduction to trace theory. In *The Book of Traces*, chapter 1, pages 1–42. World Scientific, 1995.

[MH69]   John McCarthy and Patrick J. Hayes. Some philosophical problems from the standpoint of artificial intelligence. In B. Meltzer and D. Michie, editors, *Machine Intelligence 4*, pages 463–502. Edinburgh University Press, 1969.

[MOP89]   Antoni Mazurkiewicz, A. Ochmanski, and Wojciech Penczek. Concurrent systems and inevitability. *Theoretical Computer Science*, 64:281–304, 1989.

[MP92]   Manna and Pnueli. *The Temporal Logic of Reactive and Concurrent Systems*. Springer-Verlag, 1992.

[Nie97]   Peter Niebert. *A Temporal Logic for the Specification and Verification of Distributed Behavior*. PhD thesis, Universität Hildesheim, 1997.

[Pen93]   Wojciech Penczek. Temporal logics for trace systems: on automated verification. *International Journal of Foundations of Computer Science*, 4(1):31–68, 1993.

[Ram96a]   R. Ramanujam. Local knowledge assertions in a changing world (extended abstract). In *Proceedings Theoretical Aspects of Rationality and Knowledge*, pages 1–17. Morgan Kaufmann, 1996.

[Ram96b]   R. Ramanujam. Locally linear time temporal logic. In *LICS*, volume 11, pages 118–127, New Jersey, 1996. IEEE.

[Rei90]   Raymond Reiter. On asking what a database knows. In *Proceedings of the Symposium on Computational Logic*, pages 95–113. Springer, 1990.

[Rei91]   Wolfgang Reisig. Concurrent temporal logic. SFB-Bericht 342/7/91 B, Technische Univ. München, Inst. für Informatik, August 1991.

[Sha97]   Murray Shanahan. *Solving the Frame Problem*. The MIT Press, 1997.

[SL93]   Richard B. Scherl and Hector J. Levesque. The frame problem and knowledge-producing actions. In *Proceedings of the Eleventh National Conference on Artificial Intelligence (AAAI-93)*, pages 689–697, Washington, D.C., USA, 1993. AAAI Press/MIT Press.

[Thi95]   P.S. Thiagarajan. A trace consistent subset of PTL. In *International Conference on Concurrency Theory (CONCUR)*, Lecture Notes in Computer Science, pages 438–452, 1995.

[vdHM92]   W. van der Hoek and J.J.C. Meyer. Making some issues of implicit knowledge explicit. *International Journal on Foundations of Computer Science*, 3(2):193–223, 1992.

[Wei99]   Gerhard Weiss, editor. *Multiagent Systems*. The MIT Press, 1999.

[Woo00]   Michael Wooldridge. *Reasoning about rational agents*. The MIT Press, Massachusetts Institute of Technology, Cambridge, Massachusetts 02142, 2000.

[WSQ94]   Marianne Winslett, Kenneth Smith, and Xiaolei Qian. Formal query languages for secure relational databases. *ACM Transactions on Database Systems*, 19(4):626–662, December 1994.

Historical and Computational Aspects of Paraconsistency in View of the Logic Foundation of Databases

Hendrik Decker

Instituto Tecnológico de Informática, Universidad Politécnica de Valencia, Spain
hendrik@iti.upv.es

Abstract. We expose paraconsistent logic with regard to its potential to contribute to the foundations of databases. We do so from a historical perspective, starting at the ancient inception and arriving at the contemporary use of logic as a computational device. We show that an understanding of the logic foundations of databases in terms of paraconsistency is adequate. It avoids absurd connotations of the *ex contradictione quodlibet* principle, which in fact never applies in databases. We interpret *datalog*, its origins and some of its extensions by negation and abduction, in terms of paraconsistency. We propose a procedural definition of paraconsistency and show that many well-known query answering procedures comply with it.

1 Introduction

The field of databases has benefited, more than many other computing disciplines, from a tight correspondence between theoretical foundations in classical logic and practical realizations. With "classical logic", we refer to the Boole-Frege-Russell line of development of propositional logic and predicate calculus. A stern adherence to classical logic, however, amounts to adopting a most stringent benchmark, since it does not tolerate any deviation from correctness. That is, classical logic requires the complete absence of contradictions. With the classical *ex contradictione quodlibet* (ECQ) rule, everything, and thus nothing useful at all, can be inferred from a contradiction. In the literature, ECQ is also said to "trivialize" an inconsistent theory, or to be "explosive" (cf. [par]).

In practice, though, inconsistency is ubiquitous in computing systems. Most of them could never function if ECQ really applied. For instance, both Bush and Gore, and also you and me and everybody, would be the president of the United States, and everybody also wouldn't, at the same time, if ECQ had been applied to the collected data from the notorious election in the year 2000. Thus, having databases as classical logic theory systems in the full sense of the word would mean that hardly any of them could compute useful answers. Yet, they do, in spite of the presence of inconsistent data. In databases, inconsistencies usually manifest themselves as violations of integrity constraints, but do not lead to inoperative answers in general. Indeed, it may seem surprising that the

L. Bertossi et al. (Eds.): Semantics in Databases, LNCS 2582, pp. 63–81, 2003.

tale that logic is a rock-solid foundation of databases has survived as long as it has, when facing the complete failure of classical logic to capture the semantics of databases in practice.

An apparently easy way out of this problem is to say that the logic at work in databases is paraconsistent, since it never applies ECQ in the presence of inconsistency. This was analyzed first in ([Kw1], p. 244), in terms of relevance logic [rel], and later in the conclusion of [Kw2]. Yet, we feel inclined to have a closer look. Our main motivations are to compare database logic to conventional paraconsistent logics, and to gain a more perspicious view of what it really is that makes databases paraconsistent.

In section 2, we give a brief introductory characterization of paraconsistency in general and with regard to its relevance in computational logic (CL) and databases. In section 3, we trace some of the history of logic and paraconsistency. In section 4, we look at paraconsistent logic proper, as it emerged historically. In section 5, we discuss in which sense datalog and CL should be taken as a paraconsistent reasoning paradigm.

We assume the reader is familiar with established notions in classical and computational logic. A glossary of abbreviations is annexed.

2   What Is Paraconsistency?

Informally speaking, paraconsistency is the paradigm of reasoning in the presence of inconsistency. For example, paraconsistent reasoning means to isolate, work around or simply ignore inconsistent parts of a knowledge base, thus preserving intentionally meaningful knowledge which is meaningless according to classical logic. In general, paraconsistent logic maintains the distinction of derivability and non-derivability, even in the presence of contradictions, contradictory consequences or other forms of inconsistency. The formal definition below is (roughly) equivalent to various definitions of paraconsistency in the literature.

Definition 1. *A deductive system $\mathfrak{S}$ is* paraconsistent *if there is a formula F in the language of $\mathfrak{S}$ such that $\bot \vdash F$ does not hold (i.e., F is not derivable from contradiction), where $\vdash$ denotes the derivability relation and $\bot$ the concept of contradiction or falsity in $\mathfrak{S}$ (e.g., the empty clause in datalog).*

This definition may seem at first sight to ask for a deficiency, rather than a virtue. It requires some kind of incompleteness, in the classical sense, but nothing "positive". In fact, most paraconsistent systems can be seen as a restriction of classical logic, in the sense that their logical consequences are a proper subset of the formulas derivable in classical logic. However, a paraconsistent system may very well be complete (or at least reasonably strong in terms of deductive power) for interesting classes of consistent theories. In general, though, the consistency of a theory is undecidable. Hence, when reasoning in theories that are not known to be consistent, (e.g., set theory, or most databases in practice), paraconsistency is a desirable property. Some more "positive" requirements for paraconsistency are postulated in 4.1, 4.2 and 5.2.

Besides plain database querying, examples of applications which need to operate soundly in the potential presence of inconsistency are data warehousing and decision support. Although a primary goal of these applications is to undo inconsistency and avoid its accruement as much as possible, there almost never is evidence of a complete absence of contradictory data. Fortunately, ECQ is never applied, and it shouldn't, since no useful results would be obtained otherwise. Other applications where contradictions must be endured without having them lead to chaos, are multi-agent systems (where agents may have conflicting plans or goals), legal reasoning (where rivaling counterparties defend or attack opposed beliefs) and counterfactual reasoning (e.g., for answering what-if queries in order to pursue the ramifications of hypotheses which may contradict the actual content of a database). These and other database-related applications call for a logic foundation which explains why their non-explosive behavior in the presence of inconsistency is possible at all.

3 History

In this section, we complement the preliminary understanding of paraconsistency conveyed above by highlighting some of the history which led to paraconsistent logic. In particular, we pay attention to the use or the renouncement of certain logical formalisms and laws, by which classical, intuitionistic, paraconsistent and computational logic can be distinguished and related to each other. For classical, intuitionistic and paraconsistent logic, this has already been done in [Bb], where however CL is not considered.

3.1 From Aristotle to Frege

It is well-known that formal logic commenced with Aristotle (cf. [Bc]). With deductive mathematics as a role model for provably correct reasoning, Aristotle strived for identifying the most certain of principles, upon which all reasoning should rest. He found that in the *law of contradiction* (LoC), which roughly means that a statement cannot be both true and false at a time. Aristotle's defense against his own doubts about LoC is self-sustaining and can hardly be invalidated. Essentially, he argues that any belief which holds itself as true can be defended only if LoC is applicable to that belief. Thus, trying to defend the belief that it is true that LoC was false would have to appeal to LoC and hence contradict itself. Thus, LoC appears to be a simple and evident precondition of rational reasoning at all.

Interestingly, Aristotle's defense of LoC uses (though only informally) the principle of *reductio ad absurdum* (RaA). Thus, RaA assumes a role of similar weight as LoC. Another basic, though not quite as fundamental a principle in Aristotle's philosophy is the *law of excluded middle* (LEM). Roughly, it says that a statement cannot be anything but true or false.

Predicate logic is due to Frege (cf. [Bc,KK]). One of his main goals was a purely logical foundation of mathematics. Frege set out using LoC as the ultimate basis. In particular, he tried to establish a theory of numbers on the set which contains all propositions that do not obey LoC. That set, being the empty set, represents the number 0. Given a set N representing the number n $(n > 0)$, $n+1$ is then represented by $N' = N \cup \{N\}$. However, while Frege succeeded to establish predicate logic as a broadly accepted formalization of rational reasoning, he failed in formalizing mathematics, as witnessed by Russell's paradox [Ru]. That resulted in different lines of research, each of which was motivated by formalizing the foundations of mathematics while trying to avoid any logical paradox, i.e., any violation of LoC.

3.2   Logicism, Formalism, Intuitionism

Among various attempts to provide a foundation of mathematics without violating LoC, Russell & Whitehead's logicism, Hilbert's formalism and Brouwer's intuitionism stick out as major lines of research (cf. [Bc,KK]). Mathematics is not of concern in this paper. However, each of logicism, formalism and intuitionism brought forward logic systems of axioms and inference rules. We are going to discuss these systems with regard to their relationship to paraconsistency.

3.2.1   Logicism

Principia Mathematica [RW] contains an axiomatization of classical logic. Without reproducing all details (which would involve characterizations of underlying alphabets, the substitution rule etc), the propositional core of this axiomatization, as re-formulated in [Kl], is shown in figure 1. It consists of the modus ponens (MP) as its singular inference rule and axioms 1 - 10. (For simplicity, we do not refer to quantified terms, but stick to propositional logic, in sections 3 and 4, unless explicitly stated otherwise; axioms associated to set theory and mathematics, e.g., the axiom of choice, we do not consider at all.) Both LoC and ECQ are valid in the system of fig. 1. RaA is explicitly postulated by axiom 9. The derivability of $\neg p \rightarrow (p \rightarrow q)$ in this system is demonstrated in [RW]. The interpretation of $\neg p \rightarrow (p \rightarrow q)$ as an expression of ECQ has first been given by Post [Pt]. The discovery of ECQ as a consequence of aristotelian logic has been ascribed in [KK] to the medieval philosopher Pseudo-Scotus.

Note that the proof of ECQ, i.e., its formal derivation in the system of fig. 1, requires the use of at least one of the *laws of disjunctive thinning (LDT)*, i.e., axioms 6 and 7, and cannot be accomplished without them. The trivializing effect of ECQ can be seen as follows: From the first of two contradictory premises p and $\neg p$, $p \vee q$ is obtained, for any q, by applying axiom 6. From $\neg p$ and $p \vee q$, MP then infers q. A fitting remark in this context is also that relevance logic outlaws LDT. By avoiding logical deductions in which the consequent contains predicates that do not appear in the antecedent, only formulas that bear some relevance to the given premises of an argument are inferred. As already indicated in section 1, query answering procedures of datalog share this aspect of relevance with relevance logic. So, the circumstance that LDT is necessary in classical logic

Inference rule *modus ponens*: $p \rightarrow q, \ p \vdash q$

Axioms:

1. $p \rightarrow (q \rightarrow p)$
2. $(p \rightarrow q) \rightarrow ((p \rightarrow (q \rightarrow r)) \rightarrow (p \rightarrow r))$
3. $p \rightarrow (q \rightarrow (p \wedge q))$
4. $(p \wedge q) \rightarrow p$
5. $(p \wedge q) \rightarrow q$
6. $p \rightarrow (p \vee q)$
7. $q \rightarrow (p \vee q)$
8. $(p \rightarrow r) \rightarrow ((q \rightarrow r) \rightarrow ((p \vee q) \rightarrow r))$
9. $(p \rightarrow q) \rightarrow ((p \rightarrow \neg q) \rightarrow \neg p)$
10. $\neg\neg p \rightarrow p$

Fig. 1.

for deriving ECQ (and thus enabling its applicability), and that LDT is not used (and thus ECQ is not applicable) in datalog, testifies the paraconsistent behavior of datalog.

3.2.2 Formalism

Hilbert and Ackermann generalized Post's result about the trivializing effect of ECQ in propositional logic to predicate logic [HA]. Hilbert insisted that mathematics, or any formal theory, must be logically axiomatized such that the axioms be consistent, i.e., do not violate LoC and therefore never could give reason to apply ECQ. Initially, he also demanded that formal theories should be complete. But that was shown to be impossible even for the basic mathematical theory of arithmetics, in [Goe]. Gödel also showed that predicate logic alone is complete, but Church showed its undecidability [Ch], i.e., there is no general way to tell, for a formula F, which of F or $\neg F$ (i.e., which of the disjuncts of LEM) holds.

Hilbert and Bernays presented a deductive calculus [HB] which is equivalent to Russell & Whitehead's system (fig. 1) It contains a system called *positive logic* (*PL*) which is equivalent to axioms 1 - 8 and MP in fig. 1, and three more axioms (two halves of LDN and LCP) for characterizing negation. A similar calculus was proposed earlier in [Hi], with the following two axioms for negation. N1 is one way

N1. $\ p \rightarrow (\neg p \rightarrow q)$ N2. $(p \rightarrow q) \rightarrow ((\neg p \rightarrow q) \rightarrow q)$

Fig. 2.

to represent ECQ (cf. the glossary). Hilbert interpreted N2 as a representation of LEM. He also pointed out that PL coincides with the positive part (sans occurrence of negation) of an axiomatization of intuitionistic logic (cf. 3.2.3). PL is also the core of conventional paraconsistent logic (cf. 4.2).

3.2.3 Intuitionism

As opposed to logicists and formalists, Brouwer himself resented to recur on formal logic. In particular, he rejected the use of LEM and the law of double negation (LDN, formally: $p \leftrightarrow \neg\neg p$, one half of which is axiom 10 above). He also ruled out indirect proofs using RaA. It was Kolmogorov [Kg] and Heyting [Hy] who proposed axiomatic systems for making intuitionism accessible to formal treatment (at the expense of some of Brouwer's basic philosophical beliefs).

Replacing $\neg\neg p \to p$ (axiom 10) by $\neg p \to (p \to q)$ (ECQ) in fig. 1 yields Heyting's system. In the resulting schema, neither LDN nor LEM are derivable any more. The non-validity of LDN and LEM also invalidates the law of material implication (LMI) and effectively coerces proofs to be constructive. That is, from $(p \to q)$, q can be derived if and only if p could be derived (not withstanding any proof of q by other axioms and rules). However, LoC continues to hold. Also, note that RaA, in the form of axiom 9, remains in place, even though Brouwer did not want to admit indirect proofs.

Before Heyting, Kolmogorov proposed a deductive system, resulting from the one in [Hi] (cf. fig. 2) by keeping the positive part and replacing N1 and N2 (which he considered non-intuitive) by what has been called Kolmogorov's law, i.e., axiom 9 in fig. 1. Later, he also contributed a widely recognized operational semantics to Heyting's system, and mediated between the intuitionists and the school of logicians descending from Łukasiewicz (cf. 3.3).

The axiom $\neg p \to (p \to q)$ (i.e., ECQ) in Heyting's system is abandoned in Johansson's *minimal logic* [Jo]. The latter consists of MP and axioms $1-9$ (fig. 1) and thus is essentially the same as Kolmogorov's system. In minimal logic, only each negated sentence can be deduced in the presence of contradiction, but not necessarily each sentence whatsoever. In particular, $\neg p \to (p \to \neg q)$ can be deduced from axioms 1 - 9, but $\neg p \to (p \to q)$ (ECQ) cannot. In other words, minimal intuitionistic logic shows a first, though feeble sign of tolerance of contradiction, in that only the negative consequences of an inconsistent theory are trivialized, but not its positive consequences. So, by hindsight, it could therefore be interpreted as an anticipation of paraconsistency (and it does in fact satisfy definition 1).

3.3   Precursors of Paraconsistent Logic

Łukasiewicz recalled that Aristotle had discussed several different modalities of LoC, by which certain shadings of clarity and certainty of this most fundamental of principles became apparent [Lu]. Analogous to non-euklidian geometry in which the basic axiom of parallels was no longer postulated (and which revolutionized physics in those days), Vasiliev (cf. [Bz]), Łukasiewicz and Post proposed to study non-aristotelian logics in which LoC was no longer postulated. However, after re-discovering the findings of Pseudo-Scotus, and possibly influenced by [HA] and Kolmogorov, Łukasiewicz ceased to question LoC and abandoned LEM, instead. In particular, he investigated multi-valued semantics, with three or more truth values, in which LEM is not applicable. Łukasiewicz's 3-valued logic (strong traces of which re-surfaced later in [Kl]) was elaborated by Słupecki [Sl]. In the latter's calculus, LDN does not hold and ECQ applies only in case each of p, $\neg p$, $\neg\neg p$ is derivable, for some proposition p, but not if only p and $\neg p$, or only $\neg p$ and $\neg\neg p$ are derivable.

Even though LEM does not hold in multi-valued logics, ECQ usually does, i.e. the derivability of a contradiction still entails trivialization (cf. [Ur]). In general, the introduction of more than two truth values opens up the possibility that some formulas which are classically interpreted as contradictions no longer evaluate to false. A well-known example is the three-valued database completion $p \leftrightarrow \sim p$ of the definition $p \leftarrow \sim p$ of a predicate p in [Fi,Ku], which is based on [Kl] and Łukasiewicz's three-valued systems. It clearly is inconsistent in classical logic, but it is not in the semantics of [Fi,Ku].

Another precursor of paraconsistency is Lewis' modal logic, built upon the operator $\diamond$ ($\diamond p$ means "it is possible that p") and *strict implication* $\mapsto$ [LL]. The latter is defined as $\neg\diamond(p\wedge\neg q)$, i.e., "it is not possible that both p and $\neg q$". When $\mapsto$ replaces $\rightarrow$ in ECQ, the resulting formula does not hold. Unfortunately, the number of derivable sentences using only strict instead of material implication turned out to be "very limited" [Ja2]. Thus, material implication was not abandoned but used alongside strict implication in [LL], and ECQ remains valid for the former.

Although each of intuitionistic, modal and paraconsistent logic has been initiated or spurred on by the discovery of Russell's paradox, a notable distinction persists. While paraconsistent logic accepts the presence of contradictions and avoids the trivializing effect of ECQ, the proponents of intuitionistic and modal logic have not been interested in finding ways to reason in contradictory theories, but rather in avoiding set-theoretic or mathematical anomalies, possibly caused by axioms such as LEM or LDT. Yet, we are going to see that excluding such axioms is also conductive to get closer to achieving paraconsistency.

4   Paraconsistent Logic

We are going to recapitulate paraconsistent logic as propelled by its originators. We do not deal with dialectic nor dialetheic logics nor with logics of vagueness, which are sometimes subsumed under paraconsistency (cf. [par]). Also, we do not assess relevance logic any more, except to mention that it started out as an approach to relax the limitations of modal logic's strict implication, but its further development has come to be seen as a particular approach to paraconsistency. In [par], non-adjunctive, non-truth-functional, many-valued and "relevant" logic approaches are distinguished.

Two mutually independent beginnings of paraconsistent logic have been identified, originated by Jaśkowski and da Costa. The former initiated the non-adjunctive, the latter the non-truth-functional line of development. Before each of the two researchers set out to explore the revolutionary idea that manifest inconsistencies should possibly be tolerated, rather than letting them trivialize a theory, they were involved in investigations of intuitionistic logic.

4.1   Stanisław Jaśkowski

Jaśkowski took Łukasiewicz's doubts concerning LoC a step further, by working out his *discursive logic*, "a propositional calculus for contradictory deductive systems", in 1948 [Ja2]. He postulated a logic system which would satisfy the following three requirements (note that the first one roughly corresponds to definition 1):

($j1$) Contradiction should not always lead to trivialization.
($j2$) "Practical inference" should be enabled.
($j3$) The postulated system should have "an intuitive justification".

Discursive logic satisfies ($j1$) – ($j3$). It uses a new modality of possibility for introducing *discursive implication*. For the latter, neither ECQ nor LoC remains valid. An important difference between strict and discursive implication is that MP is applicable for the latter but not for the former. Thus, discursive implication does not have to be seconded by material implication, and it is closer to the constructive implication of intuitionistic logic than strict implication (cf. 3.3).

Finally, a more general aspect merits our attention. None of Jaśkowski's three requirements above would postulate that LoC be invalid. Thus, his primary concern was not to ban LoC, but to avoid the trivializing effect of ECQ.

4.2   Newton da Costa

Da Costa's first expression of dissatisfaction with the absolute intolerance of mathematical logic against contradictions in [dC1] was amplified and resolved in [dC2]. Later, he adopted the terminology of 'paraconsistent' logic, choosing the prefix para (meaning alongside, besides), "for indicating a notion that does not necessarily go beyond, and challenges, the classical setting, but that somehow could go hand in hand with it" [Bb]. This may also serve to dispel any allusions of "para-" to other, possibly more suspicious notions (e.g., in para-normal, meaning not scientifically explainable).

In [dC2], da Costa proposed a hierarchy C_n ($1 \leq n \leq \omega$) of axiomatic schemata for formalizing paraconsistency. They all satisfy Jaśkowski's postulate ($j1$) and, arguably, also ($j2$) and ($j3$). However, da Costa's intentions are probably captured better by the following two conflicting requirements: Paraconsistent logic should

($dc1$) tolerate manifest inconsistency as much as possible,
($dc2$) conservatively approximate classical logic as much as possible.

That is, the set of formulas derivable by paraconsistent logic from consistent premises should be as close as possible to what is derivable from classical logic, while inconsistent premises should lead as rarely as possible to an explosive behavior.

By Gentzen's and Gödel's results (cf. 5.1), intuitionistic logic satisfies ($dc2$) very well, though not at all ($dc1$) nor ($j1$). On the other hand, minimal logic, which at least satisfies ($j1$), falls short of satisfying ($dc2$). Hence, it is plausible that da Costa's systems do not abandon LEM (as intuitionistic and minimal logic do), but LoC.

C_ω consists of MP and PL, (i.e., the inference rule and axioms 1 - 8 in fig. 1), plus axiom 10 (i.e., one half of LDN) and LEM. Additionally, C_1 needs two more axioms, as shown in figure 3. RaA (axiom 9, fig. 1) is missing in each C_i $(1 \leq i \leq \omega)$. Axioms 9' and 9" of C_1 amount to a weakening (in terms of deductive power) of axiom 9.

Rules and axioms of C_1: MP, 1. - 8., 10. (as in fig. 1) and
9'. $q^o \rightarrow ((p \rightarrow q) \rightarrow ((p \rightarrow \neg q) \rightarrow \neg p))$
9". $(p^o \wedge q^o) \rightarrow (((p \rightarrow q)^o \wedge (p \wedge q)^o \wedge (p \vee q)^o)$
11. $p \vee \neg p$

Fig. 3.

Axiom 9' is obtained from RaA by prefixing the antecedent q^o. For an expression e, e^o stands for $\neg(e \wedge \neg e)$, i.e., e "behaves classically", i.e. LoC applies for e. Thus, axiom 9" expresses that, if two propositions p, q behave classically, then so also do their implicative, conjunctive and disjunctive compositions. Thus, similar to [Ja2], two modes can be distinguished in da Costa's system, one which behaves classically with regard to ECQ and one which does not.

C_i $(1 < i < \omega)$ is obtained by replacing each expression e^o in axioms 9' and 9" with $e^{(n)} = e^1 \wedge e^2 \wedge ... \wedge e^n$, where $e^1 = e^o$ and $e^{n+1} = (e^n)^o$.

In terms of deductive power, C_1 is closer to classical logic, and thus satisfies $(dc2)$ better than any other C_n $(1 < n \leq \omega)$. In general, for $1 \leq i < \omega$, C_i satisfies $(dc2)$ better than C_{i+1}, and the latter still satisfies it better than C_ω. However, the latter is still stronger in deductive power than PL (which vacuously satisfies $(dc1)$). Conversely, C_ω satisfies $(dc1)$ better than all C_i, and C_{i+1} satisfies it better than C_i. In general, it can be shown that neither LoC nor ECQ nor RaA holds in any of the C_n, but LEM (axiom 11) obviously holds in each.

An arguable disadvantage of C_i $(1 \leq i < \omega)$ is the lack of clarity and simplicity, as introduced by the implicit case distinctions of the antecedents of axioms 9' and 9". After all, to determine if a sentence behaves classically or not amounts to impose an informal semantic interpretation on its constituents. Being devoid of 9' and 9", C_ω is the 'nicest' system in da Costa's hierarchy, in terms of structural clarity and simplicity. In general, it can be argued that properties such as clarity, certainty and simplicity should be added to $(j1)$ - $(j3)$ and $(dc1, 2)$ as further desiderata, even though they are hardly measurable. However, it is the complete absence of RaA (axiom 9, fig. 1) in C_ω which accounts for its weakness with regard to $(dc2)$.

Related to the C_i, Sette (and later also da Costa and Alves) defined the "maximal paraconsistent" system P^1 [Se] (called F in [dA]). It arguably satisfies $dc2$ better than each C_i, in the following sense: When adding any sentence which is valid in classical logic but not in P^1, then the so-extended system becomes equivalent to classical logic. Together with MP, P^1 consists of the following axioms.

Similar to the C_i $(i < \omega)$, P^1 may be criticized because of its implicit and explicit case distinctions in axioms 9', 9" and 9"', respectively. The explicit case

Axioms of P^1 :   1. - 8., 9', 10., 11. (as in fig. 1, 3)
9"'. $\neg(p \land \neg p)$ if p is not atomic

Fig. 4.

distinction of axiom 9"' even destroys the desirable property of *substitutivity*, which is enjoyed by both classical and intuitionistic logic. Roughly, substitutivity means that a derivable formula remains derivable if a propositional variable occurring in it is consistently substituted by any sentence of the language.

Da Costa's original version of F was P^1 diminished by axiom 9' (later added by Alves). However, the gain of clarity and simplicity is traded off to a weaker position with respect to (*dc2*) The authors of [dBB] argue to have diminished (though not removed) that weakness in their system C_1^+, obtained from C_1 by replacing axiom 9" (fig. 3) with $(p^o \lor q^o) \to ((p \to q)^o \land (p \land q)^o \land (p \lor q)^o)$, i.e., by weakening the conjunctive premise $(p^o \land q^o)$ of 9" to the disjunctive premise $(p^o \lor q^o)$. Anyway, non-substitutivity remains to be a major flaw, since axiom 9"' remains in place.

5   Computational Logic

CL, as it is commonly understood, is related to intuitionistic logic or resolution or both, comprising a wide range of subjects (cf, e.g., [tcl]). The scope of CL in this paper, however, is limited according to the following subsections.

In 5.1, we briefly address proof theory as the roots of CL, and reformulate the definition of paraconsistent systems in section 2 in terms of proof-procedures. In 5.2 – 5.5, we show that well-known resolution proof procedures used in databases are paraconsistent. We do not deal with extensions such as discussed in [ABK, BS,Pl,DP,DVW] and many others. Rather, we limit our attention to the more basic question to which extent normal logic database systems are already para-consistent.

5.1   Proof Theory

It is well-known that Tarski succeeded to give a model-theoretic semantics to classical logic, without getting into conflict with set-theoretic violations of LoC [Ta]. However, the resulting complications, which seem to be opposed to the basic quest of logic for clarity, certainty and simplicity, have caused many logicians to abandon model theory and turn to proof theory (cf. [Ro], Historical Notes). There are proof- and model-theoretic approaches also in paraconsistent logic. However, we are going to continue to emphasize proof-theoretic (i.e., syntactic and operational) aspects, skirting most of model theory.

Gentzen, the main proponent of proof theory, introduced the procedural "sequent calculus" [Ge]. He showed that intuitionistic logic is almost equivalent to classical logic, in that, for each formula F derivable in classical logic, there is an intuitionistic natural deduction proof of $\neg\neg F$. Gödel, who independently found essentially the same result, also observed that intuitionistic and classical logic are equi-consistent, i.e., LoC is violated in a theory subjected to classical logic if and only if LoC is violated when the same theory is subjected to intuitionistic logic. A little earlier, Jaśkowski came up with an intuitionistic proof calculus based on hypothetical assumptions ("suppositions") [Ja1], which, for the propositional case, is essentially equivalent to Gentzen's. In view of paraconsistency and computational logic, it is interesting to note that both calculi are *constructive* (cf. [Av]), i.e., mechanizable as proof procedures, and make explicit use of LoC and RaA.

In accord with Hilbert's requirements (cf. 3.2.2), proof theory primarily is about procedures for determining the consistency or inconsistency of given theories and finding proofs for given formulas in such theories. Hence, the definition of paraconsistency in section 2 can be rephrased as follows.

Definition 2. *A proof procedure $\wp$ is paraconsistent if there is a formula F and an inconsistent theory T such that there is no proof of F in T using $\wp$.*

Roughly, this definition is equivalent to the one in section 2. In the remainder, we are going to see that some well-known resolution-based proof procedures are paraconsistent in the sense of definition 2, and also in a sense which is more desirable than that.

5.2   Resolution

On the propositional level, the resolution inference rule is a generalization of both MP and the symmetrical modus tollens (MT), as exhibited in the annex. As shown in [Ro], resolution is equivalent to classical logic, i.e., each of LoC, ECQ, LEM, RaA, LDN (or, rather, their clausal representation) has a resolution proof. Moreover, RaA plays a crucial role in refutation-based resolution proof procedures, in that, from a refuted clause (i.e., the addition of that clause to a theory has lead to a contradiction), its negation is derived as a theorem. That seems to suggest that there is nothing in resolution which would indicate any potential of paraconsistency. However, specific refinements of the general resolution proof procedure amount to a restrained use of the available deductive resources, which may yield paraconsistency.

For instance, let us consider *SL resolution* [KuK], i.e., the proof procedure that attempts to find an SL refutation of a given root clause G in a finite set D (say) of clauses. G is the negation of a formula which is queried for theoremhood in the theory embodied by D. In disjunctive databases, SLI resolution [MZ] (the query answering procedure which draws answer substitutions from SL refutations), is well-known. To find a refutation of G, SL resolution only needs to consider those clauses from D on which G depends. Hence, an inconsistent subset of clauses in D on which G does not depend, will not cause the computation of unexpected answers. Moreover, SL avoids to derive any irrelevant consequences which would not be related to the goal of refuting G. In particular, LDT is never applied. Hence, SL resolution is paraconsistent, in the sense of definition 2.

Quite obviously, definition 2 requires less than what can be expected. In fact, it only asks for a single instance of some inconsistent theory and some formula for which the proof procedure is required to fail to comply with the predictions of ECQ. In practice, however, one would hope that, for *each* inconsistent theory, no unexpected consequences are derived from a root when the inconsistency is confined to a part of the theory on which the root does not depend. The following definition formalizes this thought. Notice that it is as general as to properly include the cases that D is inconsistent or that $D(F)$ is consistent, without requiring any of these.

Definition 3. *For a set D of clauses and a formula F, let $D(F)$ be the set of clauses in D on which the clausal representation of $\neg F$ depends (where dependency be defined as usual, by virtue of a dependency graph based on matching predicates or unifying atoms of opposed polarity).*

A proof procedure $\wp$ is fairly paraconsistent *if, for each finite set of clauses D and each formula F such that there is no proof of F using $\wp$ in $D(F)$, then there is also no proof of F using $\wp$ in D.*

Obviously, a fairly paraconsistent proof procedure is paraconsistent in the sense of definition 2, but the reverse does not necessarily hold. Clearly, SL resolution is fairly paraconsistent.

5.3   Datalog

Basically, there are two kinds of resolution proof procedures used for query answering, view materialization, updating or satisfiability checking in datalog: bottom-up and top down.

SLD resolution, which proceeds top-down, is essentially a specialization of SL resolution for computing answers to queries in denial form in definite Horn clause theories. So, it inherits the property of being a fairly paraconsistent proof procedure. But, since definite theories can never be inconsistent, paraconsistency may not seem to be a noteworthy issue at all for SLD. However, inconsistency can arise in the presence of integrity constraints, which usually are added as denial clauses to the theory. When checking integrity by querying each constraint, the fair paraconsistency of SLD ensures that none of the constraints which do not contribute to any inconsistency will ever be refuted. In other words, SLD identifies violated constraints in a given database by refuting them and not refuting those that are not violated, although ECQ would justify the refutation of each.

As pointed out in [De1] and utilized in most view update procedures (e.g., [De2,De3]), a more detailed analysis of refutations of integrity constraints and their computed answer substitutions even allows to identify the facts and rules which participate in causing integrity violation. This would not be possible as easily in a proof procedure equipped with deductive resources including LEM, ECQ and inference rules that would go beyond input resolution. Hence, the fairly paraconsistent behavior of SLD is responsible for nice and well-known features that are usually taken for granted.

Also the bottom-up inference of database facts is a paraconsistent proof procedure. Besides facts of the minimal model of all definite clauses in the database, the only further consequence that can be derived in case of violated constraints is the fact that integrity is violated, but nothing at all beyond that, although ECQ would sanction each consequence whatsoever. Hence, the usual bottom-up procedures in databases can be used to materialize something which might be called the *paraconsistent minimal model* (or, more generally, *paraconsistent stable models*) of a database, in case its integrity is violated.

5.4 Negation as Failure and the Completion

For expressing negated conditions in the body of clauses and deriving negative answers and facts from databases, *negation-as-failure* (NaF) has been introduced [Cl]. The associated completion semantics [Cl,Ll] interprets each database as a classical theory, i.e., negation is classical and LoC and LEM are valid. The completion may be inconsistent. With regard to that, query answering with SLDNF is a fairly paraconsistent proof procedure, since, similar to SLD, it does not compute unexpected answers for queries which do not depend on clauses causing the completion to be inconsistent. Independent of the consistency status of the completion, SLDNF also is fairly paraconsistent when inconsistency arises because of violated integrity constraints, much the same way as SLD is.

5.5 Abductive Extensions of Datalog

Not many would plainly deny that, in general, NaF is an inference rule which lacks clarity, certainty and simplicity. Arguably, a more logical way to extend datalog with negation in the body of clauses is the abductive approach, as initiated in [EK]. Instead of implicitly completing the database, abductive logic programming (ALP) extends SLD by dynamically augmenting the database with axioms for characterizing negation as well as with consistent hypotheses about negative facts, which are made explicit in order to justify computed answers. Axioms and hypotheses about negative literals are added (and possibly discarded again) while running the proof procedure. (Comparable approaches to dynamic, argumentation-based reasoning, which in fact could be traced back to Platon's dialogues, have been described, e.g., in [Lo] and [Ba].) The assumed axioms and hypotheses always depend on the particular query and, similar to SLD and SLDNF, disregard possibly inconsistent subsets of the database which would neither depend on the query nor the hypotheses. In this sense, also ALP is fairly paraconsistent.

With regard to the paraconsistency of ALP, the dynamically assumed axioms and hypothetical facts deserve some more attention. Due to space limitations, we only sum up the main points, below. For further details, we refer the reader to [EK,Du,KKT].

The axioms assumed about negated literals in ALP are explicit instances of LoC. For example, p is inferred from the clause $p \leftarrow \sim q$ and the hypothesis $\sim q$ if the sub-query $\leftarrow q$ could not be refuted, where the attempts to refute $\leftarrow q$

need to be consistent with the integrity constraint $\leftarrow q \wedge \sim q$, i.e., a particular instance of LoC. For nested negations, the proof procedure in [EK] also reasons with instances of LEM. However, it has been pointed out in [EK,Du] that, in general, this may be contradictory. The procedure proposed in [Du] avoids such contradictions. It distinguishes itself from the procedure in [EK] by reasoning without potentially harmful instances of LEM.

With abductive extensions of datalog that allow for more general hypotheses than just negated literals, many other tasks of knowledge-based reasoning can be accomplished (cf. [De4,KKT]). Those procedures are also fairly paraconsistent, similar to SLD and its abductive extension upon which they build. In general, they reason with application-dependent integrity constraints of arbitrary generality (not just instances of LoC), but still without computing unexpected arguments or answers, in case any of those constraints is violated.

6   Conclusion

We have taken a look at some of the history of logic which led via intuitionistic logic and various forerunners to paraconsistency. While the historic development was largely fueled by attempts to cope with problematic deficiencies of axiomatizations of set theory, our concern has been to see in which sense paraconsistent logic can contribute to the foundation of databases.

Common to all approaches to paraconsistency is the goal of avoiding the explosive behavior of ECQ. In databases, it would mean that, in case of integrity violation, each conceivable substitution of free variables in the body of a query would pass as a correct answer. Although that fortunately never happens in practice, it is disturbing that such a behavior is predicted by classical logic. Therefore, an interpretation of the logical foundations of databases in terms of paraconsistency appears to be much more adequate than the "classical" point of view. However, a closer look at traditional paraconsistent logic systems has revealed that none of them could simply be adopted as a substitute foundation for databases.

On the other hand, we have confirmed what had already been observed in [Kw1,Kw2]: CL behaves paraconsistently as long as it does not apply LDT and focuses on relevant input clauses. Also, we have reformulated the usual declarative definition of paraconsistency in procedural terms, and have shown that many CL procedures comply with it. In fact, they even comply with more demanding requirements, e.g., as expressed in definition 3.

The historical perspective has helped to bring out some other aspects: CL is more conservative with regard to classical and intuitionistic logic than any other approach to paraconsistency (perhaps except the quasi vacuously paraconsistent PL and minimal logic). In particular, CL does not need new axioms or inference rules for achieving paraconsistency (recall that NaF has not been introduced for that purpose). SL resolution is even refutation-complete in the classical sense, and SLD resolution is a Turing-complete programming language [Ll]. Thus, CL arguably satisfies traditional requirements (*j1-3*) (4.1) and (*dc1, 2*) (4.2) for paraconsistency better than the systems of those who postulated them. We have seen that (even without extensions by annotations, additional truth values, probabilistic or other constructs), datalog and its extensions embody a practically viable kind of paraconsistency. Much unlike the traditional approaches to paraconsistency, datalog is not as subversive as to mitigate or even dispense with LoC. On the contrary, abductive extensions of datalog reason explicitly with LoC while managing to behave sensibly in the presence of contradictions. We have seen this with database integrity checking, where only those constraints that have been identified as relevant to a particular update, are refuted in case of integrity violation. In general, ECQ would justify as correct the refutation of any query or constraint in a database where integrity is violated, but the paraconsistency of datalog enables a much smarter behavior. Also with regard to LEM, ALP and da Costa's systems differ considerably. The latter embrace LEM unconditionally. Akin to intuitionistic logic, abductive extensions of datalog in general do not recur on LEM when reasoning with negation.

The paraconsistency of CL also enfeebles Minsky's criticism [My]. He wrote: "The consistency that logic absolutely demands is not otherwise usually available - and probably not even desirable! - because consistent systems are likely to be too 'weak'." More generally, though, it must be conceded that all programming paradigms, including CL, are often too weak, in the sense that they fail to be robust and tolerant against plain bugs and faults. A proposal frequently given for solving the problem of fault intolerance is redundancy. In databases, redundancy can be a blessing and a curse (cf. [De3]), but is not in conflict with logic. Moreover, redundancy has proven to be a successful paradigm for coping with faults and failures, in societal and biological systems as well as computing networks (e.g., the semantic web, neural networks, distributed databases; for the latter, cf. [M+]). However, when some of the copies of a data object undergo changes while others may not, redundancy can easily lead to inconsistency. We conclude with the hope that, perhaps, there is a yet undiscovered logic by which paraconsistent and biological aspects of evolving systems could be reconciled.

Acknowledgement. The paper has benefited from critical comments by Paco Marqués and several reviewers.

References

[ABK]   M. Arenas, L. Bertossi, M. Kifer: Applications of Annotated Predicate Calculus to Querying Inconsistent Databases. *Proc. Computational Logic 2000*, Springer LNCS 1861, 926–941, 2000.

[Av]   J. Avigad: Classical and Constructive Logic.
http://www.andrew.cmu.edu/~avigad/Teaching/classical.pdf, 2000.

[Ba]   D. Batens: A survey of inconsistency-adaptive logics. In D. Batens et al (eds): *Frontiers of Paraconsistent Logic*. King's College Publications, 49–73, 2000.

[Bb]   A. Bobenrieth: *Inconsistencias ¿por qué no? Un estudio filosófico sobre la lógica paraconsistente*. Colcultura, 1996.

[Bc]   I. M. Bocheński: *Formale Logik*, 2nd edition. Verlag Karl Arber, Freiburg, München, 1956.

78   H. Decker

[BS]   H. A. Blair, V. S. Subrahmanian. Paraconsistent Logic Programming. *Theoretical Computer Science* 68(2), 135–154, 1989.

[Bz]   V. A. Bazhanov: Toward the Reconstruction of the Early History of Paraconsistent Logic: The Prerequisites of N. A. Vasiliev's Imaginary Logic. *Logique et Analyse* 41(161-163), 17–20, 1998.

[Ch]   A. Church: A Note on the Entscheidungsproblem. *J. Symbolic Logic* 1(40–41, 101–102), 1936.

[Cl]   K. Clark: Negation as Failure. In H. Gallaire, J. Minker (eds): *Logic and Data Bases*, Plenum Press, 293–322, 1978.

[dA]   N. da Costa, E. Alves: On a Paraconsistent Predicate Calculus. In O. T. Alas, N. da Costa, C. Hönig (eds): *Collected papers dedicated to Professor Edison Farah on the occasion of his retirement*. Instituto de Matematica e Estatistica, Universidade de São Paulo, 83–90, 1982.

[dBB] N. da Costa, J.-Y. Béziau, O. Bueno: Aspects of Paraconsistent Logic. *IGLP Bulletin* 3(4), 597–614, 1995.

[dC1] N. da Costa: Nota sobre o conceito de contradicao. *Anuario da Sociedade Paranaense de Mathematica (2a. Serie)* 1, 6–8, 1958.

[dC2] N. da Costa: Sistemas Formais Inconsistentes. Universidade do Paraná, 1963.

[De1] H. Decker: Integrity Enforcement on Deductive Databases. In L. Kerschberg: *Expert Database Systems*. Benjamin Cummings, 381–395, 1987.

[De2] H. Decker: Drawing Updates From Derivations. *Proc. 3rd ICDT*, Springer LNCS 470, 437–451, 1990.

[De3] H. Decker: Some Notes on Knowledge Assimilation in Deductive Databases. *In Transactions and Change in Logic Databases*, Springer LNCS 1472, 249–286, 1998.

[De4] H. Decker: An Extension of SLD by Abduction and Integrity Maintenance for View Updating in Deductive Databases. *Proc. JICSLP*. MIT Press, 157–169, 1996.

[DP]   C. Damasio, L. Pereira: A Survey of Paraconsistent Semantics for Logic Programs. In D. Gabbay, P. Smets (eds): *Handbook of Defeasible Reasoning and Uncertainty Management Systems*, Vol. 2, Kluwer, 241–320, 1998.

[Du]   P. Dung: An argumentation-theoretic foundation for logic programming. *J. Logic Programming* 22(2), 151–171, 1995.

[DVW] H. Decker, J. Villadsen, T. Waragai (eds): *Paraconsistent Computational Logic (Proc. FLoC'02 Workshop)*. Datalogiske Skrifter, Vol. 95, Roskilde University, 2002. Electronic proceedings, edited by D. Goldin et al, available at http://www.imm.dtu.dk/jv/pcl.html.

[EK]   K. Eshghi, R. Kowalski: Abduction compared with negation by failure. *Proc. 6th ICLP*, MIT Press, 234–254, 1989.

[Fi]   M. Fitting: A Kripke-Kleene semantics for logic programs. *J. Logic Programming* 2(4), 295–312, 1985.

[Ge]   G. Gentzen: Untersuchungen über das logische Schließen. *Mathematische Zeitschrift* 39, 176–210, 405–431, 1934, 1935.

[Goe] K. Gödel: Über formal unentscheidbare Sätze der Principia Mathematica und verwandter Systeme I. *Monatshefte für Mathematik und Physik* 38, 173–198, 1931. Reprinted in [Hj].

[HA]   D. Hilbert, W. Ackermann: Grundzüge der theoretischen Logik. Springer, 1928.

[Hi]   D. Hilbert: Die logischen Grundlagen der Mathematik. *Mathematische Annalen* 88, 151–165, 1923.

[HB]   D. Hilbert, P. Bernays: *Grundlagen der Mathematik I, II*. Springer, 1934, 1939.

[Hj]   J. van Heijenoort: *From Frege to Gödel.* Harvard University Press, 1967.

[Hy]   A. Heyting, Die intuitionistische Grundlegung der Mathematik. *Erkenntnis* 2, 106-115, 1931.

[int]   http://plato.stanford.edu/entries/logic-intuitionistic 1999.

[Ja1]   S. Jaśkowski: The theory of deduction based on the method of suppositions. *Studia Logica* 1, 5–32, 1934. Reprinted in St. McCall (ed): *Polish Logic 1920–1939*, North-Holland, 232–258, 1963.

[Ja2]   S. Jaśkowski: Rachunek zdan dla systemow dedukcyjnych sprzecznych. *Studia Societatis Scientiarun Torunesis, Sectio A*, 1(5), 55-77, 1948. Translated as: Propositional Calculus for Contradictory Deductive Systems. *Studia Logica* 24, 143–157, 1969.

[Jo]   I. Johansson: Der Minimalkalkül, ein reduzierter intuitionistischer Formalismus. *Compositio Mathematica* 4(1), 119–136, 1936.

[Kg]   A. N. Kolmogorov: O principie tertium non datur. *Matematiceskij Sbornik (Recueil Mathématique)* 32, 1924/25. Translated as: On the principle of excluded middle, in [Hj].

[KK]   W. Kneale, M. Kneale: *The Development of Logic.* Clarendon Press, 1962.

[KKT]   A. Kakas, R. Kowalski, F. Toni: The Role of Abduction in Logic Programming. *Handbook of Logic in Artificial Intelligence and Logic Programming*, Oxford University Press, 235–324, 1995.

[KL]   M. Kifer, E. Lozinskii: RI: A Logic for Reasoning with Inconsistency. *Proc. 4th LICS*, 253–262, 1989.

[Kl]   S. C. Kleene: *Introduction to Metamathematics.* North-Holland, 1952.

[Ku]   K. Kunen: Negation in Logic Programming. *J. Logic Programming* 4(4), 289–308, 1987.

[KuK]   R. Kowalski, D. Kuehner: Linear Resolution with Selection Function. *Artificial Intelligence* 2(3/4), 227–260, 1971.

[Kw1]   R. Kowalski: *Logic for Problem Solving.* North-Holland, 1979.

[Kw2]   R. Kowalski: Logic without Model Theory. In D. Gabbay (ed): *What is a logical system?* Oxford University Press 1994, pp 35–71.

[LL]   C. Lewis, C. Langford: *Symbolic Logic.* Century Co., 1932.

[Ll]   J. Lloyd: Foundations of Logic Programming, 2nd edition. Springer, 1987.

[Lo]   P. Lorenzen: *Einführung in die operative Logik und Mathematik*, 2nd edition. Springer, 1969.

[Lu]   J. Łukasiewicz: *O zasadzie sprzecznosci u Arystotelesa: Studium Krytyczne* (On Aristotle's Principle of Contradiction: A Critical Study), 1910. Translated as: On the Principle of Contradiction in Aristotle, *Review of Metaphysics* 24, 485–509, 1971.

[My]   M. Minsky: A Framework for Representing Knowledge. Tech. Report 306, MIT AI Lab, 1974. Reprinted (sans appendix) in P. Winston (ed): *The Psychology of Computer Vision*, McGraw-Hill, 177–211, 1975.

[MZ]   J. Minker, G. Zanon: An Extension to Linear Resolution with Selection Function. *Information Processing Letters* 14(4), 191–194, 1982.

[M+]   F. Muñoz, L. Irún, P. Galdámez, J. Bernabéu, J. Bataller, C. Bañuls, H. Decker: Flexible Management of Consistency and Availability of Networked Data Replications. *Proc. 5th FQAS*, Springer LNCS 2522, 289–300, 2002.

[par]   http://plato.stanford.edu/entries/logic-paraconsistent/, 2000.

[Pl]   D. Poole: Logic Programming, Abduction and Probability. *Proc. Fifth Generation Computer Systems '92*, 530–538, 1992.

[Pt]   E. Post: Introduction to a General Theory of Propositions. *American Journal of Mathematics* 43, 1921. Reprinted in [Hj].

[Ro]   J. A. Robinson: Logic: *Form and Function – The Mechanization of Deductive Reasoning*. Edinburgh University Press, 1979.

[rel]   http://plato.stanford.edu/entries/logic-relevance/, 1998.

[Ru]   B. Russell: Letter to Frege, 1902, in [Hj].

[RW]   B. Russell, A. Whitehead: *Principia Mathematica*. Cambridge University Press, 1910–1913, reprinted 1962.

[Se]   A. Sette: On the Propositional Calculus P1. *Mathematica Japonicae* 18, 173–180, 1973.

[Sl]   J. Słupecki: Der volle dreiwertige Aussagenkalkül. *Comptes Rendus Séances Société des Sciences et Lettres Varsovie* 29, 9–11, 1936.

[Ta]   A. Tarski: Der Wahrheitsbegriff in den formalisierten Sprachen. *Studia Philosophica* 1, 261–405, 1935.

[tcl]   http://www.acm.org/pubs/tocl/scope.html, 2002.

[Ur]   A. Urquhart: Many-Valued Logic. In Gabbay, Guenthner (eds): *Handbook of Philosophical Logic*. Kluwer, 71–116, 1986.

Annex: Glossary

Acronyms and technical terms are listed alphabetically on the left. Explications, synonyms, formalizations and comments are on the right.

ALP   Abductive Logic Programming

CL   Computational Logic

datalog Horn clause logic, possibly extended with NaF. Its declarative interpretation yields a specification language for databases, and its procedural interpretation yields the corresponding query language.

ECQ   Ex Contradictione Quodlibet; ex contradictione sequitur quodlibet; ex falso quodlibet; ex absurdo quodlibet; Duns Scotus' law.
        Either one of the following four formulas can be found in the literature as a propositional formalization of ECQ:
        $(p \wedge \neg p) \rightarrow q$, $(\neg p \wedge p) \rightarrow q$, $p \rightarrow (\neg p \rightarrow q)$, $\neg p \rightarrow (p \rightarrow q)$,
        occasionally prefixed with $\vdash$, or in parentheses-free notation, or with other symbols for $\wedge$, $\neg$, $\rightarrow$. The four versions are mutually equivalent in classical, intuitionistic, minimal and da Costa's logics C_n $(1 \leq n \leq \omega)$ and F. That can be shown by applying the law of commutativity of conjunction and LEI, which are also valid in each of the logic systems just mentioned.

LCP   Law of Counter-Position: $(p \rightarrow q) \rightarrow (\neg q \rightarrow \neg p)$

LDN   Law of Double Negation: $\neg \neg p \leftrightarrow p$
        Sometimes, LDN and LEM are taken to be equivalent, which is all

right for classical and intuitionistic logic. However, in da Costa's systems C_i $(1 \leq i \leq \omega)$ and F, one does not follow from the other.

LDT Laws of Disjunctive Thinning; laws of disjunctive weakening; laws of addition: $p \to (p \vee q)$, $p \to (q \vee p)$

LEI Law of Importation and Exportation: $((p \wedge q) \to r) \leftrightarrow (p \to (q \to r))$
'Importation' is the *if* ($\leftarrow$), 'exportation' the *only-if* ($\to$) half of LEI.

LEM Law of Excluded Middle; tertium non datur: $p \vee \neg p$

LMI Law of Material Implication: $(p \to q) \leftrightarrow (\neg p \vee q)$

LoC Law of Contradiction; law of noncontradiction: $\neg(p \wedge \neg p)$
Sometimes, LoC and ECQ are inadvertently identified (e.g., in [int]). That may be all right for classical, intuitionistic and da Costa's paraconsistent logic. In CL, however, they are different: LoC holds in datalog and is explicitly reasoned with in ALP, while ECQ is never applied.

LP Logic Programming

MP Modus Ponens: $p \to q, p \vdash q$

MT Modus Tollens: $p \to q, \neg q \vdash \neg p$

NaF Negation as Failure
(names the connective $\sim$ as well as the associated inference rule)

PL Positive Logic (axioms 1–8 in fig. 1)

RaA Reductio ad Absurdum; Kolmogorov's law:
$(p \to q) \to ((p \to \neg q) \to \neg p)$

SL Selection-controlled Linear Resolution;
linear input resolution with selection function

SLD Selection-controlled Linear Definite Resolution

SLDNF SLD extended with NaF

Soft Constraints and Heuristic Constraint Correction in Entity-Relationship Modelling

Sven Hartmann

Dept. of Information Systems & Information Science Research Centre
Massey University, Palmerston North, New Zealand

Abstract. In entity-relationship modelling, cardinality constraints impose restrictions on the number of occurrences of objects in relationships. If violations may appear, cardinality constraints should be treated as soft constraints rather than as integrity constraints. Nevertheless one often expects them to be satisfied at least in average or up to a small number of exceptions. These expectations may compete each other and cause new kinds of inconsistencies. We discuss how these inconsistencies can be detected and repaired.

1 Introduction

Conceptual modelling is widely accepted as an inherent phase of database design. It aims on determining and describing the structure, organization and interaction of data to be stored in the database under development. It results in a conceptual database schema which provides an abstract description of the target of the database using high-level terms and acts as a basis for further design steps. In addition, a conceptual database schema happens to be a useful tool for tasks like database integration or reverse engineering.

In database design great attention is devoted to the modelling of semantics. During requirements engineering database designers usually come across a variety of actual and desirable properties of the data to be stored. Integrity constraints specify the way by that objects in the domain of interest are associated to each other and, thus, describe those database instances which plausibly represent the target of the database. Whenever an integrity constraint is violated, this indicates that the present database instance is wrong, i.e., does not reflect any possible state of the domain of interest and, therefore, should not occur during the lifetime of the database.

In addition, there exist properties which are desirable or normally satisfied, but actually violations may occur. These properties do not yield integrity constraints. Nevertheless, they represent valuable information which should be collected. Desirable properties of the data give rise to soft constraints which are used to express ideal situations or preferences. Good database specification languages allow to specify them with the conceptual schema.

The most popular approach towards conceptual modelling, the entity-relationship model (ER), was introduced in the 1970s and has been considerably

L. Bertossi et al. (Eds.): Semantics in Databases, LNCS 2582, pp. 82–99, 2003.

extended and widely used ever since. Two kinds of integrity constraints are almost always defined for conceptual database schemas using the ER approach: keys and cardinality constraints. A cardinality constraint is a restriction that bounds the number of relationships an object from a given object set participates in. For example, if A is a set of professors and B a set of lectures, we may require that every lecture in B is taught by exactly one professor and every professor in A teaches between 3 and 5 lectures.

Cardinality constraints are good candidates for soft constraints. In practice it is often difficult to specify exact bounds for the occurrence of objects in relationships. In many cases we only have a vague idea of what might be or what should be. When cardinality constraints turn out to be soft, we may not expect them to be satisfied by all objects in every state of the target of the database. However, when specifying a soft constraint one often has certain ideas up to which degree the constraint corresponds to reality. In our example above, we could expect that professors in average teach between 3 and 5 lectures or that the number of professors teaching less than 3 or more than 5 lectures is considerably small.

It is well-known that cardinality constraints collected during requirements engineering may compete each other. While every set of cardinality constraints is satisfiable by the empty database, it may happen that some of the object sets stay invariably empty during the lifetime of the database. Usually, this indicates an error in the conceptual design. Constraint sets observing this property are said to be inconsistent [16,22]. Inconsistent sets of cardinality constraints were characterized in [2,9,10,16,21]. A polynomial-time algorithm for consistency checking was presented in [8], and in [11] we discussed a catalogue of strategies to repair inconsistent sets of cardinality constraints.

Of course, considering constraints as soft does not prevent them from competing each other. Conflicts of this kind are usually accepted as they may actually occur in the target of the database. For a discussion of this issue, the interested reader is referred to [19]. However, if cardinality constraints are treated as soft, but expected to be satisfied in average or with a small number of exceptions, this may lead to new inconsistencies which should be repaired.

The objective of this paper is to extend the investigation in [11] to soft cardinality constraints and to discuss strategies to revise expectations in case of conflicts among soft cardinality constraints. The paper is organized as follows. In Section 2, we assemble basic notions of the ER approach to conceptual modelling. In Section 3, we give a formal definition of cardinality constraints. Section 4 addresses the consistency problem for cardinality constraints treated as integrity constraints. In Section 5, we introduce soft cardinality constraints and study possible inconsistencies arising from expectations on their satisfaction. Finally, in Section 6, we survey strategies for constraint correction.

2   Entity-Relationship Modelling

The entity-relationship approach to conceptual design considers the target of a database as consisting of entities and relationships. In this section, we shall briefly review basic ideas of this approach. For a recent comprehensive survey on entity-relationship modelling, we refer to [22].

Entities and relationships are objects that are stored in a database. Intuitively, entities may be seen as basic objects in the domain of interest, whereas relationships are derived objects representing connections between other objects. Usually, a database contains lots of objects with common properties. By classifying them and pointing out their significant properties, we obtain *object types* that are used to model the objects under discussion.

More formally, we specify for every object type $\underline{v}$ its set of components $Co(\underline{v})$, its set of attributes $Attr(\underline{v})$ and its set of primary keys $Id(\underline{v})$. Roughly speaking, components are again object types, while attributes are intended to store additional properties of the objects modelled by $\underline{v}$. For some applications it is useful to allow some object type to occur several times as a component of the fixed object type $\underline{v}$. To avoid confusion, *roles* are associated with the different occurrences. As an example consider an object type DESCENDENT whose components are both of type PERSON. The roles of the two components are Parent and Child. Though not always explicitly set out, all our considerations apply to this case, too.

The *arity* of an object type $\underline{v}$ is the number of its components. Object types without components are *entity types*. Designing a database usually starts with specifying entity types to model the basic real-world objects in the target of the database. Conversely, object types with components are *relationship types*. In practice, unary object types are often used to model specializations of other object types, while object types of arity 2 or larger usually reflect real-world aggregations or relationships between objects of the types in $Co(\underline{v})$.

2.1   Database Schemas

All object types declared for the domain of interest, collectively, form a database schema. In particular, with every object type all its components belong to the database schema, too. For our investigation we found it convenient to consider a *database schema* as a finite digraph $S = (V, L)$. Its vertices are the *object types* of the schema, its arcs are said to be *links*. S contains a link $(\underline{r}, \underline{c})$ just when $\underline{c}$ is a component of $\underline{r}$.

The *order* of an object type $\underline{v}$ is the maximum length of a path in S with initial vertex $\underline{v}$. Clearly, entity types have order 0. A relationship type $\underline{r}$ of order k has components of order $k-1$ or smaller. Most textbooks on entity-relationship modelling do not allow cyclic specifications of object types, that is, S is supposed to be an acyclic digraph. In this case, all object types are of finite order. As a consequence, database schemas without dicycles possess a hierarchical structure: Every object type is declared on the basis of types already specified, i.e., types of smaller order. Object-oriented extensions of the entity-relationship approach, however, do allow dicycles. As an example consider an object type PERSON possessing itself as a component. This provides an alternative way of modelling Parent-Child associations. Again all our considerations apply to this case, too.

Example. Suppose we want to store data on a weekly university schedule. Each lecture offered to students yields a relationship between certain objects such as professors, courses, time slots and halls. This is modelled by an object type LECTURE with four components PROFESSOR, COURSE, TIME SLOT and HALL, see Fig. 1. Herein, LECTURE is a relationship type of order 1, while all its components are entity types.

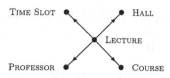

Fig. 1. A database schema modelling a university schedule.

2.2 Populations and Database Instances

All objects modelled by an object type $\underline{v}$ in a certain state of the target of the database form an *object set* $\underline{v}^t$. Its members are *instances of type* $\underline{v}$. Entities are instances of entity types, while relationships are instances of relationship types.

For every attribute A, let $dom(A)$ denote its domain, that is, the set of feasible values of A. If A is an attribute of $\underline{v}$, every object v of type $\underline{v}$ assigns a value $v(A) = a$ from $dom(A)$ to the attribute A. Similarly, if $\underline{c}$ is a component of $\underline{v}$, the object v assigns an object $v(\underline{c}) = c$ of type $\underline{c}$ to the component $\underline{c}$. Herein, c is said to *be involved* in v or to *participate* in v.

Usually an object set $\underline{v}^t$ is considered to be a finite subset of the cartesian product $\underline{c}_1^t \times \ldots \times \underline{c}_n^t \times dom(A_1) \times \cdots \times dom(A_m)$, where $\underline{c}_1, \ldots, \underline{c}_n$ are the components of $\underline{v}$ and $A_1, \ldots, A_m$ are its attributes. Hence, an object set $\underline{v}^t$ may be regarded as a table whose columns correspond to the components and attributes of $\underline{v}$ and whose rows correspond to the objects in $\underline{v}^t$. The number of objects in an object set is its *size*. The object sets $\underline{c}_i^t$ over the components of $\underline{v}$ are called the *codomains* of the object set $\underline{v}^t$.

In order to distinguish the objects in an object set, primary keys are specified for every object type. A non-empty subset K of $Co(\underline{v}) \cup Attr(\underline{v})$ is a *key* for a given object set $\underline{v}^t$ if the restrictions $v[K]$ are mutually distinct for all objects v in the set, that is, if for any two objects v_1 and v_2 there exists some component $\underline{c}$ (or some attribute A) such that $v_1(\underline{c}) \neq v_2(\underline{c})$ (or $v_1(A) \neq v_2(A)$) holds. Some subsets K may be keys by accident, while others are keys for all real-world object sets over $\underline{v}$. The latter ones are used within the identification mechanism in entity-relationship databases. For every object type $\underline{v}$ a set of *primary keys* $Id(\underline{v})$ is specified containing certain subsets $K \subseteq Co(\underline{v}) \cup Attr(\underline{v})$. An object set $\underline{v}^t$ over $\underline{v}$ is called a *population* if each $K \in Id(\underline{v})$ is a key for $\underline{v}^t$.

Example. Possible primary keys for the object type LECTURE in Fig. 1 are the subsets {TIME SLOT, PROFESSOR}, {TIME SLOT, COURSE} and

{TIME SLOT, HALL}, while {TIME SLOT} and {PROFESSOR, HALL} are rather unlikely to be keys, and should not be specified as primary keys.

A *database instance* S^t contains a population $\underline{v}^t$ for each object type $\underline{v}$ in the database schema S. It suggests itself that in a database instance S^t the codomains of an object type $\underline{v}$ are just the populations $\underline{c}_i^t$ over its components $\underline{c}_i \in Co(\underline{v})$.

3   Integrity Constraints in Entity-Relationship Modelling

3.1   Cardinality Constraints

Cardinality constraints are regarded as one of the basic constituents of the entity-relationship model. Consider a relationship population $\underline{r}^t$. For every object c in the codomain $\underline{c}^t$, let $\deg(\underline{r}^t, c)$ denote the number of all those relationships r in $\underline{r}^t$ that the object c participates in. We call $\deg(\underline{r}^t, c)$ the *degree* of c. A *cardinality constraint* on $\underline{r}$ is a statement $card(\underline{r}, \underline{c}) = M$ where M is a set of non-negative integers. The cardinality constraint *holds* in a population $\underline{r}^t$ if, for every object c in the codomain $\underline{c}^t$, the degree $\deg(\underline{r}^t, c)$ lies in M.

For most applications, the sets M are intervals, i.e., of the form $(a, b) = \{a, a+1, \ldots, b\}$ or $(a, \infty) = \{a, a+1, \ldots\}$. Throughout, let $cmin(\underline{r}, \underline{c})$ denote the minimum value in M, and $cmax(\underline{r}, \underline{c})$ the maximum value in M. If M is infinite, we put $cmax(\underline{r}, \underline{c}) = \infty$. If no cardinality constraint is specified for a link $(\underline{r}, \underline{c})$, we may assume the *trivial* cardinality constraint $card(\underline{r}, \underline{c}) = (0, \infty)$. It is easy to see that this does not impose any restriction, but is just a technical agreement.

For surveys on cardinality constraints, the interested reader is referred to [17, 22]. These monographs also furnish a great deal of information on generalized versions of cardinality constraints.

Example. In our weekly university schedule, objects do not appear arbitrarily often. Every professor teaches only a bounded number of lectures, say between 3 and 5. Further there are never too many lectures in parallel, say every time slot is used up to 4 times. These conditions may be expressed by cardinality constraints $card(\text{LECTURE}, \text{PROFESSOR}) = (3, 5)$ and $card(\text{LECTURE}, \text{TIME SLOT}) = (0, 4)$. Cardinality constraints are often reflected graphically by labelling the corresponding links, see Fig. 2

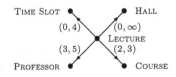

Fig. 2. A database schema with labels for the cardinality constraints.

3.2 Legal Database Instances

If a constraint σ holds in a population, then this population *satisfies* σ. A population satisfies a constraint set Σ if every constraint from Σ holds in it. Similarly, a database instance S^t satisfies a constraint set Σ specified on S if the constraints in Σ hold in the corresponding populations in S^t. Databases are used to store and manipulate data representing the target of the database. The typical situation in database design is that we generate a database schema S together with an domain-specific constraint set Σ declared on it. A database instance S^t is only meaningful if it satisfies Σ. Such a database instance is *legal* for Σ.

An object type is *superfluous* for a constraint set Σ if its population is empty in every database instance satisfying Σ. Further, Σ is called *consistent* if it does not force any object type to be superfluous, and *inconsistent* otherwise. A single constraint σ *competes* a consistent constraint set Σ if the union $\Sigma \cup \{\sigma\}$ is inconsistent.

3.3 Implications

Given a database schema S some constraints hold in every database instance over S, as for example the trivial cardinality constraints. Moreover, every primary key $K = \{\underline{c}\}$ consisting of a single component of an object type $\underline{r}$ gives rise to the cardinality constraint $card(\underline{r}, \underline{c}) = (0, 1)$. Thus, whenever a 1-element subset of $Co(\underline{r})$ is among the primary keys of $\underline{r}$, we immediately have the corresponding cardinality constraint, too.

Example. Consider the database schema in Fig. 3 which models information on the formation of project teams within a research institute. Suppose every project team may be identified by its manager, that is, {MANAGER} is a primary key for PROJECT TEAM. Clearly, this provides the cardinality constraint $card(\text{PROJECT TEAM}, \text{MANAGER}) = (0, 1)$, too.

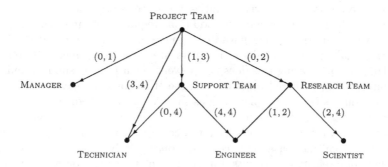

Fig. 3. A database schema modelling the formation of project teams.

In addition, the constraints satisfied by database instances over a given schema are usually not independent. A constraint set Σ semantically implies a further constraint σ if σ holds in every database instance $\underline{S}^t$ which satisfies Σ. In [10] a list of implication rules is given that may be used to derive new cardinality constraints from given ones. This list is proved to be a sound and complete system of implication rules.

Example. In the database schema in Fig. 3 a straightforward calculation shows that the number of research teams will always be exactly four times the number of engineers. Consequently, we may conclude the cardinality constraint $card(\text{PROJECT TEAM, SUPPORT TEAM}) = (1,1)$ which is implied by the original constraints and provides additional semantic information.

4   Consistency of Cardinality Constraints

During requirements engineering semantic information is gathered from domain experts and potential users of the database under development. In general, they express their knowledge and requirements in a language they are familiar with, that is, in almost all cases in a natural language. It is the task of the database designer to understand the domain-specific terminology, to extract semantic information and to translate this information into integrity constraints.

4.1   Problems during Constraint Acquisition

In practice, the acquisition of integrity constraints in conceptual modelling is a rather complex task. Though this task covers only a small proportion of the overall database development effort, it has a profound impact on the final result. Errors that occur during later phases of the design process or when developing applications on top of the database system are often the result of inaccurate requirements engineering.

There are a number of potential sources for mistakes during constraint acquisition. First, domain experts and users may have different viewpoints and preferences as well as different abilities and experience in formulating their domain knowledge and demands. In addition, requirements may change over the time. Moreover, translating natural language into formal language is far from being trivial. Even if database designers are assisted by methods from natural language processing this does not prevent them from misunderstanding and misinterpreting gathered requirements, see [1]. Problems may also occur when further sources come into play such as sample databases which may be used to extract semantic information.

Second, database designers have to decide which information is essential to guarantee legal database instances and, consequently, has to be captured with the help of integrity constraints. Finally, for the formalization of semantic information a deep understanding of logic is required. This is essential for choosing appropriate integrity constraints to capture actual requirements. Some requirements may even be too complicated to be modelled directly in the specification language used for database design. Then the quality of design decisions depends on the skill and experience of the individual designer.

Often designers are confronted with 'too little' or 'too much' information about the target of the database. The problem of incomplete information has been discussed mostly with respect to bottlenecks during normalization [12,18]. On the other hand, 'too much' information may lead to inconsistent constraint sets.

Example. Consider the database schema in Fig. 3 and suppose we are confronted with the additional requirement that every research team participates in at least 3 project teams. Obviously, this competes the present cardinality constraint $card$(PROJECT TEAM, RESEARCH TEAM) $= (0, 2)$. This kind of problem has been discussed in [15]. Conflicts arising from constraints specified for the same link are easy to detect by scanning the schema.

On the other hand, inconsistencies may be caused by conflicting constraints which are specified for different links. Suppose we gather the additional requirement that every research team participates in at most one project team. At the first glance, this goes with the present cardinality constraint $card$(PROJECT TEAM, RESEARCH TEAM) $= (0, 2)$ which should be strengthened to $card$(PROJECT TEAM, RESEARCH TEAM) $= (0, 1)$. However, this new constraint forces every legal database instance to contain neither research teams nor project teams and, thus, indicates some kind of global inconsistency caused by conflicts among distributed semantic information.

In view of these problems it is desirable to have support for the acquisition of semantic constraints as suggested e.g. in [1,11]. When conflicts occur during requirements engineering this is not necessarily a problem as long as the database designer knows how to proceed. There are at least two major tasks where database designers need assistance: consistency checking and constraint correction in case of inconsistencies.

4.2   Consistency Checking

Consistency is a basic requirement for the correctness of the chosen conceptual schema. In our example, the research institute wants to know in advance whether it is possible to form teams satisfying all the specified rules, that is, whether legal database instances exist. Though the empty database is always legal, the research institute is surely not interested in solutions without any teams.

As pointed out in [9,16,21], methods from combinatorial optimization may be used to reason about cardinality constraints. Let us consider a database schema $S = (V, L)$. For every link $\ell = (\underline{r}, \underline{c})$, we define its *reverse arc* $\ell^{-1} = (\underline{c}, \underline{r})$. By L^{-1} we denote the set of all reverse arcs, and put $A = L \cup L^{-1}$. Now we turn our attention to the resultant digraph $D = (V, A)$, known as the *symmetric digraph* of S. On its arc set A, we define a weight function $\omega : A \to \mathbb{Q}_0 \cup \{\infty\}$ by

$$\omega(a) = \begin{cases} \frac{1}{cmin(\underline{r},\underline{c})} & \text{if } a \text{ is a link } \ell = (\underline{r}, \underline{c}), \\ cmax(\underline{r}, \underline{c}) & \text{if } a \text{ is the reverse arc } \ell^{-1} = (\underline{c}, \underline{r}) \text{ of a link } \ell. \end{cases}$$

The weight function may easily be extended to subgraphs in D. Special interest is devoted to walks in D. A *walk* P from a vertex v_0 to a vertex v_k is a sequence of arcs $(v_0, v_1), (v_1, v_2), \ldots, (v_{k-1}, v_k)$. Herein k denotes the length of the walk. Note that we do not claim the arcs in a walk to be pairwise distinct. The weight of a walk P is given by setting

$$\omega(P) = \prod_{i=1}^{k} \omega((v_{i-1}, v_i)).$$

If $k = 0$ we put $\omega(P) = 1$. A walk is a *dipath* if the vertices $v_0, v_1, \ldots, v_k$ are mutually distinct, whereas it is a *dicycle* if $v_0 = v_k$ holds and the vertices $v_1, \ldots, v_k$ are mutually distinct. A dicycle is said to be *critical* for a set Σ of cardinality constraints if its weight is less than 1. In combinatorial optimization, dicycles of this kind are also called *absorbing*.

Characterizations of consistent sets of cardinality constraints were given in [9,16,21]. The impact of critical dicycles was first pointed out in [16]. In [8] we developed a polynomial-time algorithm to recognize consistent sets. The consistency problem for cardinality constraints in the presence of primary keys declared for the object types as described in Section 2 was studied in [10]. We record the derived criterion for further reference.

Theorem 1 (cf. [10]) *Σ is consistent if and only if the symmetric digraph D contains neither critical dicycles nor arcs of weight 0.*

Example. In Fig. 3 the dicycle from PROJECT TEAM via SUPPORT TEAM, ENGINEER, RESEARCH TEAM back to PROJECT TEAM has weight 1. If we replace the actual cardinality constraint for the link between PROJECT TEAM and SUPPORT TEAM by $card$(PROJECT TEAM, SUPPORT TEAM) $= (2, 3)$ the weight becomes $\frac{1}{2}$, that is, we find a critical dicycle. In fact, this new constraint does not admit any legal database instance containing a team.

In [10] we also presented a method to find all object types that are superfluous due to the inconsistency of a given set of cardinality constraints Σ. For any two object types $\underline{v}$ and $\underline{w}$ the *distance* from $\underline{v}$ to $\underline{w}$ is given by

$$dist(\underline{v}, \underline{w}) = \inf\{\omega(P) : P \text{ is a walk from } \underline{v} \text{ to } \underline{w} \text{ in } D\}.$$

An object type $\underline{w}$ is superfluous for Σ whenever there is an object type $\underline{v}$ with $dist(\underline{v}, \underline{w}) = 0$, or $\underline{w}$ has a superfluous component, or $\underline{w}$ itself is the component of some superfluous object type $\underline{r}$ and $cmax(\underline{r}, \underline{w})$ is finite. This observation enabled us to determine all superfluous object types in polynomial time.

5   Soft Constraints

Apart from integrity constraints the objects in the target of the database possess properties which are normally satisfied, but which may be violated during the lifetime of the database. These properties should not be translated into integrity constraints. Instead they can be captured by soft constraints which are used to express ideal states of the domain of interest or preferences of the potential users of the database under development. Though it is sometimes difficult to decide whether a constraint is soft or not, the occurrence of inconsistencies may indicate that soft constraints have been incorrectly specified as (hard) integrity constraints.

Designing large information systems usually involves a lot of soft constraints. For every soft constraint the responsible designer may specify what should happen in case of violation. It is possible to report the violation, to restore integrity or to take some other compensation action. In recent years deontic logic has become increasingly popular as a tool to model soft constraints. Deontic logic is a modal logic in which the modal operator O is used to express that something is obliged. For a survey on deontic logic in database design, we refer to [19]. Sometimes all integrity constraints per se are regarded as soft [14].

In general, cardinality constraints are good candidates for soft constraints, in particular if their *cmax*-value is larger than 1. In many situations domain experts and potential users are not able to give exact bounds for the occurrence of objects in relationships.

Example. Recall the database schema in Fig. 3. Most of the constraints express properties that are desired by the management of the research institute. According to the cardinality constraint $card($RESEARCH TEAM, SCIENTIST$) = (2, 4)$ every scientist is a member of 2, 3 or 4 research teams. However it is likely that nobody will bother if some scientist participates in 5 research teams, as long as this is an exception and not the normal case. Under this assumption the cardinality constraint under discussion should be treated as a soft constraint.

Conversely, $card($PROJECT TEAM, MANAGER$) = (0, 1)$ states that every manager cares for only one project team. This constraint holds since the component MANAGER forms a primary key for the object type PROJECT TEAM and may not be considered as a soft constraint. Generally, cardinality constraints with *cmax*-value equal to 1 are less suited as soft constraints since they are frequently used within the identification mechanism of the database system.

In practice, one is usually prepared to accept violations of the ideal situation described by a soft constraint. In particular, considering constraints as soft does not prevent them from competing each other. In this case, of course, violations are not only possible but obligatory. Nevertheless, in almost all cases we expect soft constraint to be satisfied at least 'up to a certain

degree'. In the sequel we survey two possible approaches to formalize relaxed versions of satisfaction.

5.1   Satisfaction in Average

Let $\underline{r}$ be a relationship type in the database schema S, and let $\underline{c}$ be one of its components. A population $\underline{r}^t$ satisfies a cardinality constraint $card(\underline{r}, \underline{c}) = (a, b)$ *in average* if the average degree

$$\deg_{average}(\underline{r}^t, \underline{c}^t) = \frac{1}{|\underline{c}^t|} \sum_{c \in \underline{c}^t} \deg(\underline{r}^t, c)$$

over all objects in the codomain $\underline{c}^t$ lies in the interval (a, b).

We call a set of cardinality constraints Σ *consistent in average* if there is no object type whose population is empty in every database instance satisfying all constraints in Σ in average. Surprisingly, consistency in average turns out to be equivalent to consistency.

Lemma 2 *Let Σ be a set of cardinality constraints, and suppose that the symmetric digraph D contains an arc of weight 0 or a critical dicycle for Σ. Then there exists an object type $\underline{w}$ whose population is empty in every database instance S^t which satisfies all constraints in Σ in average.*

Proof. Given a database instance S^t, let $g(\underline{v})$ denote the size of its population $\underline{v}^t$. If the population $\underline{r}^t$ satisfies a cardinality constraint $card(\underline{r},\underline{c}) = (a,b)$ in average we immediately conclude the inequality $a\,|\underline{c}^t| \leq \sum_{c \in \underline{c}^t} \deg(\underline{r}^t, c) \leq b\,|\underline{c}^t|$ and thus $a\,g(\underline{c}) \leq g(\underline{r}) \leq b\,g(\underline{c})$.

Suppose S^t satisfies every constraint in Σ in average. Due to the definition of the weight function we derive $g(\underline{w}) \leq g(\underline{v})\omega(a)$ for every arc $a = (\underline{v}, \underline{w})$ in the symmetric digraph D of the schema. By induction, this provides the inequality $g(\underline{w}) \leq g(\underline{v})\omega(P)$ for every walk P from $\underline{v}$ to $\underline{w}$. Suppose there is an arc $a = (\underline{v}, \underline{w})$ of weight 0. This gives us $g(\underline{w}) = 0$, that is, the population over $\underline{w}$ is empty. Otherwise, suppose there is a dicycle C with weight smaller than 1. For every object type $\underline{w}$ on C we have $g(\underline{w}) \leq g(\underline{w})\omega(C)$ which is valid only if $g(\underline{w}) = 0$ holds, that is, the population over $\underline{w}$ is again empty. $\square$

The previous result shows that for sets of cardinality constraints consistency in average implies consistency. Conversely, consistency trivially allows us to conclude consistency in average.

Theorem 3 *Let Σ be a set of cardinality constraints. Σ is consistent in average if and only if Σ is consistent.*

5.2   Satisfaction at a Certain Percentage

Even if a cardinality constraint may be violated by some objects in a population we often expect that the number of exceptions is small in comparison with the total size of the actual population.

Let $\underline{r}$ be a relationship type with component $\underline{c}$, and let p be a real number with $0 \leq p \leq 100$. A population $\underline{r}^t$ satisfies a cardinality constraint $card(\underline{r}, \underline{c}) = M$ *at p percent* if at least p percent of the objects in the codomain $\underline{c}^t$ have their degree $\deg(\underline{r}^t, c)$ in the preferred set M.

Obviously, if we expect a cardinality constraint to be satisfied at 100 percent this means that we treat the constraint as a (hard) integrity constraint. Conversely, if we expect satisfaction at less than 100 percent this means that we consider the cardinality constraint as a soft constraint.

It is worth mentioning that cardinality constraints which are considered to be soft but expected to be satisfied at a certain percentage still give rise to conflicts. We call a set of cardinality constraints Σ *p-consistent* if there is no object type whose population is empty in every database instance satisfying all constraints in Σ at p percent. Clearly, 100-consistency is equivalent to consistency. For $p < 100$ this is usually not true. In order to study satisfaction at a certain percentage, we replace the original weight function ω by a revised weight function $\omega_p : A \to \mathbb{Q}_0 \cup \{\infty\}$ with

$$
\omega_p(a) = \begin{cases} \frac{100}{p \, cmin(\underline{r},\underline{c})} & \text{if } a \text{ is a link } \ell = (\underline{r}, \underline{c}), \\ 1 & \text{if } a \text{ is a reverse arc } \ell^{-1} = (\underline{c}, \underline{r}) \text{ and } \{\underline{c}\} \in Id(\underline{r}), \\ \infty & \text{otherwise.} \end{cases}
$$

Using this new weight function we again derive a criterion which tells us whether there are object types having invariably empty population.

Lemma 4 *Let Σ be a set of cardinality constraints, let $p < 100$ be a given percentage and suppose that the symmetric digraph D contains a critical dicycle for Σ with respect to the revised weight function ω_p. Then there exists an object type $\underline{w}$ whose population is empty in every database instance S^t which satisfies all constraints in Σ at p percent.*

Proof. For a fixed link $(\underline{r}, \underline{c})$ let Z denote the set of objects c in the codomain $\underline{c}^t$ which satisfy the inequality $cmin(\underline{r}, \underline{c}) \leq deg(\underline{r}^t, c)$. If the population $\underline{r}^t$ satisfies $card(\underline{r}, \underline{c}) = M$ at p percent, the ratio $|Z|/|\underline{c}^t|$ is at least $\frac{p}{100}$. This yields

$$
\frac{p}{100} g(\underline{c}) \, cmin(\underline{r}, \underline{c}) \leq |Z| \, cmin(\underline{r}, \underline{c}) \leq \sum_{c \in Z} deg(\underline{r}^t, c) \leq g(\underline{r}).
$$

With our new weight function we again obtain $g(\underline{w}) \leq g(\underline{v})\omega_p(a)$ for every arc $a = (\underline{v}, \underline{w})$ in the symmetric digraph D of the schema. By induction, this provides the inequality $g(\underline{w}) \leq g(\underline{v})\omega_p(P)$ for every walk P from $\underline{v}$ to $\underline{w}$. Suppose there is a dicycle C with weight smaller than 1. For every object type $\underline{w}$ on C we have $g(\underline{w}) \leq g(\underline{w})\omega_p(C)$ which is valid only if $g(\underline{w}) = 0$ holds, that is, the population over $\underline{w}$ must be empty. $\qquad\qquad\square$

Example. Consider the database schema in Fig. 3, but with the constraint $card(\text{PROJECT TEAM}, \text{SUPPORT TEAM}) = (2, 3)$ for the link between PROJECT TEAM and SUPPORT TEAM. The weight of this link is $\omega_p(ProjectTeam, SupportTeam) = \frac{50}{p}$ and the weight of the dicycle C from PROJECT TEAM via SUPPORT TEAM, ENGINEER, RESEARCH TEAM back to PROJECT TEAM amounts to $\omega_p(C) = \infty$. Hence, we do not have a critical dicycle. In fact, it is always possible to find a database instance satisfying the constraints under discussion up to a fixed percentage $p < 100$. For $p \leq 50$ we may construct a database instance where the number of project teams equals the number of support teams with some support teams participating in less than 2 project teams. For $50 < p < 100$, we must allow some engineers to appear in more than 2 research teams or some research teams to appear in more than 2 project teams. Thus, the constraint set is inconsistent, but p-consistent for every $p < 100$.

Similar to the case where cardinality constraints are treated as integrity constraints, see [10], we may derive the following characterization of p-consistent sets of cardinality constraints.

Theorem 5 *Let Σ be a set of cardinality constraints and $p < 100$ a given percentage. Σ is p-consistent if and only if the symmetric digraph D contains no critical dicycles with respect to the revised weight function ω_p.*

6   Strategies for Constraint Correction

When the criteria derived in the previous sections fail, the constraint set Σ under discussion is inconsistent (inconsistent in average, p-inconsistent) and we may conclude the existence of superfluous object types. Resolving inconsistencies usually means relaxing or removing one or more constraints. If no additional information is available, any constraint is a candidate for this. However most constraints do not have any impact on the conflict. Thus we have to localize the conflict before we can fix it in an appropriate way.

In [11] we suggested several strategies to correct inconsistent sets of cardinality constraints. We shall now discuss how these approaches may be used when cardinality constraints are treated as soft constraints but expected to be satisfied in average or at a certain percentage. In the case of inconsistency in average, the situation is fairly easy: since consistency in average is equivalent to consistency we may use the very same rules to correct cardinality constraints. In case of p-inconsistency, however, we shall apply the revised weight function ω_p. In the sequel let p always be a fixed percentage strictly less than 100.

6.1   Incremental Consistency Checking

In general, the occurrence of superfluous object types goes back to a mistake made during the acquisition of constraints. If we do not fix this mistake, it will be a potential source for complications in future design steps or even failures during the lifetime of the database.

A simple strategy to avoid constraint conflicts is *incremental consistency checking*, see [23]. Whenever a new constraint is gathered and should be added to the schema, we first test whether it forces some object type to become superfluous. This enables us to decide whether the new constraint contributes to a p-inconsistency, and to react accordingly. We say that a cardinality constraint σ *p-competes* a p-consistent set of cardinality constraints Σ if the union $\Sigma \cup \{\sigma\}$ is p-inconsistent.

To investigate when a constraint is p-competing we again use the revised weight function ω_p. Based on this function we derive modified distances

$$dist_p(\underline{v}, \underline{w}) = \inf\{\omega_p(P) : P \text{ is a walk from } \underline{v} \text{ to } \underline{w} \text{ in } D\}$$

between any two object types $\underline{v}$ and $\underline{w}$. Given a database instance S^t, let $g(\underline{v})$ denote the size of its population $\underline{v}^t$. Suppose S^t satisfies all cardinality constraints in Σ at p percent. As mentioned above we obtain the inequality $g(\underline{w}) \leq g(\underline{v})\omega_p(P)$ for every walk P between $\underline{v}$ to $\underline{w}$ in the symmetric digraph D and, consequently,

$$g(\underline{w}) \leq g(\underline{v})dist_p(\underline{v}, \underline{w})$$

holds for any two object types $\underline{v}$ and $\underline{w}$.

Clearly, critical dicycles with respect to ω_p can only be caused by cardinality constraints with $cmin$-value larger than 1, or by cardinality constraints derived from 1-element primary keys. However, cardinality constraints from primary keys should not be corrected as primary keys are essential for the identification mechanism in databases. In fact, the latter constraints do never compete each other. Hence, our concentration is on the $cmin$-value of cardinality constraints.

Theorem 6 *Let σ be a cardinality constraint $card(\underline{r}, \underline{c}) = M$ defined for the link $\ell = (\underline{r}, \underline{c})$ such that $M \neq \emptyset$ and $\{\underline{c}\}$ is not a primary key for $\underline{r}$. Then σ p-competes Σ if and only if we have $dist_p(\underline{c}, \underline{r}) < \frac{p}{100} cmin_\sigma(\underline{r}, \underline{c})$.*

If the cardinality constraint to be added causes a conflict, it should be either refused or corrected. The previous result immediately proposes an appropriate correction:

Rule 1. If $dist_p(\underline{c}, \underline{r}) < \frac{p}{100} cmin_\sigma(\underline{r}, \underline{c})$ then put $cmin_\sigma(\underline{r}, \underline{c}) = \lfloor \frac{100}{p} dist_p(\underline{c}, \underline{r}) \rfloor$.

The discussion above suggests a possible strategy to avoid p-inconsistencies. We simply start with an initial constraint set containing only the cardinality constraints given by the 1-element primary keys, and successively add all other cardinality constraints under inspection. Whenever a new constraint is p-competing, it has to be refused or corrected before it is added to the constraint set. With this, the generated constraint set always stays p-consistent.

It goes without saying that result of this procedure depends on the order in which the constraints are handled: Constraints added later are more likely to be corrected or refused. Often some constraints are more significant or preferred than others. These ones should be considered first.

Further, incremental constraint checking can also be applied when constraint sets have to be updated due to new requirements or domain knowledge. Constraints may change over time. To avoid undesired constraint violations by database operations, the specified semantics has to be revised, cf. [5]. Deleting constraints will never cause a new conflict, while adding a new constraint can be critical. In this case we can again apply the strategy presented above.

6.2   Minimal Inconsistent Constraint Sets

Suppose we observe that a constraint set is p-inconsistent. Naturally, we are interested in detecting the source of this inconvenience. We call a p-inconsistent constraint set Σ_0 *minimal* if every proper subset of Σ_0 is p-consistent. A good strategy towards conflict resolution is the following: For every minimal p-inconsistent subset Σ_0 of Σ we choose a suitable constraint $\sigma \in \Sigma_0$ and remove it from Σ. for this, we first need to detect all minimal p-inconsistent subsets.

Consider a dicycle C in the symmetric digraph D of the schema. Each arc in this dicycle carries a cardinality constraint. Let $\Sigma(C)$ consist of the constraints $card(\underline{r}, \underline{c}) = M$ where either the link $\ell = (\underline{r}, \underline{c})$ or its reverse arc $\ell^{-1} = (\underline{c}, \underline{r})$ belongs to C. We say that $\Sigma(C)$ is *carried* by the dicycle C.

Theorem 7 *A set of cardinality constraints Σ_0 is minimal p-inconsistent if and only if it is carried by a critical dicycle with respect to the weight function ω_p.*

It is easy to see, that every link in a critical dicycle carries a cardinality constraint with *cmin*-value larger than 1, while every reverse arc carries a cardinality constraint derived by a 1-element primary key. The polynomial-time algorithm to decide the existence of a critical dicycle [8] uses single-root-shortest-path methods. This idea may be reused after some slight modification. Since there can be exponentially many minimal p-inconsistent subsets, we cannot expect a polynomially time algorithm for their detection. Our goal is rather to obtain a method which determines these subsets quickly with respect to their number. In this section, we study a method for this task which works fairly well in practice.

To begin with we introduce a new root vertex $\underline{s}$ to the symmetric digraph D and insert arcs $a = (\underline{s}, \underline{v})$ connecting $\underline{s}$ to all object types $\underline{v}$. Moreover we put $\omega_p(a) = 1$ such that all the new arcs have weight 1. For every object type $\underline{v}$, we maintain its potential $\pi(\underline{v})$ and its parent $f(\underline{v})$. Initially we have $\pi(\underline{v}) = \infty$ and $f(\underline{v}) = nil$ for every $\underline{v}$. For the root $\underline{s}$ we declare $\pi(\underline{s}) = 1$ which stays constant during the whole procedure. At each step, the method selects an arc $a = (\underline{v}, \underline{w})$ such that $\pi(\underline{v}) < \infty$ and $\pi(\underline{w}) > \pi(\underline{v})\omega_p(a)$ and puts $\pi(\underline{w}) = \pi(\underline{v})\omega_p(a)$ and $f(\underline{w}) = \underline{v}$. If no such arc exists, the algorithm terminates. The resultant potential function π is always positive rational.

Efficient implementations of this method such as the Bellman-Ford-Moore algorithm use priority queues to decide in which order the arcs are inspected. In the literature, several variations of this algorithm have been suggested to compute the shortest paths from a root to all the other vertices. For surveys on this topic, see [7,4].

At each step, the *parent digraph* D_p is the subgraph of D which contains the arcs $(f(\underline{w}'), \underline{w}')$ for all object types $\underline{w}'$ with $f(\underline{w}') \neq nil$. The parent digraph has a nice property which is profitable for our purpose here: We have $\pi(\underline{w}') \geq \pi(\underline{v}')\omega_p(a)$ for every arc $(\underline{v}', \underline{w}') = (f(\underline{w}'), \underline{w}')$ in D_p. Hence we conclude that

$$\pi(\underline{w}') \geq \pi(\underline{v}')\omega_p(P)$$

holds for every dipath P from a vertex $\underline{v}'$ to a vertex $\underline{w}'$ in the parent digraph D_p. Suppose the algorithm selects in a certain step an arc $a = (\underline{v}, \underline{w})$ such that $\pi(\underline{v}) < \infty$ and $\pi(\underline{w}) > \pi(\underline{v})\omega_p(a)$ hold. Then this arc will be inserted into the parent digraph. Assume now that the arc a produces a dicycle $C = P \cup \{a\}$, where P is some dipath from $\underline{w}$ to $\underline{v}$ which has already been in D_p. We have $\pi(\underline{v}) \geq \pi(\underline{w})\omega_p(P)$, which implies

$$1 > \omega_p(a)\omega_p(P) = \omega_p(C).$$

This shows that C is a critical dicycle.

Lemma 8 *If a dicycle occurs in D_p then it is critical with respect to the weight function ω_p.*

Consequently, we modify the algorithm as follows: After every step, when a new arc has to be inserted into P, we look for a dicycle in D_p. This is fairly easy and can be done in linear time $O(|A|)$: Every vertex lies on at most one incoming and at most one outgoing arc in the parent digraph. It is not difficult to see that this search will be successful after some time:

Theorem 9 *If Σ is p-inconsistent, then D_p will contain a dicycle after a finite number of steps. Clearly, this dicycle is critical with respect to ω_p.*

Computational aspects of classical shortest-path algorithms in the absence and presence of critical dicycles were studied in [4,3,13]. Heuristic improvements of the algorithms were discussed in [6,20], some of which also analyze the parent digraph. The best-known time bound $O(|V||L|)$ is achieved by the Bellman-Ford-Moore algorithm as long as there are no critical dicycles.

Corollary 10 *If Σ is p-consistent, the algorithm above stops in time $O(|V||L|)$. Otherwise it finds a minimal p-inconsistent subset of Σ in finite time.*

Whenever we find a critical dicycle C, it carries a constraint set $\Sigma(C)$ which is minimal p-inconsistent. At least one of the constraints in this set has to be removed or revised in order to resolve the p-inconsistency. Recall that all cardinality constraints carried by a critical dicycle either have *cmin*-value larger than 1 or correspond to a 1-element primary key. In particular, there must be at least one constraint of the first kind. Since cardinality constraints derived from primary keys should not be corrected, the database designer must select one constraint of the first kind to be corrected or deleted. It suffices to relax the chosen constraint $card(\underline{r}, \underline{c}) = M$ as follows:

Rule 2. If $\ell = (\underline{r}, \underline{c})$ lies in C then put $cmin(\underline{r}, \underline{c}) = \lfloor \frac{100}{p} \frac{\omega_p(\ell)}{\omega_p(C)} \rfloor$.

Afterwards, the algorithm has to be restarted with the modified constraint set. By iterated application of the algorithm we are able to find all critical dicycles. Note that the minimal p-inconsistent subsets are usually not disjoint: Thus the correction of a constraint may eliminate more than just a single minimal p-inconsistent subset. The essential factor determining the overall running time is the number of mutually disjoint critical dicycles. This number is unknown in advance, but expected to be reasonably small in most situations.

7 Conclusion

In the present paper, we pointed out the emergence of soft cardinality constraints. Often conflicts among constraints turn out to be caused by contradicting desires, interpretations or expectations, which may actually be violated in the target of the database. We studied cardinality constraints which are treated to be soft, but expected to be satisfied in average or with a small number of exceptions. Both approaches may still lead to inconsistencies which should be repaired. We discussed strategies to handle these inconsistencies. The idea is to offer suggestions which constraints should be refused or relaxed. These strategies may be used to implement algorithms assisting the database designer in the process of conflict resolution. The final selection of the constraint to be corrected or removed, however, is naturally at the discretion of the individual designer.

References

1. M. Albrecht, E. Buchholz, A. Düsterhöft, and B. Thalheim. An informal and efficient approach for obtaining semantic constraints using sample data and natural language processing. *LNCS*, 1358:1–11, 1996.
2. D. Calvanese and M. Lenzerini. On the interaction between ISA and cardinality constraints. In *Proc. of Tenth Int. Conf. on Data Engin.*, pp. 204–213, 1994.
3. B.V. Cherkassy and A.V. Goldberg. Negative cycle detection algorithms. *Math. Programming*, 85:277–311, 1998.
4. B.V. Cherkassy, A.V. Goldberg, and T.Radzik. Shortest path algorithms: theory and experimental evaluation. *Math. Programming*, 73:129–174, 1996.
5. E. Di Nitto and L. Tanca. Dealing with deviations in DBMSs: an approach to revise consistency constraints. *Integrity in Databases, FMLDO96*, pp. 11–24. 1996.
6. A.V. Goldberg and T.Radzik. A heuristic improvement of the Bellman-Ford algorithm. *Appl. Math. Letters*, 6:3–6, 1993.
7. M. Gondran and M. Minoux. *Graphs and algorithms*. Wiley, Chichecter, 1990.
8. S. Hartmann. Graphtheoretic methods to construct entity-relationship databases. *LNCS*, 1017:131–145, 1995.
9. S. Hartmann. On the consistency of int-cardinality constraints. *LNCS*, 1507:150–163, 1998.
10. S. Hartmann. On the implication problem for cardinality constraints and functional dependencies. *Annals Math. Artificial Intell.*, 33:253–307, 2001.
11. S. Hartmann. Coping with inconsistent constraint specifications. *LNCS*, 2224:241–255, 2001.
12. W. Kent. Consequences of assuming a universal relation. *ACM Trans. Database Syst.*, 6:539–556, 1981.
13. S.G. Kolliopoulos and C. Stein. Finding real-valued single source shortest paths in $o(n^3)$ expected time. In *Proc. 5th Int. Prog. Combin. Opt. Conf.* 1996.
14. K. Kwast. A deontic approach to database integrity. *Annals Math. Artificial Intell.*, 9:205–238, 1993.
15. M.L Lee and T.W. Ling. Resolving constraint conflicts in the integration of entity-relationship schemas. *LNCS*, 1331:394–407, 1997.
16. M. Lenzerini and P. Nobili. On the satisfiability of dependency constraints in entity-relationship schemata. *Inform. Systems*, 15:453–461, 1990.
17. S.W. Liddle, D.W. Embley, and S.N. Woodfield. Cardinality constraints in semantic data models. *Data Knowledge Engrg.*, 11:235–270, 1993.
18. H. Mannila and K. Räihä. *The design of relational databases*. Addison-Wesley, Reading, 1992.

19. J.-J.Ch. Meyer, R.J. Wieringa, and F.P.M. Dignum. The role of deontic logic in the specification of information systems. In J. Chomicki and G. Saake, editors, *Logics for databases and information systems*, pages 71–116. Kluwer, Dordrecht, 1998.
20. R.E. Tarjan. *Data structures and network algorithms*. SIAM, Philadelphia, 1983.
21. B. Thalheim. Foundations of entity-relationship modeling. *Annals Math. Artificial Intell.*, 6:197–256, 1992.
22. B. Thalheim. *Entity-relationship modeling*. Springer, Berlin, 2000.
23. D. Theodorates. Deductive object oriented schemas. *LNCS*, 1157:58–72, 1996.

Characterization of Type Hierarchies with Open Specification

Stephen J. Hegner

Umeå University
Department of Computing Science
SE-901 87 Umeå, Sweden
hegner@cs.umu.se
http://www.cs.umu.se/~hegner

Abstract. Type hierarchies which arise in applications are often described incompletely; this missing information may be handled in a variety of ways. In this work, such incomplete hierarchies are viewed as *open specifications*; that is, descriptions which are sets of constraints. The actual hierarchy is then any structure satisfying these constraints. For such specifications, two forms of characterization are provided. The first is algebraic and utilizes a generalization of weak partial lattices; it provides a structure-based characterization in which optimality is characterized via an initial construction. The second is logical, an inference-based representation, in which models are characterized as products of models of propositional-based specifications.

1 Introduction

Type hierarchies play a central rôle in the foundations of database and knowledge-base systems; consequently, a vast literature surrounding them has developed. Frameworks such as description logics [3] and formal concept analysis [10], as well as formal models for object-oriented database systems themselves [2,18,19], have evolved in response to the need for a comprehensive theoretical foundation. The work reported in this paper lays the foundation for extending the ideas of these formalisms to contexts in which information about the type hierarchy is *open*, in the sense that specification is via a set of constraints rather than via a single instance, so there may be none, one, or many instances which satisfy those constraints.

1.1 Background: the relationship between subsumption, suprema and infima. Regardless of the specific formalism, the major underlying notion in a type hierarchy is that of subsumption. Type τ_1 *is subsumed* by type τ_2, (or, equivalently, τ_2 *subsumes* τ_1), written $\tau_1 \sqsubseteq \tau_2$, just in case every object of type τ_1 is also an object of type τ_2. Writing $\mathfrak{O}(\tau)$ to denote the collection of objects of type τ, this subsumption is expressed as $\mathfrak{O}(\tau_1) \subseteq \mathfrak{O}(\tau_2)$.

Subsumption is often visualized using lattice like diagrams, such as that of Fig. 1, which depicts information about a simple hierarchy for university people.

L. Bertossi et al. (Eds.): Semantics in Databases, LNCS 2582, pp. 100–118, 2003.

The symbol $\perp$ represents the empty type, with $\top$ the universal type of all such people. A line from a higher object to a lower one indicates subsumption; e.g., Grad $\sqsubseteq$ Student.

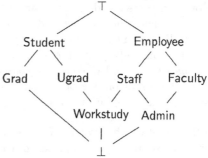

Fig. 1: Visualization of a simple type hierarchy

It is tempting to make further use of the lattice-like properties of the infimum and supremum operations on the ordering induced by type subsumption. For example, suppose that it is known that Mary is both an undergraduate (Mary $\in$ $\mathfrak{O}$(Ugrad)) and that she is a member of the staff (Mary $\in$ $\mathfrak{O}$(Staff)). Since inf({Ugrad, Staff}) = Workstudy in this hierarchy, one might conclude that she is a workstudy (Mary $\in$ $\mathfrak{O}$(Workstudy)). More generally, from the above hierarchy, one might conclude that $\mathfrak{O}$(Ugrad)$\cap\mathfrak{O}$(Staff) = $\mathfrak{O}$(Workstudy). In [16], this has been called the *natural meet semantics*. However, it is not appropriate to presume this semantics universally; rather such a decision must be based upon further information. Regardless of whether or not it is named and represented explicitly, there is a type τ which embodies precisely those people who are both undergraduates and staff members, i.e., $\mathfrak{O}(\tau) = \mathfrak{O}$(Ugrad) $\cap$ $\mathfrak{O}$(Staff). If this type τ is represented explicitly in the hierarchy, then it must have the same objects as Workstudy. However, it is quite possible that the hierarchy of Fig. 1 is just a partial representation of the total state of affairs, with τ not explicitly represented. For example, undergraduate students could be hired to do grounds maintenance, which might not be classified as workstudy. The point is that, from the subsumption relationships, it can only be concluded that Workstudy $\sqsubseteq \tau$; to conclude that Workstudy = τ, further information is needed.

The ideas described above are central to an operation known as *(conjunctive) type unification*, which may be described as follows. Given that it is known that an object a is of both type τ_1 and of type τ_2, determine the most specific type τ for which a is of type τ. Such unification is of fundamental importance in parsing of natural language, particularly within formalisms such as HPSG [17], which employ type hierarchies to classify linguistic objects. Indeed, in parsers built using systems designed to support such parsing, which include ACQUILEX [4], CUF [7], and TFS [20], the underlying database is effectively a large type hierarchy, which embodies both the lexicon and the grammar, without recourse to traditional phrase-structure grammars. Even in systems such as ALE [5], which does incorporate an underlying context-free grammar, type unification plays a central rôle.

There is a dual question, which looks at infima rather than suprema. Again referring to Fig. 1, suppose that it is known that Mary is a student (Mary $\in$ Student). The fact that Student = sup({Grad, Ugrad}) in the hierarchy suggests that Mary must be either an undergraduate or a graduate student.

However, there could easily be another class of student, such as nondegree students, which is not represented explicitly in the hierarchy. The conclusion that $\mathfrak{O}(\text{Student}) = \mathfrak{O}(\text{Grad}) \cup \mathfrak{O}(\text{Ugrad})$ is not automatic; it depends upon further information. In [16], this property is called the *natural join semantics*. Although not as widely used in constraint-based problem solving as conjunctive unification, the corresponding operation of *disjunctive unification* has seen use in the management of solution search [8].

1.2 Focus of this work. The work reported in this paper was motivated by an interest in coming to grips, in a formal manner, with the semantics of the infimum and supremum operations in type hierarchies, particularly in applications involving unification. More often than not, in the current literature, such hierarchies are described in terms of subsumption only, and while infimum (and often supremum as well) is used in the computational process, the semantics of these operators is not described explicitly. The reader is thus left to reverse engineer the framework in order to determine these important details.

The starting point of the work presented here is that all constraints on the hierarchy are to be expressed via formulas of the forms $(\bigsqcap\{\tau_1, \tau_2, .., \tau_n\} = \tau)$ and $(\bigsqcup\{\tau_1, \tau_2, .., \tau_n\} = \tau)$, with $\bigsqcap$ denoting a generalized meet operation, and $\bigsqcup$ a generalized join operation. The semantics of these operations are $(\bigcap\{\mathfrak{O}(\tau_1), \mathfrak{O}(\tau_2), .., \mathfrak{O}(\tau_n)\} = \mathfrak{O}(\tau))$ and $(\bigcup\{\mathfrak{O}(\tau_1), \mathfrak{O}(\tau_2), .., \mathfrak{O}(\tau_n)\} = \mathfrak{O}(\tau))$, respectively, with $\mathfrak{O}(\omega)$ denoting the set of objects of type ω. It is important to note that operations of extended arity are required. It is quite possible for $\bigsqcap\{\tau_1, \tau_2, \tau_3\}$ to be defined without any stated definition for $\bigsqcap\{\tau_1, \tau_2\}$, $\bigsqcap\{\tau_1, \tau_3\}$, or $\bigsqcap\{\tau_2, \tau_3\}$. The subsumption $\tau_1 \sqsubseteq \tau_2$ is expressed via the constraint $(\bigsqcap\{\tau_1, \tau_2\} = \tau_1)$. A description based upon constraints of this form is called an *open specification* because it does not state the complete properties of a single hierarchy; rather, it provides constraints which any *model* hierarchy must obey. There may be none, one, or several such models for a given set of constraints.

In this work, there is no explicit notion of attribute; thus, it differs fundamentally from work in description logics and formal concept analysis. This decision, however, was not made due to any lack of belief in the importance of these concepts. Clearly, a meaningful notion of attribute, and its behavior under order-induced operations, is central to any comprehensive theory of type hierarchies. At the same time, it seems clear that any attempt to introduce attributes into the work at this initial stage would only complicate matters and cloud development of the fundamental issues. A useful formalism of simple attribute-free types under open specification must precede the development of a more complex formalism which incorporates attributes. The focus here is upon how pure types behave in the presences of open specification. Once this is understood, the results should be combined with the more general notions embodied in other formalisms.

1.3 Content and organization of this paper. The work reported here is based, to a substantial degree, on the papers [13], [16], and [15]. Nonetheless, the method of presentation, and even some of the results, particularly with respect to algebraic characterization, are entirely new.

The primary emphasis of this paper is the *characterization* of type hierarchies under open specifications; that is, the development of mathematical principles necessary to model and compute upon such hierarchies. Two major avenues of characterization are presented and contrasted. In Section 3, an algebraic characterization, based upon a generalized form of bounded partial lattice, and couched in a general framework which makes use of universal constructions, is provided. In Section 4, a logical characterization, based upon propositional logic, is given. Each of these approaches has its strengths and weaknesses, as are exposed in the presentation.

An additional topic of great importance in this area is that of computational complexity and algorithms for the manipulation of and inference on such hierarchies. Unfortunately, space limitations preclude a thorough treatment. However, the most important results are summarized in Section 5.

Finally, Section 2 provides a common backdrop of basic definitions and results, in support of both of the characterizations, while Section 6 provides some conclusions and further directions.

1.4 Prerequisites and notation. Of course, it is presumed that the reader has an appreciation for the central rôle of type hierarchies in computer science in general and database systems in particular. Beyond that, a knowledge of standard propositional logic is expected. It is also assumed that the reader has a basic knowledge of the theory of partial orders and lattices; the necessary background may be obtained from standard references such as [12] and [6]. To minimize the possibility of confusion with the corresponding logical symbols $\wedge$ and $\vee$, the order relation and the lattice-like operations of join, and meet in such structures will always be denoted with the squared symbols $\sqsubseteq$, $\sqcup$, and $\sqcap$, respectively. A *bounded lattice* is denoted as $\mathbf{L} = (L, \sqcup, \sqcap, \top, \bot)$, with $\top$ the greatest element and $\bot$ the least element. In particular, the boldface symbol (e.g., $\mathbf{L}$) denotes the entire structure, while the normal symbol L denotes just the underlying set. This font convention will also be used with other algebraic structures (e.g., the generalized bounded weak partial lattices of Section 3).

A very small amount of category theory background will prove helpful. Specifically, *isomorphisms* are always characterized as morphisms which have both left and right inverses. In addition, some familiarity with the ideas of free and initial objects would prove helpful, although full definitions are always given. A suitable reference is [1].

The following specific notation, which may be a bit less than standard, is used. If f is a partial function, then $f(s)\!\downarrow$ (resp. $f(s)\!\uparrow$) means that f is defined (resp. undefined) on argument s. If A is a set, then 2^A (resp. 2^A_f) denotes the set of all subsets (resp. finite subsets) of A, while $\mathsf{Card}(A)$ denotes the cardinality of A.

2   Constraints and Models

In this section, the formal foundations for the description of type hierarchies via open specification are presented. The presentation follows most closely that of [15]. However, some of the ideas, particularly those related to morphisms and completions, are new and were developed in support of the algebraic characterization of the next section.

2.1 Elementary positive constraints. A *clean set* is any finite set P which does not contain either of the special symbols $\perp$ and $\top$. Define $\mathsf{Aug}(P) = P \cup \{\perp, \top\}$. The elements $\perp$ and $\top$ are called the *extreme types*; the elements of P are called the *base types*. As shall soon be formalized, $\top$ (resp. $\perp$) represents the greatest (resp. least) type in the hierarchy.

The most fundamental class of constraint is the *elementary positive constraint*, of which there are two basic forms.[1] An *elementary positive meet constraint* has the form $(\bigsqcap\{\tau_1, \tau_2, .., \tau_n\} = \tau)$, with the τ_i's and τ members of $\mathsf{Aug}(P)$. The set of all such constraints over P is denoted $\mathsf{ElemConstr}_{\sqcap}^{+}$. Dually, an *elementary positive join constraint* has the form $(\bigsqcup\{\tau_1, \tau_2, .., \tau_n\} = \tau)$, with the set of all such constraints denoted by $\mathsf{ElemConstr}_{\sqcup}^{+}$. Combining these two classes, $\mathsf{ElemConstr}^{+}(P)$ denotes $\mathsf{ElemConstr}_{\sqcap}^{+}(P) \cup \mathsf{ElemConstr}_{\sqcup}^{+}(P)$.

2.2 Interpretation of types. Let P be a clean set of types. An *interpretation* over P is a pair $S = (\mathfrak{U}, \mathfrak{D})$, in which $\mathfrak{U}$ is a finite nonempty set, called the *universe of objects*, and $\mathfrak{D} : \mathsf{Aug}(P) \to 2^{\mathfrak{U}}$ is a function which associates a subset of $\mathfrak{U}$ to each type in $\mathsf{Aug}(P)$, subject to the conditions that $\mathfrak{D}(\top) = \mathfrak{U}$ and $\mathfrak{D}(\perp) = \emptyset$. Think of $\mathfrak{D}(\tau)$ as the set of all objects of type τ.

An interpretation $S = (\mathfrak{U}, \mathfrak{D})$ is to be viewed as the specification for a unique, complete type hierarchy over P, in which all infima and suprema have their natural semantics. Specifically, define $\mathsf{Lat}(S)$ to be the smallest set of subsets of $\mathfrak{U}$ which contains every member of $\{\mathfrak{D}(\tau) \mid \tau \in \mathsf{Aug}(P)\}$, and which is closed under union and intersection. It is easy to see that $\mathsf{Lat}(S)$ admits the structure of a finite, bounded distributive lattice, with union as join, intersection as meet, $\mathfrak{U}$ as top element and $\emptyset$ as least element. This lattice is denoted $\mathbf{Lat}(S) = (\mathsf{Lat}((S), \cup, \cap, \mathfrak{U}, \emptyset)$.

The *size* of an interpretation $S = (\mathfrak{U}, \mathfrak{D})$ is the cardinality of $\mathfrak{U}$; an interpretation of size m is often called an m-element interpretation. In particular, a *one-element interpretation* is of the form $S = (\{a\}, \mathfrak{D})$.

Let $S = (\mathfrak{U}, \mathfrak{D})$ be an interpretation for P. For each nonempty subset $\mathfrak{B} \subseteq \mathfrak{U}$, define the $\mathfrak{B}$-*projection* of S to be $S_{|\mathfrak{B}} = (\mathfrak{B}, \mathfrak{D}_{|\mathfrak{B}})$, with $\mathfrak{D}_{|\mathfrak{B}} : \mathsf{Aug}(P) \to 2^{\mathfrak{B}}$ given by $\mapsto X\mathfrak{D}(X) \cap \mathfrak{B}$. Conversely, let $S_i = (\mathfrak{U}_i, \mathfrak{D}_i)$ be interpretations for P for $i = 1, 2$. Assume further that $\mathfrak{U}_1 \cap \mathfrak{U}_2 = \emptyset$. The *product interpretation* $S_1 \times S_2$ is given by $(\mathfrak{U}_1 \cup \mathfrak{U}_2, \mathfrak{D}_1 \times \mathfrak{D}_2)$, with $\mathfrak{D}_1 \times \mathfrak{D}_2\mathsf{Aug}(P) \to 2^{\mathfrak{U}_1 \cup \mathfrak{U}_2}$ given

---

[1] The definition here differs slightly from that found in [15]; in which equality is replaced by subsumption. This is of no major consequence, since the two formalisms are equivalent in expressive power. See 2.7 for a further discussion.

by $X \mapsto \mathfrak{O}_1(X) \cup \mathfrak{O}_2(X)$. This definition extends easily to any finite number of interpretations. Note that, for any interpretation $S = (\mathfrak{U}, \mathfrak{O})$, $\prod_{a \in \mathfrak{U}} S_{|\{a\}}$ is just S, up to a renaming of $\mathfrak{O}$.

2.3 Satisfaction of constraints. The constraint $(\sqcap\{\tau_1, \tau_2, .., \tau_n\} = \tau)$ is *satisfied* by the interpretation $S = (\mathfrak{U}, \mathfrak{O})$ if $\mathfrak{O}(\tau_i) \cap \mathfrak{O}(\tau_2) \cap .. \cap \mathfrak{O}(\tau_n) = \mathfrak{O}(\tau)$. Similarly, $(\bigsqcup\{\tau_1, \tau_2, .., \tau_n\} = \tau)$ is *satisfied* by S if $\mathfrak{O}(\tau_i) \cup \mathfrak{O}(\tau_2) \cup .. \cup \mathfrak{O}(\tau_n) = \mathfrak{O}(\tau)$. In general, if $\varphi \in \mathsf{ElemConstr}^+(P)$ and S is an interpretation, then $M \models \varphi$ denotes that φ is satisfied by S; in this case, S is called a *model* of φ. If $\Phi \subseteq \mathsf{ElemConstr}^+(P)$, $S \models \Phi$ holds iff $S \models \varphi$ for each $\varphi \in \Phi$. The set of all S for which $S \models \varphi$ (resp. $S \models \Phi$) holds is denoted $\mathsf{Mod}_P(\varphi)$ (resp. $\mathsf{Mod}_P(\Phi)$). Two sets Φ_1 and Φ_2 of constraints over P are *equivalent* if $\mathsf{Mod}_P(\Phi_1) = \mathsf{Mod}_P(\Phi_2)$.

Following standard notation from mathematical logic, if $\varphi \in \mathsf{ElemConstr}^+(P)$ (resp. $\Psi \subseteq \mathsf{ElemConstr}^+(P)$), and $\Phi \subseteq \mathsf{ElemConstr}^+(P)$, then $\Phi \models \varphi$ (resp. $\Phi \models \Psi$) holds just in case $\mathsf{Mod}_P(\Phi) \subseteq \mathsf{Mod}_P(\varphi)$ (resp. $\mathsf{Mod}_P(\Phi) \subseteq \mathsf{Mod}_P(\Psi)$). Although this assigns double duty to the symbol $\models$, no confusion can result, since the nature of the first argument (interpretation or constraint) identifies the usage unambiguously.

2.4 Abbreviations and variants. In addition to these elementary constraints, there are a few notational variants which are important enough to warrant their own notation. First of all, $(\tau_1 \sqsubseteq \tau_2)$ is an abbreviation for $(\sqcap\{\tau_1, \tau_2\} = \tau_1)$. Second, $(\tau_1 = \tau_2)$ is an abbreviation for $(\sqcap\{\tau_1\} = \tau_2)$.

In addition to these definitions, the notational variant of infix (as opposed to prefix) notation is allowed for all forms of constraints. Thus, for example, $(\tau_1 \sqcap \tau_2 \sqcap \tau_3 = \tau)$ is a perfectly acceptable (and obvious) abbreviation for $(\sqcap\{\tau_1, \tau_2, \tau_3\} = \tau)$. This notation in no way implies that meets for subsets (e.g., $\tau_1 \sqcap \tau_2$) are defined.

2.5 Open specification Let P be a clean set of types. An *elementary positive open specification* is a pair (P, Φ) in which $\Phi \subseteq \mathsf{ElemConstr}^+(P)$.

2.6 Examples. In each of the cases below, let (Q, Ω) be the elementary positive open specification with $Q = \{\tau_a, \tau_b, \tau_c\}$ and $\Omega = \{(\tau_a \sqcup \tau_c = \top), (\tau_c \sqsubseteq \tau_b)\}$.
(a) Let $\mathfrak{B}_1 = \mathfrak{B}_2 = \mathfrak{B}_3 = \mathfrak{B}_4 = \{a\}$, and define $\mathfrak{O}_1\mathsf{Aug}(Q) \to 2^{\mathfrak{B}_1}$ by $\tau_a \mapsto \{a\}$, $\tau_b \mapsto \emptyset$, $\tau_c \mapsto \emptyset$; $\mathfrak{O}_2\mathsf{Aug}(Q) \to 2^{\mathfrak{B}_2}$ by $\tau_a \mapsto \{a\}$, $\tau_b \mapsto \{a\}$, $\tau_c \mapsto \emptyset$; $\mathfrak{O}_3\mathsf{Aug}(Q) \to 2^{\mathfrak{B}_3}$ by $\tau_a \mapsto \emptyset$, $\tau_b \mapsto \{a\}$, $\tau_c \mapsto \{a\}$; $\mathfrak{O}_4\mathsf{Aug}(Q) \to 2^{\mathfrak{B}_4}$ by $\tau_a \mapsto \{a\}$, $\tau_b \mapsto \{a\}$, $\tau_c \mapsto \{a\}$. Then $\mathfrak{S}_i = (\mathfrak{B}_i, \mathfrak{O}_i)$ for $i = 1 \dots 4$ are models for (Q, Ω). It is not difficult to see that they are the only one-element models, up to renaming of elements.
(b) Let $\mathfrak{B}_5 = \{a, b\}$, and define $\mathfrak{O}_5\mathsf{Aug}(Q) \to 2^{\mathfrak{B}_5}$ by $\tau_a \mapsto \{a\}$, $\tau_b \mapsto \{b\}$, $\tau_c \mapsto \{b\}$. Then $\mathfrak{O}_5 = (\mathfrak{B}_5, \mathfrak{O}_5)$ is a model for (Q, Ω). This model also satisfies positive constraints which are not embodied in Ω, including $(a \sqcap c = \bot)$ and $(b = c)$.
(c) Let $\mathfrak{B}_6 = \{a, b, c\}$, and define define $\mathfrak{B}_6 : Q \to 2^{\mathfrak{B}_6}$ by $\tau_a \mapsto \{a\}, \{b\}$, $\tau_b \mapsto \{b\}, \{c\}$, $\tau_c \mapsto \{c\}$. $\mathfrak{S}_5 = (\mathfrak{B}_5, \mathfrak{O}_5)$ is also a model for (Q, Ω). It still satisfies a positive constraint not embodied in Ω, namely $(a \sqcap c = \bot)$.

(d) Let $\mathfrak{B}_7 = \{a, b\}, c\}, \mathfrak{d}\}, e\}$, and define $\mathfrak{O}_7 Q \to 2^{\mathfrak{B}_7}$ by $\tau_a \mapsto \{a, b\}, c\}, \mathfrak{d}\}, e\}$, $\tau_b \mapsto \{b, c, \mathfrak{d}\}$, $\tau_c \mapsto \{c, \mathfrak{d}\}$. $\mathfrak{S}_7 = (\mathfrak{B}_7, \mathfrak{O}_7)$ is a model for (Q, Ω). While this model does not satisfy any positive constraints not embodied in Ω, as do $\mathfrak{S}_4$ and $\mathfrak{S}_5$ above, it is redundant in that the elements a and e are not distinguishable.

(e) Let $\mathfrak{B}_8 = \mathfrak{B}_8 = \{a, b, c, \mathfrak{d}\}$, and define $\mathfrak{O}_8 Q \to 2^{\mathfrak{B}_8}$ by $\tau_a \mapsto \{a, b, \mathfrak{d}\}$, $\tau_b \mapsto \{b, c, \mathfrak{d}\}$, $\tau_c \mapsto \{c, \mathfrak{d}\}$. $\mathfrak{s}_8 = (\mathfrak{B}_8, \mathfrak{O}_8)$ is a model for (Q, Ω). This model is "optimal" in the sense that it neither enforces unnecessary constraints, as do $\mathfrak{S}_4$ and $\mathfrak{S}_5$, nor does it introduce superfluous elements, as does $\mathfrak{S}_6$. See 3 for a further discussion of this idea.

2.7 Subsumption-based constraints. In [15], the basic forms of the constraints are taken to be $(\sqcap\{\tau_1, \tau_2, , \ldots \tau_n\} \sqsubseteq \tau)$ and $(\tau \sqsubseteq \sqcup\{\tau_1, \tau_2, \ldots, \tau_n\})$; i.e., equality is replaced by subsumption. This choice was made because such *subsumption constraints* are more suitable to the computational algorithms of that paper. On the other hand, the *equality constraints* used in this paper are more natural and intuitive when characterization — particularly algebraic characterization — is the principal topic. These two representations are effectively equivalent. The equality constraint $(\sqcap\{\tau_1, \tau_2, , \ldots \tau_n\} = \tau)$ is equivalent to the subsumption constraint $(\sqcap\{\tau_1, \tau_2, , \ldots \tau_n\} \sqsubseteq \tau)$ together with the set $\{(\tau \sqsubseteq \tau_1), (\tau \sqsubseteq \tau_2), \ldots, (\tau \sqsubseteq \tau_n)\}$. To represent the subsumption constraint $(\sqcap\{\tau_1, \tau_2, , \ldots \tau_n\} \sqsubseteq \tau)$ using equality constraints, use $(\sqcap\{\tau_1, \tau_2, , \ldots \tau_n\} = \sigma)$ together with $(\sqcap\{\sigma, \tau\} = \sigma)$. Here σ is a new type symbol not used previously. The join constraints are represented similarly.

3   Algebraic Characterization

While an open specification is not a lattice, it is nonetheless apparent that the constraints look very much like constraints on a lattice. A major difference is that, with an open specification, there is no guarantee that $\tau_1 \sqcap \tau_2$ and $\tau_1 \sqcup \tau_2$ will be defined for an arbitrary pair $\{\tau_1, \tau_2\} \subseteq P$. Thus, a natural question to ask is whether there is some sort of algebraic characterization of open specifications and their models, using a lattice-like notion with partial operations. In this section, this question is answered in the affirmative.

The author's initial report on open specification [13] contains a rather detailed development of their algebraic properties. While these results remain valid, simpler and more concise characterizations, which have not been previously published, have been discovered since the appearance of [13]. In this section, many of these newer results are presented.

3.1 Generalized Bounded Weak Partial Lattices. In [12, Ch. I, Sec. 5], Grätzer describes two notions, the *partial lattice* and the *weak partial lattice*. Roughly speaking, a partial lattice may be characterized as a subset of a lattice under the induced operations, while a weak partial lattice is a set with partial lattice-like operations. Grätzer also provides an example of a weak partial lattice which cannot be embedded in a partial lattice.

Neither of these concepts is directly applicable to the modelling of open specifications. Most importantly, the definitions given in [12] have their meet and join operations restricted to pairs of elements, while the constraints described here allow these operations to take finite subsets as arguments. This is an essential difference, as it is quite possible to have $\bigsqcap\{\tau_1, \tau_2, \ldots, \tau_n\}$ and/or $\bigsqcup\{\tau_1, \tau_2, \ldots, \tau_n\}$ defined without having the these operations defined on any proper subset of $\{\tau_1, \tau_2, \ldots, \tau_n\}$ of cardinality greater than one. On the other hand, it is quite possible to extend the definitions of [12] to this more general context. The appropriate one to generalize for the context at hand is the weak partial lattice. The definition given here, which originally appeared in [13, 1.2.3], generalizes that of [12] to meet and join operations with more than two arguments, and adds universal bounds as well.

A *generalized bounded weak partial lattice* (*GBWPL*) is a five-tuple $\mathbf{L} = (L, \bigsqcup, \bigsqcap, \top, \bot)$ in which the following nine conditions are satisfied.

(gbwpl:1) L is a set (the *underlying set*).

(gbwpl:2) $\bigsqcup \to 2_f^L \to L$ is a partial operation, called the *generalized join*.

(gbwpl:3) $\bigsqcup \to 2_f^L \to L$ is a partial operation, called the *generalized meet*.

(gbwpl:4) $\bot, \top \in L$ with $\bot \neq \top$.

The operations $\bigsqcup$ and $\bigsqcap$ are subject to the following conditions.

(gbwpl:5) $\bigsqcup \emptyset = \bot$ and $\bigsqcap \emptyset = \top$.

(gbwpl:6) For all $a \in L$, $\bigsqcup\{a\} = a$, $\bigsqcap\{a\} = a$, $\bigsqcup\{a, \bot\} = a$, $\bigsqcap\{a, \top\} = a$.

(gbwpl:7) If $S_1, S_2, \ldots, S_n$ are finite subsets of L with $\bigsqcup S_i \downarrow$ for all i, $1 \leq i \leq n$, then $\bigsqcup(\bigcup_{i=1}^n S_i) \downarrow$ iff $\bigsqcup\{\bigsqcup S_1, \bigsqcup S_2, \ldots \bigsqcup S_n\} \downarrow$, and $\bigsqcup\{\bigsqcup S_1, \bigsqcup S_2, \ldots \bigsqcup S_n\} = \bigsqcup(\bigcup_{i=1}^n S_i)$ in this case.

(gbwpl:8) If $S_1, S_2, \ldots, S_n$ are finite subsets of L with $\bigsqcap S_i \downarrow$ for all i, $1 \leq i \leq n$, then $\bigsqcap(\bigcup_{i=1}^n S_i) \downarrow$ iff $\bigsqcap\{\bigsqcap S_1, \bigsqcap S_2, \ldots \bigsqcap S_n\} \downarrow$, and $\bigsqcap\{\bigsqcap S_1, \bigsqcap S_2, \ldots \bigsqcap S_n\} = \bigsqcap(\bigcup_{i=1}^n S_i)$ in this case. The constraints of (gbwpl:7) and (gbwpl:8) are called the *generalized associativity laws*.

(gbwpl:9) If S is a finite subset of L with $\bigsqcup S \downarrow$, then for all $a \in S$, $\bigsqcap\{a, \bigsqcup S\} \downarrow$ with $\bigsqcap\{a, \bigsqcup S\} = a$. Dually, for all $a \in S$, $\bigsqcup\{a, \bigsqcap S\} \downarrow$ with $\bigsqcup\{a, \bigsqcap S\} = a$. These are called the *generalized absorption identities*.

In (gbwpl:5), the condition $\bigsqcup\{a\} = a$ is a restatement of the idempotency of the join, which becomes $a \vee a = a$ in a lattice. The condition $\bigsqcup\{a, \bot\} = a$ states that $\bot$ is the least element. The other two new conditions are dual. In (gbwpl:9), The condition $\bigsqcap\{a, \bigsqcup S\} = a$ for $a \in S$ and $\bigsqcup S \downarrow$ is a generalization of the absorption identity $a \wedge (a \vee b) = a$ of an ordinary lattice [12, Cond. (L4), p. 5]. The other condition is dual.

Let $\mathbf{L_1} = (L, \bigsqcup, \bigsqcap, \top, \bot)$ and $\mathbf{L_2} = (L, \bigsqcup, \bigsqcap, \top, \bot)$ be GBWPL's. A *morphism* $f : \mathbf{L_1} \to \mathbf{L_2}$ is a function $f : L_1 \to L_2$ subject to the following constraints.

(gbwpl:mor1) For a finite $S \subseteq L_1$, if $\bigsqcup S \downarrow$, then $\bigsqcup(f(S)) \downarrow$ and $f(\bigsqcup S) = \bigsqcup(f(S))$. Dually, if $\bigsqcap S \downarrow$, then $\bigsqcap(f(S)) \downarrow$ and $f(\bigsqcap S) = \bigsqcap(f(S))$.

(gbwplmor:mor2) $f(\bot) = \bot$ and $f(\top) = \top$.

It is easy to see that $f : \mathbf{L}_1 \to \mathbf{L}_2$ is an isomorphism iff the underlying function is a bijection which preserves and reflects $\top$, $\bot$, and all meets and joins.

Any bounded lattice may be viewed as a GBWPL in which all of the operations are total; that is, for any finite $S \subseteq L$, $\sqcap S$ and $\sqcup S$ are defined. Conversely, any GBWPL for which all such operations are total is a bounded lattice; this is easily verified from the axioms.

Given a GBWPL $\mathbf{L} = (L, \sqcup, \sqcap, \top, \bot)$ and a subset $M \subseteq L$ containing $\{\top, \bot\}$, the *restriction* of $\mathbf{L}$ to M is the GBWPL $\mathbf{L}_{|M} = (M, \sqcup, \sqcap, \top, \bot)$. The operations of $\mathbf{L}_{|M}$ are just those of $\mathbf{L}$, restricted to M. Note that this means that all components of an operation must lie in M. For example, if $\sqcap\{\tau_1, \tau_2, \ldots, \tau_n\} = \tau$ holds in $\mathbf{L}$, then for it to apply to $\mathbf{L}_{|M}$, the entire subset $\{\tau_1, \tau_2, \ldots, \tau_n, \tau\}$ must lie in M. It is easy to see that such a restriction of a GBWPL is always a GBWPL; the conditions (gbwpl:5)-(gbwpl:9) never generate any new elements, other than $\top$ and $\bot$, so it is never necessary to add any additional elements to M to maintain the GBWPL properties.

Finally, the notion of the product $\prod_{i \in I}$ of a family $\{\mathbf{L}_i \mid i \in I\}$ of GBWPL's is well defined, with the obvious coordinatewise operations.

3.2 Notational convention. For the rest of this section, unless noted otherwise, fix (P, Φ) to be an elementary positive open specification.

3.3 Representations of open specifications. A *GBWPL representation* of (P, Φ) is a pair $(\mathbf{L}, f)$ in which $\mathbf{L}$ is a GBWPL and $f : \mathsf{Aug}(P) \to L$ is a function which satisfies the following rules.

(rep:1) $f(\top) = \top$ and $f(\bot) = \bot$.

(rep:2) For $\tau_1, \tau_2, \ldots, \tau_n, \tau \in \mathsf{Aug}(P)$, if $(\sqcap\{\tau_1, \tau_2, \ldots, \tau_n\} = \tau) \in \Phi$, then $(\sqcap\{f(\tau_1), f(\tau_2), \ldots, f(\tau_n)\} = f(\tau))$ holds in $\mathbf{L}$.

(rep:3) For $\tau_1, \tau_2, \ldots, \tau_n, \tau \in \mathsf{Aug}(P)$, if $(\sqcup\{\tau_1, \tau_2, \ldots, \tau_n\} = \tau) \in \Phi$, then $(\sqcup\{f(\tau_1), f(\tau_2), \ldots, f(\tau_n)\} = f(\tau))$ holds in $\mathbf{L}$.

A *morphism* $h : (\mathbf{L}, f) \to (\mathbf{M}, g)$ of GBWPL representations is a GBWPL morphism $h : \mathbf{L} \to \mathbf{M}$ with the property that that the diagram to the right commutes. It is easy to see that h is an isomorphism of representations iff it is an isomorphism of the underlying GBWPL's.

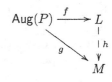

A representation $(\mathbf{L}, f)$ is *surjective* precisely in the case that f is a surjective function. It is always possible to derive a surjective representation from an arbitrary one; just replace $\mathbf{L}$ with $\mathbf{L}_{|f(\mathsf{Aug}(P))}$.

Finally, if $\{(\mathbf{L}_i, f_i) \mid i \in I\}$ is a family of representations of (P, Φ), so too is $(\prod_{i \in I} \mathbf{L}_i, \prod_{i \in I} f_i)$, with $\prod_{i \in I} f_i$ sending τ to the tuple whose i^{th} entry is $f(\tau)$.

Perhaps the most fundamental example of a representation of an open specification is one which arises from a model of the specification. The following proposition is immediate.

3.4 Examples. The examples of 2 may easily be transformed to GBWPL representations. Indeed, it is easy to see that, for any model $\mathfrak{S}_i = (\mathfrak{V}_i, \mathfrak{O}_i)$ of (Q, Ω), $(\textbf{Lat}(\mathfrak{S}_i), \mathfrak{f}_i)$ is a GBWPL representation. Here $\mathfrak{f}_i : \textsf{Aug}(Q) \to \textsf{Lat}(\mathfrak{S}_i)$ is the function which is identical to $\mathfrak{O}_i$, save that its codomain is restricted to $\textsf{Lat}(\mathfrak{S}_i)$. Of course, these representations are in fact bounded lattices, with all meets and joins defined.

To obtain a "best" GBWPL representation of (Q, Ω), define $\mathfrak{L}_8 = (\mathfrak{L}_8, \bigsqcup, \sqcap, \top; \bot)$ to be the GBWPL in which $\mathfrak{L}_8 = Q \cup \{\bot, \top\}$; the constraints of this GBWPL are those which include $(\tau_a \sqcup \tau_c = \top)$ and $(\tau_c \sqsubseteq \tau_b)$, together with those constraints implied by these under the rules given in 3. Note that the closure of this set of constraints includes, in particular, $(\tau_a \sqcup \tau_b = \top)$. Define $\mathfrak{f}_8 \textsf{Aug}(Q) \to \mathfrak{L}_8$ to be the natural identity embedding $\tau_x \mapsto \tau_x$. Then $(\mathfrak{L}_8, \mathfrak{f}_8)$ is a GBWPL representation for (Q, Ω). This representation is best in the sense that it includes neither unnecessary constraints nor superfluous information. The precise meaning of "best" is developed in 3-3 below.

It might seem that one could always obtain a best GBWPL representation of (P, Φ) as a GBWPL whose elements are precisely the members of $\textsf{Aug}(P)$, and whose constraints are a natural closure of the obvious translation of Φ. However, this is not always the case, since Φ may force the collapsing of certain elements. For example, let $R = \{\kappa_i \mid 1 \leq i \leq 6\}$, and let $\Gamma = \{(\bigvee\{\kappa_1, \kappa_2\} = \kappa_5), (\bigvee\{\kappa_2, \kappa_3\} = \kappa_5), (\bigvee\{\kappa_1, \kappa_3\} = \kappa_4), (\bigwedge\{\kappa_2, \kappa_3\} = \kappa_6), (\bigwedge\{\kappa_1, \kappa_6\} = \bot), ((\kappa_5 \leq \kappa_4)\}$. Then, as shown in [16, 5.5], any model $(\mathfrak{U}, \mathfrak{O})$ of (R, Γ) has the property that $\mathfrak{O}(\kappa_4) = \mathfrak{O}(\kappa_5)$. This means that, when constructing a GBWPL representation from R, the elements $\{\kappa_4, \kappa_5\}$ must be collapsed into a single equivalence class. This idea is expanded below; in particular in the development of the canonical representation 3.

3.5 Initial representations. In view of the preceding examples, there are many representations of an open specification. The next task is to characterize formally the notion of a best representation, as illustrated by the final two examples above. Such a representation must be optimal in the sense that it contains just the right amount of information, without adding superfluous elements, and without imposing unnecessary constraints. In the world of arrows, the standard way to characterize such entities is via a free construction [1, Chap. 7]; or, equivalently, via an initial object.

Formally, an *initial GBWPL representation* for (P, Φ) is a GBWPL representation $(\textbf{L}, \eta)$ with the property that, for any other GBWPL representation $(\textbf{M}, g)$, there is a unique GBWPL morphism $h : (\textbf{L}, \eta) \to (\textbf{M}, g)$. Such a representation is unique, up to isomorphism [1, 7.3].

3.6 Canonical representations. The notion of an initial representation is abstract, and says nothing about how to construct one. The canonical representation, described next, provides a concrete realization of an initial representation.

Assume that (P, Φ) has at least one representation. Define the equivalence relation $\equiv_{(P, \Phi)}$ on $\textsf{Aug}(P)$ by $\tau_1 \equiv_{(P, \Phi)} \tau_2$ iff $f(\tau_1) = f(\tau_2)$ for all representations

$(\mathbf{L}, f)$ of (P, Φ). For $\tau \in \mathsf{Aug}(P)$, $[\tau]_{\equiv_{(P,\Phi)}}$ denotes the equivalence class of τ in $\equiv_{(P,\Phi)}$. When no confusion can result, the subscript will be dropped; i.e., $[\tau]$. Further, for $A \subseteq \mathsf{Aug}(P)$, $[A]$ denotes $\{[\tau] \mid \tau \in A\}$. For $\varphi \in \Phi$, let $[\varphi]$ denote the constraint on $[\mathsf{Aug}(P)]$ obtained by replacing all members of $\mathsf{Aug}(P)$ with their respective equivalence classes. For example, $(\sqcap\{\tau_1, \tau_2, \ldots, \tau_n\} = \tau)$ becomes $(\sqcap\{[\tau_1], [\tau_2] \ldots, [\tau_n]\} = [\tau])$. Define $[\Phi] = \{[\varphi] \mid \varphi \in \Phi\}$.

Next, define $\mathsf{GBWPL}(P, \Phi) = ([\mathsf{Aug}(P)]\sqcup, \sqcap, [\top], [\bot])$. The constraint $(\sqcap\{[\tau_1], [\tau_2] \ldots, [\tau_n]\} = [\tau])$ holds in $\mathsf{GBWPL}(P, \Phi)$ iff for every GBWPL representation $(\mathbf{L}, f)$ of (P, Φ), $(\sqcap\{f(\tau_1), f(\tau_2) \ldots, f(\tau_n)\} = f(\tau))$ holds in $\mathbf{L}$. Similarly, the constraint $(\sqcup\{[\tau_1], [\tau_2] \ldots, [\tau_n]\} = [\tau])$ holds in $\mathsf{GBWPL}(P, \Phi)$ iff for every GBWPL representation $(\mathbf{L}, f)$ of (P, Φ), $(\sqcup\{f(\tau_1), f(\tau_2) \ldots, f(\tau_n)\} = f(\tau))$ holds in $\mathbf{L}$.

The pair $\mathsf{CanRep}(P, \Phi) = (\mathsf{GBWPL}(P, \Phi), [-]_{\equiv_{(P,\Phi)}})$, with $[-]_{\equiv_{(P,\Phi)}} : \mathsf{Aug}(P) \to [\mathsf{Aug}(P)]$ given by $\tau \mapsto [\tau]$, is called the *canonical representation* of (P, Φ). Of course, it remains to be shown that $\mathsf{GBWPL}(P, \Phi) = ([\mathsf{Aug}(P)]\sqcup, \sqcap, [\top], [\bot])$ is indeed a GBWPL; this is the next topic.

3.7 Proposition *If (P, Φ) has a GBWPL representation, then it has an initial GBWPL representation, which is given by $\mathsf{CanRep}(P, \Phi)$.*

Proof Assume that (P, Φ) has a GBWPL representation. Begin with the class A of all GBWPL representations of (P, Φ), and pare this collection down in two ways. First of all, limit it to just the surjective representations; in view of the discussion at the end of 3, each original element of A will give rise to such a surjective relative. Next, pare the resulting collection down by choosing just one representative of each isomorphism class. Note that this resulting collection B is not only a set, but a finite one, since P is finite.

Denote B explicitly as $\{(\mathbf{M}_j, f_j) \mid j \in J\}$, and form the product $\prod B = (\prod_{j \in J} M_j, \prod_{j \in J} f_j)$. The diagram to the right shows that $(\prod B)_{\mid(\prod_{j \in J} f_j)(\mathsf{Aug}(P))}$ is an initial GWBPL representation for (P, Φ). Let $(\mathbf{L}, g)$ be any GBWPL representation of (P, Φ) whatever. The first and fourth

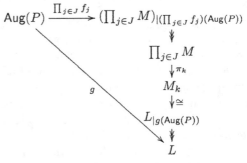

vertical morphisms are just the natural injections. The second morphism π_k is the obvious projection; and the third morphism $\cong$ is the isomorphism between $(\mathbf{L}_{\mid g(\mathsf{Aug}(P))}, g)$ and its representative $(\mathbf{M}_k, f_k)$ in B. Finally, it is immediate from the way in which $\mathsf{CanRep}(P, \Phi)$ has been defined that $\prod B$ is isomorphic to $\mathsf{CanRep}(P, \Phi)$. $\square$

3.8 Examples. $\mathfrak{L}_8$ with the natural embedding of 3 is already a canonical representation of (Q, Ω). To obtain the canonical representation of (R, Γ) of 3, the types κ_4 and κ_5 must be identified in the equivalence relation $\equiv_{(R,\Gamma)}$.

As illustrated by the conversions in 3 of the example models in 2, every model $S = (\mathfrak{U}, \mathfrak{O})$ of (P, Φ) gives rise to a GBWPL representation of same which in fact a bounded lattice; namely, **Lat**(S). This leads naturally to the converse question, which asks whether an arbitrary representation (L, f) in which L is a bounded lattice identifies a model. The answer is a qualified yes. Two additional conditions must be met are, first, the lattice must be distributive, and, second, the mapping f must be dense in **L**, so that no superfluous elements arise. The details follow.

3.9 Dense mappings and completions. Let $\mathbf{L} = (L, \sqcup, \sqcap, \top, \bot)$ be a bounded lattice, let A be any set, and let $f : A \to L$ be a function. f is said to be *dense* in **L** if the smallest sublattice of **L** containing $f(A)$ is **L** itself.

A GBWPL representation $(\mathbf{L}, f)$ for (P, Φ) in which **L** is a lattice is said to be *dense* just in case $f : \text{Aug}(P) \to L$ is dense in **L**. By construction, the representation $(\textbf{Lat}(L), \mathfrak{O})$ associated with an interpretation $S = (\mathfrak{U}, \mathfrak{O})$ is dense. Similarly, if $h : (\mathbf{L}_1, f_1) \to (\mathbf{L}_2, f_2)$ is a morphism of GBWPL's, with $\mathbf{L}_2$ furthermore a lattice, then h is said to be *dense* if $h(L_1)$ dense in $\mathbf{L}_2$.

A representation $(\mathbf{L}, f)$ of (P, Φ) in which **L** is a bounded distributive lattice and f is dense is called a *completion* of (P, Φ). Note in particular that for every model $S = (\mathfrak{U}, \mathfrak{O})$ of (P, Φ), the representation $(\textbf{Lat}(S), \mathfrak{O})$ is a completion. The following proposition shows that, up to isomorphism, all completions arise in this fashion.

3.10 Proposition *Let $(\mathbf{L}, f)$ be a completion of (P, Φ). Then there is a model $S = (\mathfrak{U}, \mathfrak{O})$ for (P, Φ) with the property that $(\textbf{Lat}(S), \mathfrak{O})$ is isomorphic to $(\mathbf{L}, f)$. In particular, $\mathbf{L}$ must be finite.*

Proof In view of the characterization theorem of Birkhoff and Stone [12, Ch. II, Sec. 1, Thm. 19], **L** may be taken to be a ring of sets over a (finite) set A. Define $\mathfrak{U} = A$, and take $\mathfrak{O}$ to be identical to f, save that the codomain is extended to be all of $\to 2^{\mathfrak{U}}$. It is immediate that $S = (\mathfrak{U}, \mathfrak{O})$ is a model for (P, Φ). Furthermore, since f is dense, it follows immediately from the definition of **Lat**(S) that Lat(S) and L are identical. Thus, the pair $(\mathfrak{U}, \mathfrak{O})$ is the desired model. $\square$

3.11 Theorem: algebraic characterization of models *There is a natural bijective correspondence between completions $(\mathbf{L}, f)$ of (P, Φ) and dense GBWPL morphisms h from $(\text{GBWPL}(P, \Phi), [-]_{\equiv(P,\Phi)})$ to distributive lattices, as illustrated in the diagram to the right. Thus, every $S \in \text{Mod}(\Phi)$ is represented, up to isomorphism, by a dense GBWPL morphism from $\textsf{CanRep}(P, \Phi)$ to a distributive lattice.*

$$\text{Aug}(P) \xrightarrow{[-]_{\equiv(P,\Phi)}} \text{GBWPL}(P, \Phi)$$

with f and h mapping to L.

Proof The proof follows immediately from the fact that the canonical representation $\textsf{CanRep}(P, \Phi)$ is in fact initial, as shown in 3. $\square$

3.12 The central rôle of distributivity. In view of the above theorem, as well as the very definition of completion in 2, it is clear that any completion of an open specification **must** be distributive. If a diagram such as that of Fig. 1 (which is easily shown not to be distributive) is presented as a description of the hierarchy under consideration, then it is certain that some information is missing. Either natural semantics do not hold for meet or for join (in which case the diagram is a partial description of an open specification), or else there are some types which are equivalent (in which case the actual hierarchy would take the form of a distributive quotient of the one presented).

3.13 Initial and canonical models. In general, a satisfiable open specification (P, Φ) will have many models. All of these models must satisfy every constraint in Φ, but some may satisfy other constraints as well. It turns out that there is a unique (up to isomorphism) model, called the *canonical model*, which satisfies as few constraints as possible; in fact, it satisfies only those constraints which are a consequence of Φ.

Space limitations do not permit presentation of the full construction here; rather, the reader is referred to [16, 3.3-3.8] for details. However, a brief discussion is in order. A set A of subsets of P is called a *crown* of P if whenever $C_1, C_2 \in A$ and $C_1 \subseteq C_2$, then $C_1 = C_2$. Crown(P) denotes the set of all crowns of P, while $\text{Crown}_\top(P)$ denotes $\text{Crown}(P) \cup \{\top\}$. The set $\text{Crown}_\top(P)$ has the structure of a bounded distributive lattice, with $\top$ the upper bound and $\{\emptyset\}$ the least element. The join operation is union followed by removal of elements which are proper subsets of others, while the meet operation is pairwise intersection.

In general, the canonical model is a quotient of this lattice of crowns. The key point to be observed here is that this lattice can be *very* large. If Φ does not contain any constraints, then the canonical model is the lattice of crowns itself. Note that this is a super-exponential construction. In general, there is no way to represent it explicitly when P contains more than a few elements. The bottom line is that any explicit representation of the canonical model is impossible for all but the most trivial of situations.

3.14 Hierarchies with complements. In some applications, it is desired that types in the hierarchy have complements. It is possible to extend the results given here to such hierarchies; the key idea is to replace bounded distributive lattices with Boolean algebras in the characterization. The rôle of GBWPL's does not change. Some results along these lines are presented in [13].

3.15 Limitations. Despite the elegance of this algebraic characterization, there are some inherent limitations. First of all, only positive, equality constraints are readily characterizable. It is possible to check for satisfaction of other types of constraints, such as type inequality constraints (e.g., $(\tau_1 \neq \tau_2)$ indirectly, but there is no direct construction of canonical models which satisfy such constraints.

Another major limitation is that the characterization is not computational; that is, there are no obvious tools of any efficiency for computing whether models exist, or for answering queries about existing models. Therefore, it is important to have an alternate characterization without these weaknesses. The logical characterization, presented in the next section, is a step in this direction.

4   Logical Characterization

In this section, a logical characterization of open specifications is developed. The approach consists of two distinct branches. First, a syntax for formulas, as well as a semantics, for constraints over a clean system P of types is developed from the elementary constraints introduced in Section 2. Second, a mapping from such formulas to a special propositional logic, denoted $\mathcal{L}_P$, is introduced. It is shown that the two logics are equivalent for *one-element models*; that is, models for which the underlying universe contains only one element.

The advantage of tying the constraint logic to propositional logic is that there exists a wealth of knowledge on the theoretical and practical aspects of solving satisfiability and model identification problems within propositional logic. Thus, with this approach, computational tools are ready at hand.

Unfortunately, one element models are not adequate to model all situations. Therefore, additional results are needed which show how to combine one-element models (*qua* propositional models) to obtain general models of sets of constraints. Much of this section is devoted to the presentation of such results.

Some of the basic ideas in this section are based upon the initial part of [15]. However, the presentation given here emphasizes logical characterization in a general form, while the emphasis of that of [15] is upon laying the foundations for the development of efficient inference algorithms on certain classes of constraints. (See 5).

4.1 Notational convention. Throughout this section, P denotes a clean set of types.

4.2 General constraints and their semantics. The set $\mathsf{UnresConstr}(P)$ of *unrestricted constraints over* P is built up from $\mathsf{ElemConstr}^+(P)$ using the usual logical connectives $\wedge, \vee, \neg$. The semantics of such constraints is the obvious one; if S is an interpretation over P and $\varphi, \varphi_1, \varphi_2 \in \mathsf{UnresConstr}(P)$, then $S \models (\varphi_1 \wedge \varphi_2)$ iff $S \models \varphi_1$ and $S \models \varphi_2$; $S \models (\varphi_1 \vee \varphi_2)$ iff $S \models \varphi_2$ or $S \models \varphi_2$; $S \models (\neg\varphi)$ iff $S \not\models \varphi$.

The *positive (unrestricted) constraints*, denoted $\mathsf{UnresConstr}^+(P)$, is the subset of $\mathsf{UnresConstr}(P)$ which is constructed using only the connectives $\wedge$ and $\vee$; without negation. Given an arbitrary set $\Phi \subseteq \mathsf{UnresConstr}(P)$, the subset $\Phi \cap \mathsf{UnresConstr}^+(P)$ is denoted Φ^+.

The definition of open specification is also extended to this more general context. Specifically, an *open specification* (resp. *positive open specification*) is a pair (P, Φ) in which P is a clean set of types and $\Phi \subseteq \mathsf{UnresConstr}(P)$ (resp. $\Phi \subseteq \mathsf{UnresConstr}^+(P)$).

4.3 The propositional logic of one-element models. Define the propositional logic $\mathcal{L}_P$ to have as proposition symbols the set $\{\mathfrak{r}_\tau | \tau \in \mathsf{Aug}(P)\}$. Also, within this context, **false** and **true** will be used to denote special propositions which always have the values false and true, respectively. For $\varphi \in \mathsf{ElemConstr}^+(P)$, associate a formula $\mathfrak{F}(\varphi)$ according to Table 1.

Table 1. Translation from constraints to propositional logic.

Constraint φ	Associated Logical Formula $F(\varphi)$
$(\sqcap\{\tau_1, \tau_2, .., \tau_n\} = \tau)$	$(r_{\tau_1} \wedge r_{\tau_2} \wedge \ldots \wedge r_{\tau_n}) \Leftrightarrow r_\tau$
$(\sqcup\{\tau_1, \tau_2, .., \tau_n\}) = \tau$	$(r_{\tau_1} \vee r_{\tau_2} \vee \ldots \vee r_{\tau_n}) \Leftrightarrow r_\tau$

$\mathfrak{F}$ is extended to formulas in $\mathsf{UnresConstr}(P)$ in the obvious manner: $\mathfrak{F}((\varphi_1 \wedge \varphi_2)) = (\mathfrak{F}(\varphi_1) \wedge \mathfrak{F}(\varphi_2))$; $\mathfrak{F}((\varphi_1 \vee \varphi_2)) = (\mathfrak{F}(\varphi_1) \vee \mathfrak{F}(\varphi_2))$; $\mathfrak{F}((\neg\varphi)) = (\neg\mathfrak{F}(\varphi_1))$. $\mathfrak{F}$ may furthermore be extended to one-element interpretations. Let $S = (\{a\}, \mathfrak{O})$ be any one-element interpretation for P. Define the interpretation (i.e., truth assignment) $\mathfrak{F}(S)$ in $\mathcal{L}_P$ to be true on the propositions in the set $\{r_\tau \in \mathsf{Aug}(P) \mid \mathfrak{O}(\tau) = \{a\}\}$, and false otherwise.

Using this translation to propositional logic, it is possible to characterize the one-element models of an arbitrary family of constraints in a simple fashion. The proof of the following is immediate from the definition of $\mathfrak{F}$.

4.4 Theorem — propositional models as models of open specifications. *Let (P, Φ) be an open specification. A one-element interpretation $S = (\{a\}, \mathfrak{O})$ for P is a model of (P, Φ) iff $\mathfrak{F}(S)$ is a model of $\mathfrak{F}(\Phi)$.* $\square$

In order to use the above result in a more general context, it is necessary to understand how one-element models combine to form arbitrary models. The next result shows that, at least for positive constraints, there is a direct componentwise decomposition.

4.5 Proposition *Let (P, Φ) be a positive open specification, and let $(\mathfrak{U}, \mathfrak{O})$ be an interpretation for P. Then S is a model for (P, Φ) iff for each $\varphi \in \Phi$ and every $a \in \mathfrak{U}$, $S_{|\{a\}} \models \varphi$.*

Proof For $\Phi \subseteq \mathsf{ElemConstr}^+(P)$, the result is immediate, since the basic set operations involved ($\cup$, $\cap$, and $=$) are all defined pointwise. For $\varphi = \varphi_1 \wedge \varphi_2 \wedge \ldots \varphi_n \in \mathsf{UnresConstr}^+(P)$, with each $\varphi_i \in \mathsf{ElemConstr}^+(P)$, it suffices to note that $M \models \varphi$ iff $M \models \varphi_i$ for each i, and then to replace Φ with $(\Phi \cup \{\varphi_1, \varphi_2, \ldots, \varphi_n\}) \setminus \{\varphi\}$. Thus, the results holds whenever $\Phi \subseteq \mathsf{UnresConstr}^+(P)$. Next, suppose that $\varphi \in \mathsf{UnresConstr}(P)$ is of the form $(\varphi_1 \vee \varphi_2 \vee \ldots \varphi_n)$ for $\varphi_1, \varphi_2, \ldots, \varphi_n \in$

UnresConstr$^+(P)$, By definition, there must be some $i,$, $1 \leq i \leq n$, for which $S \models \varphi_i$. The above result then applies. The most general case now follows easily by induction on the structure of the formula. $\square$

4.6 Some issues with negative constraints. Unfortunately, the above characterization does not extend to contexts with negation in the constraints. For example, consider an open specification which includes the three constraints $(\tau_1 \neq \tau_2)$, $(\tau_1 \neq \tau_3)$, and $(\tau_2 \neq \tau_3)$. Any model of these constraints must be based upon a universe with at least two distinct elements, since three distinct types are required. In general, a collection of constraints which mandates at least n distinct elements requires a universe of at least n elements. The following two paragraphs establish some more general results in the direction of model size.

4.7 Conjunctive normal form for constraints. An ElemConstr(P)-*clause* is an element of UnresConstr(P) of the form $\varphi_1 \vee \varphi_2 \vee \ldots \vee \varphi_n$, with $\varphi_i \in$ ElemConstr(P) for $1 \leq i \leq n$. In this context, each $\varphi_i \in$ ElemConstr(P) is called a ElemConstr(P)-*literal*. A formula $\varphi \in$ UnresConstr(P) is said to be in UnresConstr(P)-*CNF* (or just *CNF*, if the context is clear) if it is the conjunction of ElemConstr(P)-literals. It is easy to see that any constraint in UnresConstr(P) may be converted to an equivalent one which is in CNF. The procedure is analgous to that used for ordinary propositional formulas. This leads to the following results.

4.8 Theorem – characterization of model size. *Let (P, Φ) be an open specification with each member of Φ in CNF. If (P, Φ) is satisfiable, then it has a model of cardinality no greater than the total number of ElemConstr(P)-clauses in all formulas of Φ.*

Proof Assume that (P, Φ) is satisfiable, and let $S = (\mathfrak{U}, \mathfrak{O})$ be a model. Without loss of generality, Φ may be taken to be a single constraint ψ in CNF. If it is not of that form, just replace it with the conjunction of its members. Now, let $\varphi = \ell_1 \vee \ell_2 \vee \ldots \vee \ell_n$ be a conjunct of ψ. By assumption, φ is an ElemConstr(P)-clause. Furthermore, it must be the case that $S \models \varphi$, so that $S \models \ell_i$ for some i. If $\ell_i \in$ ElemConstr$^+(P)$, then every one-element projection $S_{|\{a\}}$ is a model of ℓ_i, in view of 4. On the other hand, if $\ell_i \in$ ElemConstr$^-(P)$, then $\neg \ell_i$ must fail to be a model for some one-element projection $S_{|\{a\}}$, again in view of 4. In any case, there is a element $a_\varphi \in \mathfrak{U}$ with the property that $S_{|a_\varphi} \models \varphi$. Define $A = \{a_\varphi \mid \varphi$ is a conjunct of $\psi\}$. Then, $S_{|A} \models \psi$, and Card(A) is, by construction, no larger than the number of clauses in Φ. $\square$

This bound on model size also provides a simple way of establishing decidability of satisfaction of open specification.

4.9 Corollary – decidability. *The question of whether or not an arbitrary open specification (P, Φ) is satisfiable is decidable.*

Proof First, convert Φ to CNF. Then, compute the bound on model size n stipulated by the above theorem. Up to isomorphism, there can be only a finite number of models of size n or less, which can be tested, in turn. $\square$

5   Complexity and Algorithms

The major focus of this paper is characterization. However, it is important to give a flavor for the computational implications of adopting an open-specification modelling strategy. Therefore, a brief overview of the most important aspects are provided here.

5.1 Important problems on open specifications. Let $\Omega = (P, \Phi)$ be an open specification.

Satisfiability: Does Ω have a model? In other words, does there exist an interpretation S such that $S \models \Phi$?

Query solution: Given $\varphi \in \mathsf{Constraints}(P)$, does $\Phi \models \varphi$ hold?

It is important to point out that query solution is a particularly important problem, in light of the combinatorial explosion which can easily result when attempting to construct a canonical model.

5.2 Theorem — complexity results. *Let* $\mathbf{S} = (P, \Phi)$ *be an open specification.*

(a) *The satisfiability problem for* Ω *is NP-complete.*

(b) *The query solution problem for* Ω *is co-NP-complete.*

These results remain valid even under the following circumstances:

(i) *The context is limited to positive constraints.*

(i) *The context is limited to positive equality constraints.*

(iii) *The context is limited to constraints with fanout at most three (that is, constraints which involve the operations* $\sqcap$ *and* $\sqcup$ *on at most three elements).*

The size of an instance is the length of the associated formula. $\square$

Sketch of Proof (a) Given the result 4 of the previous section, it should be no surprise that these problems are closely related to the satisfaction problem for propositional formulas, which is known to be NP-compete [11, Sec. 2.6]. However, a formal proof of the NP-completeness of satisfiability for open specifications must present a reduction in the opposite direction, showing that satisfiability for propositional formulas can be reduced to satisfiability for open specifications. The proof is quite nontrivial; complete details may be found in [13, Sec. 2].

(b) This is immediate from (a), since $\Phi \models \varphi$ iff $\Phi \cup \{\neg\varphi\}$ is unsatisfiable.

5.3 Efficient inference over a class of constraints. Despite the intractability results described above, it is possible to develop efficient inference algorithms over a useful subset of constraints. Results in this direction are reported in [15]. These results are based upon the fundamental observation that inference on Horn clauses is computable in linear time [9], [14].

6   Conclusions and Further Directions

Two distinct forms of characterization of openly specified type hierarchies have been provided. The *algebraic* characterization provides insight into the structure of such hierarchies, and in particular provides a canonical representation, in the form of a generalized bounded weak partial lattice, which recaptures in algebraic form a structure which embodies exactly those constraints which the specification mandates. The *logical* characterization provides a representation, in terms of propositional logic, which provides the basis for effective inference on the properties of such hierarchies.

For open specification to become a useful tool, it is clear that substantial further advances must be made in the tractability of inference on such structures. Classes of constraints which are at the same time useful in modelling those hierarchies which arise in practice and amenable to tractable inference algorithms must be identified. Although significant first steps in this direction appear in [15], significant further steps remain to be taken. Finally, in many situation there is further structure associated with the hierarchy. The computational results must be extended to representations which involve attributes, such as description logics [3] and formal concept analysis [10], as well as associated methods, as are commonly found in the object-oriented database context [2,18,19].

References

[1] J. Adámek, H. Herrlich, and G. Strecker. *Abstract and Concrete Categories.* Wiley-Interscience, 1990.

[2] C. Beeri. A formal approach to object-oriented databases. *Data and Knowledge Engineering*, 5:353–382, 1990.

[3] A. Borgida. Description logics in data management. *IEEE Trans. Knowledge Data Engrg.*, 7:671–682, 1995.

[4] T. Briscoe, V. de Paiva, and A. Copestake, editors. *Inheritance, Defaults, and the Lexicon.* Cambridge University Press, 1993.

[5] B. Carpenter and G. Penn. ALE: The Attribute Logic Engine user's guide, Version 3.1 Beta. Technical report, Bell Laboratories and Universität Tübingen, 1998.

[6] B. A. Davey and H. A. Priestly. *Introduction to Lattices and Order.* Cambridge University Press, 1990.

[7] J. Döerre and M. Dorna. CUF – a formalism for linguistic knowledge representation. In J. Dörre, editor, *Computational Aspects of Constraint-Based Linguistic Description, DYANA-2 Deliverable R.1.2.A*, pages 3–22. ESPRIT, 1993.

[8] J. Dörre and A. Eisele. Feature logic with disjunctive unification. In *Proceedings of the COLING 90, Volume 2*, pages 100–105, 1990.

[9] W. F. Dowling and J. H. Gallier. Linear-time algorithms for testing the satisfiability of propositional Horn clauses. *J. Logic Programming*, 3:267–284, 1984.

[10] B. Ganter and R. Wille. *Formal Concept Analysis.* Springer-Verlag, 1999.

[11] M. R. Garey and D. S. Johnson. *Computers and Intractability.* W. H. Freeman, 1979.

[12] G. Grätzer. *General Lattice Theory.* Birkhäuser Verlag, second edition, 1998.

[13] S. J. Hegner. Distributivity in incompletely specified type hierarchies: Theory and computational complexity. In J. Dörre, editor, *Computational Aspects of Constraint-Based Linguistic Description II, DYANA-2, ESPRIT Basic Research Project 6852, Deliverable R1.2B*, pages 29–120. DYANA, 1994.

[14] S. J. Hegner. Properties of Horn clauses in feature-structure logic. In C. J. Rupp, M. A. Rosner, and R. L. Johnson, editors, *Constraints, Languages and Computation*, pages 111–147. Academic Press, 1994.

[15] S. J. Hegner. Efficient inference algorithms for databases of type hierarchies with open specification. Submitted for publication, 2001.

[16] S. J. Hegner. Computational and structural aspects of openly specified type hierarchies. In M. Moortgat, editor, *Logical Aspects of Computational Linguistics, Third International Conference, LACL '98 Grenoble, France, December 1998, Selected Papers*, in press, 2001.

[17] C. Pollard and I. A. Sag. *Head-Driven Phrase Structure Grammar*. University of Chicago Press, 1994.

[18] K.-D. Schewe and B. Thalheim. Fundamental concepts of object oriented databases. *Acta Cybernetica*, 11:49–84, 1993.

[19] J. Van den Bussche. *Formal Aspects of Object Identity in Database Manipulation*. PhD thesis, University of Antwerp, 1993.

[20] R. Zajac. Notes on the Typed Feature System, Version 4, January 1991. Technical report, Universität Stuttgart, Institut für Informatik, Project Polygloss, 1991.

Null Values in Relational Databases and Sure Information Answers

Hans-Joachim Klein

Institut für Informatik und Prakt. Mathematik
Universität Kiel, 24098 Kiel, Germany
hjk@is.informatik.uni-kiel.de

Abstract. Null values are ubiquitous in database applications. There is, however, no common agreement upon how to deal with null values, neither in practice nor in theory. In this paper, the null value problem is revisited with special emphasis on answers representing sure information with respect to possible world semantics. The focus is on the *unknown* and the *no information* interpretation of null values. A new semantics for the *no information* interpretation of nulls is proposed taking into account a problem arising in connection with the closed world assumption. Results are applied to the query language SQL.

1 Introduction

In one of his early papers on the relational data model, E.F. Codd stated: "The usual distinction between the primary key and other candidate keys (if any) is that no tuple is allowed to have an undefined value for any of the primary key components, whereas any other components may have an undefined value." ([Cod72a], p. 37). In a subsequent paper he discussed relational algebra and tuple relational calculus and showed their equivalence without mentioning undefined values ([Cod72b]). This state of "secondary importance" of incompleteness caused by missing data values in attributes persisted in the forthcoming discussion of the classical relational data model.

The record oriented processing of data by "legacy" database systems puts responsibility for the handling of missing values to the application programmer. The close connection between CODASYL-systems and the programming language COBOL, for example, allows to use the figurative constant *low-value* of COBOL for representing and processing missing values. There are strong advocates for adopting a similar solution in the relational data model by providing *defaults* (special values) in domains ([DD98]). We do not consider this approach but concentrate on facilities to support the handling of missing information by the database system.

The main difference between concrete values of a domain and undefined values is the metalevel character of the latter. Since values in a database represent observations made in the "real world", undefined values correspond to certain phenomena of "reality". Examples are the missing knowledge about the birthday

L. Bertossi et al. (Eds.): Semantics in Databases, LNCS 2582, pp. 119–138, 2003.
© Springer-Verlag Berlin Heidelberg 2003

of a person, the nonexistence of a maiden name for an unmarried person, or the missing of an expected news. The reason why such a phenomenon is not represented in the database by an ordinary data value can be well–known or unknown. Among the well–known causes is the mismatch between the granularity of facts and their representation in a database. This mismatch shows up, for example, as anomalies in connection with transitive functional dependencies or it is a consequence of special views such as the universal relation view ([MUV84]). In these cases, a value may be undefined since there is no corresponding phenomenon in "reality". This means that the knowledge about the value is complete (its nonexistence coincides with an observed fact) but not representable by an ordinary value due to the chosen information structure. Null values marking such *inapplicable properties* ([Cod86]) are also called *nonapplicable* or *nothing nulls* ([LeLi86], [Vas79]).

The uniform interpretation of missing values as *value existent but unknown* has been assumed for most of the approaches dealing with incomplete information in relational databases. Usually, no distinction is made between permanently and temporarily unknown values (for an exception see [GM90]). In [Cod79] a duplicate removal rule and the use of a three–valued logic were proposed obviously with influence on the null value concept of SQL ([ISO92]). Critique on these proposals originated in different perspectives such as the incompleteness of query evaluation ([Grt77]) and semantical problems in connection with the duplicate removal rule ([Bis83]).

A thorough investigation of representation problems in connection with the *unknown* interpretation of missing values and the evaluation of queries can be found in [IL84]. From there we know that in general one must be satisfied with approximations for the representation of answers and that working with variables (marked nulls) instead of a single null value often results in better approximations. The necessity to approximate answers even if there are no problems with their representation follows from complexity problems connected with incompleteness ([Var86]). Several proposals can be found in the literature to circumvent these problems with the consequence that query results are either incomplete ([Cod79], [Rei86], [Var86]) or may contain uncertain information ([Lak89]) not to mention erroneous proposals (see [ZP97] for comments into this regard). Another approach is to focus on a more basic kind of null value by assuming that there is no information at all if some value is missing ([Zan84]). Efficient evaluation strategies have been proposed for this interpretation.

In order to deal with different kinds of null values some approaches have focused on the introduction of special logics and operators (e.g.: [GZ88], [Ges90], [Yue91]). A deep understanding of the consequences of their application is necessary in order to be able to interpret query results correctly. This may be a major obstacle to their use.

A different level of incompleteness in relational databases is the missing of tuples in relations. Because the number of negative facts in a domain of application is usually very large it is common practice to store only positive facts and to evaluate queries under the so–called *closed world assumption* ([Rei78]). If this assumption shall be weakened in order to take some incomplete knowledge about the validity of facts into account, relations can be partitioned appropriately. Proposals into this direction can be found in [LS90] and [OO93], for example. We refer to such a partitioning only in connection with query evaluation but assume the structure of relations in database to be homogeneous.

There are a number of interesting contributions in the field of incomplete relational databases we do not cover here. Overviews can be found in recent textbooks on database systems (e.g.: [LeLo99]). A source presenting a broad survey in the area of imperfect information is [MS97].

In this paper, our focus is on sure information answers in case of missing values interpreted as *value existent but unknown* or *there is no information*. In the next section we first discuss some basic concepts in connection with the modeling of incomplete information by a single special value and introduce some notation. Then we investigate the representation problem for answers to queries in a more general form and consider some approximations to sure answers based on model–theoretic semantics of incomplete databases. Section 4 deals with existential nulls. Complexity and decidability problems are pointed out and efficient but incomplete evaluation methods are discussed. Some research in this field is revisited under the viewpoint of sure information answers. No information nulls and null values in SQL are investigated in Section 5. It is shown that the correct handling of no information nulls with respect to a model–theoretic semantics has to solve problems implied by the closed world assumption. We make a proposal how to cope with this problem and point out some peculiarities of SQL semantics in connection with null values. Then a method for the transformation of SQL queries into query formulations approximating sure information answers is sketched. We conclude with some remarks.

2   Modeling Incomplete Information

2.1   Domain Extensions

A major advantage of the classical relational data model is its simplicity. Information is structured in finite sets of finite tuples belonging to the same type. Tuples are composed of atomic values taken from specified domains. A domain is a simple not necessarily finite set possibly with an order relation on it. Its elements are treated as uninterpreted objects.

If the model has to be generalized in order to cope with incomplete information not representable in this framework and if such a generalization shall preserve the main characteristics of the model, the domain concept can be enriched. Special domains with values specifying the information content of tuples can be introduced or existing domains can be extended by distinguished elements.

Additional attributes with special domains for the specification of the information content of tuples have been considered in [Bis83] and [IL84]. Entries into these attributes are generated during query evaluation to represent knowledge about the inclusion of tuples in results. The additional attributes, however, are not used in query formulations.

Our focus is on the extension of domains by a single distinguished value with the interpretation *value existent but unknown* (*unknown*, for short) or *there is no information on the value*. Up to some comments at the end of Section 5 we consider both interpretations separately, i.e. we do not mix occurrences of different kinds of null values in databases. The *unknown* interpretation has been assumed in most of the studies dealing with incomplete information in relational databases and fits well to many applications. The *no information* interpretation may be considered as more basic than the *unknown* and the *nonapplicable* interpretation ([Zan84]). It assumes that a missing value represents the information that no corresponding value exists in "reality" or that one or more values exist which are not known. This semantics is closely related to the null value concept of SQL.

Another important approach which should be mentioned is the extension of every domain by an infinite set of *variables* (Skolem constants) which allows to model information on the equality of unknown values ([IL84]). We sometimes refer to this kind of extension but do not consider it in detail. Sure information answers taking variables into account are proposed in [Kln99a] where also efficient algorithms for approximating answers can be found.

2.2   Notation and Basics

We review first some notations for relational databases used in the sequel.

For our investigations it is sufficient to assume a single countably infinite domain of values D to be given.

Let a finite set $\alpha = \{A_1, ..., A_k\}, k \geq 1$, of *attributes* be given. A *relational database schema* σ on α is a finite set $\{(id_1, \alpha_1), ..., (id_m, \alpha_m)\}, m \geq 1$, of *relation types* $(id_i, \alpha_i), i \in \{1, ..., m\}$ where id_i is a unique identifier and α_i a subset of α. $(id_i, \alpha_i), \alpha_i = \{A_{i_1}, ..., A_{i_l}\}$, is also denoted as $id_i(A_{i_1}, ..., A_{i_l})$.

A *relation* on a subset α_i of α is a finite set of mappings $\quad t \mid \alpha_i \rightarrow D^l, l = \mid \alpha_i \mid$. Its elements are called *tuples* over α_i. A *database state* or *database* over some database schema $\sigma = \{(id_1, \alpha_1), ..., (id_m, \alpha_m)\}$ is a set of relations $\{R_1, ..., R_m\}$, with one relation R_i on α_i for every relation type $(id_i, \alpha_i), i = 1, ..., m$.

A relation is called *incomplete* if some of its tuples are partial mappings; otherwise it is called *complete*. A database is called *incomplete*, if some of its relations are incomplete; otherwise, it is called *complete*. We use *idb* to denote incomplete databases and *pdb* for complete databases.

To simplify investigations it is assumed that different occurrences of missing values in databases are unrelated.

Independent of the interpretation of partiality a so–called information order can be introduced on sets of tuples over the same set of attributes. Let t, t' be two partial tuples over α_i. t' *subsumes* t (or is *more informative* than t), written $t' \geq t, \Leftrightarrow_{def} t(A)$ defined $\Rightarrow t'(A) = t(A)$ for every $A \in \alpha_i$. If t' is a total tuple then it is called an *extension* of t.

In operations it is sometimes necessary to check for tuples having identical subsumptions. We therefore define: Let t', t be as above; t' and t are *compatible*, written $t' \Delta t$, if there is some tuple t'' over α_i such that $t'' \geq t'$ and $t'' \geq t$.

The information order on tuples can be extended to relations in several ways reflecting different semantics of incomplete relations. We distinguish the following alternatives: closed world (CWA) or open world assumption (OWA), a missing value is of type *no information (ni)* or *unknown (unk)*, and an unknown value represents exactly one ($\exists_1$) or at least one ($\exists_{\geq 1}$) concrete value. From the eight possible cases $((OWA, ni, \exists_1)$ and $(OWA, unk, \exists_1))$ can be discarded since they are covered by $(OWA, ni, \exists_{\geq 1})$ and $(OWA, unk, \exists_{\geq 1})$, resp. This leaves us with the following alternatives (up to close subsumption see [Mai83]):
Let R, S be two incomplete relations on the same set of attributes.

$(OWA, ni, \exists_{\geq 1})$: R is a *subsumption* of S if the following holds:
$$(\forall t \in S)(\exists t' \in R)(t' \geq t).$$

$(OWA, unk, \exists_{\geq 1})$: R is an *extension* of S if R is a subsumption of S and total.

$(CWA, ni, \exists_{\geq 1})$: R is a *close subsumption* of S if R can be obtained from S by replacing some occurrences of undefined values by one or more concrete values in each case.

$(CWA, unk, \exists_{\geq 1})$: R is a *close extension* of S if R is an extension of S and the following holds: $(\forall t' \in R)(\exists t \in S)(t' \geq t)$.

$(CWA, ni, \exists_1)$: R is an *augmentation* of S if R can be obtained from S by replacing some occurrences of undefined values by a concrete value in each case.

$(CWA, unk, \exists_1)$: R is a *completion* of S if R is an augmentation of S and total.

Example 1. For the incomplete relations P and Q below with attributes ranging over the natural numbers, two completions (R_1, R_2), an augmentation (R_3), two close extensions (R_4, R_5), and a close subsumption (R_6) are shown for P and Q, resp. ω denotes undefined values in P and Q; its semantics is *valuenotexistent* in R_3 and R_6.

P			R_1			R_2			R_3			Q		
A	B	C	A	B	C	A	B	C	A	B	C	A	B	C
1	1	ω	1	1	2	1	1	4	1	1	4	ω	1	3
1	2	ω	1	2	1	1	2	4	1	2	ω	ω	1	ω
1	3	3	1	3	3	1	3	3	1	3	3	2	1	ω

R_4			R_5			R_6		
A	B	C	A	B	C	A	B	C
1	1	3	2	1	3	1	1	3
3	1	4	2	1	4	ω	1	2
2	1	1				ω	1	3
2	1	3				2	1	ω

We are interested in CWA–semantics which can now be defined for an incomplete relation R as the set of close subsumptions, close extensions, augmentations, or completions depending on the interpretation of incompleteness. Under the *no information* interpretation null values can be replaced either by one (non–key attribute) or several (key–attribute) concrete values or they remain unchanged and are interpreted as *value not existent*. If null values are allowed in key–attributes of an incomplete relation R then semantics is given by the set of close subsumptions of R; otherwise, semantics of R is given by the set of augmentations of R. For the *unknown* interpretation we have to take the set of close extensions and the set of completions of R, resp. Note that close subsumptions and augmentations represent complete information although they may be considered as incomplete relations at the syntactical level. We use the notations $POSS(R)$ and $POSS(idb)$ if the concrete kind of semantics is not relevant. The elements of $POSS(idb)$ are also called *possible worlds* to idb.

3   Query Semantics

3.1   Sure and Maybe Information in Answers

Query semantics based upon possible world semantics for incomplete databases can first be discussed without referring to concrete semantics and query languages. It is sufficient to consider queries as total functions mapping databases to relations and to assume *nonexistent* values to be represented by a unique symbol of the domain D. For the following, we suppose the reader to be familiar with relational algebra and relational calculus.

In case of complete databases, a query formulated in one of the common query languages against some schema σ can be considered as a mapping $q : S_\sigma \to \mathcal{R}_\beta$, where S_σ is the set of databases (or states) for σ and $\mathcal{R}_\beta$ the set of relations (tables) on β. The attribute set β is called *type of q* and denoted $type(q)$. It is fixed by the query formulation. *POSS*–semantics as defined in the previous section suggests to define semantics of a query q posed against an incomplete database idb as the set of mappings $\{q \mid q : POSS(idb) \to \mathcal{R}_\beta\}$. For a given incomplete database idb the set $\{q(pdb) \mid pdb \in POSS(idb)\}$ is called set of *possible answers* to q for idb.

On such a view, query semantics for incomplete databases can be based on query semantics for complete databases. In case of close subsumptions and augmentations, however, it has to be defined how queries are evaluated against relations with *nonexistent* values in attributes. Furthermore, this semantics is appropriate only under the assumption that query formulations do not use constructs directly addressing incomplete information. The IS NULL predicate of SQL, for example, is such a construct.

In [Lip79] an *upper value* has been defined for selection queries against a so–called object information system. This value corresponds to the union of possible answers and is finite for finite systems. It is not well–suited for the relational data model in general since it can be infinite even for finite databases and too much information about the structure of possible answers can be lost: Consider two relations $R = \{(1, \omega)\}$ and $S = \{(1, 2), (1, 3)\}$ on $\{A, B\}$ assuming ω to represent exactly one missing value. The possible answers to $R[B \neq 1]$ are all singletons $\{(1, c)\}$ with $c \neq 1$; its union is infinite if the domain of B is infinite. The possible answers to the join of R and S are $\emptyset, \{(1, 2)\}$ and $\{(1, 3)\}$. Their union $\{(1, 2), (1, 3)\}$ does not reflect these three possibilities.

The *lower value* defined in [Lip79] is the intersection of possible answers. It is a useful approximation to sure information answers for the information system considered in [Lip79] and is often applied to the relational data model (e.g.: [Gra91], [AHV95]). In [IL84] it is used in connection with the discussion of so–called *representation systems*. Such a system exists for a given set of relational operators, a given representation form for relations (single null value or variables), and a fixed semantics if the following holds: Answers can be computed incrementally for arbitrary expressions over the given set of operators without losing information with respect to the *lower value*, i.e. the intersection of possible answers does not change. This means that intermediate results can be represented in a finite form such that no information is lost necessary for the determination of the *lower value* for the complete expression (*truth–preservation*). One of the main results in [IL84] is that for so–called *Codd–tables* (a single null value and completion semantics) such representation systems only exist for the trivial empty set of operators and sets consisting of projection and/or selection operations with equality as single comparison operator. As a consequence, we cannot expect to find a finite representation for the set of possible answers in general (see also the example above). This negative result holds for *V–tables* (relations with variables) as well if the set of operators contains non–monotone operators ([IL84]). Hence one has to look for approximations describing sets of possible answers as precise as possible. In the next subsection, we show how single answer relations representing sure information to queries can be used for this purpose.

3.2 Sure Information Answers as Approximations

For the following, it is important to note that missing values in close subsumptions and augmentations are interpreted as *valuenotexistent* and treated like constants in the information order. Let us apply the concept of a *lower value* as an approximation to the set of possible answers for relational databases.

Definition 1. *The* sure total answer *to a query q posed against some incomplete database idb is defined as the set $\bigcap_{pdb \in POSS(idb)} q(pdb)$.*

The set of possible answers is always a superset of the sure total answer.

This kind of answer is also called *f–information* ([IL84]), *certain answer* ([Gra91]), or *set of sure facts* ([AHV95]). The restriction to total tuples representing sure information makes it more suitable for theoretical investigations than for practical purposes. To give an example, for the query 'R' we get an empty sure total answer if $R = \{(1, \omega)\}$; for '$R[A]$' the answer is $\{(1)\}$.

From a practical viewpoint, partial tuples have to be allowed in answers. Since semantics of an incomplete database is given by a set of complete databases, it is reasonable to consider all those partial tuples as sure information having an extension in every possible answer. This means that the information represented by such a partial tuple is implied by every possible answer.

Definition 2. *The* set of sure answer tuples *to q posed against idb, written $S(q, idb)$, is defined as follows:*

$$S(q, idb) =_{def} \{t \mid t \text{ partial tuple on } type(q) \ \wedge$$
$$(\forall pdb \in POSS(idb))(\exists t' \in q(pdb))(t' \geq t)\}.$$

A set of sure answer tuples is always a lower semilattice with respect to the information ordering on partial tuples. Since it is uniquely defined by its maximal elements, the following kind of sure answer is of interest:

Definition 3. *The* sure max–answer *to q posed against idb is the set of all maximal elements of* $S(q, idb)$.

Obviously, the sure total answer to a query is always a subset of the sure max–answer. Both answers coincide if the set of sure answer tuples does not contain maximal tuples which are partial. Every possible answer can be obtained as an extension of the sure max–answer.

Example 2. Let an incomplete database *idb* consist of the union–compatible relations $P = \{(1,2),(1,3)\}$ and $Q = \{(1,\omega)\}$ with ω assumed to represent exactly one missing value (completion semantics). In every possible world to *idb* the relation for Q consists of exactly one tuple. Consider the query $P - Q$. Either $(1, 2)$ or $(1, 3)$ or both tuples of P qualify for the answer. Hence we get $\{(1, \omega), (\omega, \omega)\}$ as set of sure answer tuples and $\{(1, \omega)\}$ as sure max–answer. The sure total answer is empty. In case of close extension semantics the set of sure answer tuples is empty as well because there are possible worlds with Q having $(1, 2)$ and $(1, 3)$ among its elements.

Sure max–answers can be enriched by subsumed sure answer tuples in order to get better approximations for possible answers. The idea is to include a minimal number of sure answer tuples such that in case of a non–empty set of sure answer tuples every possible answer can be obtained as a close extension. Details can be found in [Kln99a] (for V–tables) and [Kln99b] (for relations with a single null value).

Null value occurrences in sure max–answers have to be distinguished from those in incomplete databases insofar as they do not represent arbitrary values in general (cf. Example 2 where ω in the answer tuple represents 2, 3, or both values). If simple efficient evaluation methods such as three–valued logic are applied, the meaning of nulls in the database and in answers is *uniform*, i.e. null value occurrences in answers represent arbitrary values. Therefore subsets of the following kind of answer are of special interest:

Definition 4. *Let q and idb be as before. The* sure uni–answer *to q posed against idb is the following subset of* $S(q, idb)$:

$$\{t \in S(q, idb) \mid (\forall A \in type(q))(t(A) = \omega \Rightarrow (\forall d \in D)(\exists t' \text{ on } type(q))$$
$$(t'(A) = d \,\wedge\, t' \geq t \,\wedge\, (\exists pdb \in POSS(idb))(t' \in q(pdb))))\}.$$

The sure uni–answer includes a tuple t of the set of sure answer tuples only if there is no restriction given for any occurrence of ω in an attribute A of t, i.e. for every element d from the corresponding domain there is a possible answer with a tuple $t' \geq t$ having d as value in A. Possible answers can be obtained as extensions of the sure uni–answer.

The sure uni–answer to $P - Q$ from Example 2 is empty since ω in the sure answer tuple $(1, \omega)$ has to be interpreted as 1 or 2.

4   Existential Nulls

4.1   Tautology and Decidability Problems

In the following, we assume the *unknown* interpretation for missing values and denote occurrences of these null values in examples by *unk*. A query against an incomplete database with this kind of missing information may state conditions which are fulfilled or not independent of the replacement of null values by concrete values. Such tautologies and contradictions may be valid independent of the given database or they may hold for a part of the database.

Example 3. Let $R = \{(5, unk, 2), (unk, 1, 3)\}$ be a relation on $\{A, B, C\}$. The query $q = R[A \geq B \vee A < B]$ formulates a condition which is true for tuples of R independent of their values in the attributes A and B.
The condition formulated in the query $q = R[A > 4 \wedge B = 0 \vee C < 3 \wedge B \neq 0]$ is false for the second tuple but true for the first tuple independent of the null value in B and A, resp.

Co–NP–completeness with respect to query complexity has been shown in [Var86]. Further complexity problems have been investigated for different representation forms of incomplete information in [AKG91]. These complexity problems force to look for approximations to all kinds of sure answers.

For close extension semantics, not only complexity is the problem: Since a null value represents an arbitrary number of missing values it follows from the undecidability of finite validity in first–order logic (theorem of Trachtenbrot) that sure information answers are not computable for query languages which are as expressive as relational calculus (for a proof see [Kln97]).

4.2   Approximations for Relational Calculi and Algebra

In [Kle38], S.C. Kleene introduced a three–valued logic for reasoning about partial recursive functions. The third truth value represents the indefiniteness of computation, a situation quite similar to the missing knowledge about values. The tables for the connectives *and*, *or*, and *not* are identical to the three–valued logic proposed earlier by J. Łukasiewicz for the investigation of modality:

$\wedge$	T	F	U
T	T	F	U
F	F	F	F
U	U	F	U

$\vee$	T	F	U
T	T	T	T
F	T	F	U
U	T	U	U

$\neg$	T	F	U
	F	T	U

Both logics differ in the definition of implication. Kleene's definition ($U \Rightarrow U$ is equivalent to U) fits well the *unknown* interpretation. Therefore, equivalences of this logic should be valid when three–valued logic is used for relational query languages.

In [Cod79], the above definitions are used for the evaluation of calculus queries under completion semantics. Comparison expressions are evaluated to *unknown* if they involve the null value. Quantifiers are assumed to behave like iterated *or* and *and*. The monotonicity property of three–valued logic guarantees that tuples in results are always sure answer tuples: If some tuple qualifies for the answer then it qualifies even if at evaluation some occurrences of *unknown* are replaced by *true* or *false*. This means that *true* (and *false*) results are independent of null value occurrences. One problem has to be observed, however: If the identification of tuples is stated indirectly, i.e. without use of comparison expressions, as in $(x)/Rx \wedge Sx$, then the calculus interpretation has to avoid the identification of independent partial tuples. Tuple identification in set operations in connection with the removal of duplicates is regarded in [Cod79] as "an operation at a lower level of detail than equality testing in the evaluation of retrieval conditions". This view is exactly what we find in SQL where three–valued logic is used for the evaluation of WHERE–clauses but independent occurrences of the null value may be identified in set operations ([ISO92]). As a consequence, set operations may yield results not representing sure information.

If operations of relational algebra shall be defined in accordance with possible world semantics, the independence of different occurrences of null values in a database has to be guaranteed. In [Bis83], the need to change completion semantics into close extension semantics in results of operations is argued. Tuples in results are classified as *sure* or *maybe* depending on their status with respect to evaluation. Definitions for relational algebra operations are given and proven to guarantee the correct status of tuples in results. Expressions of relational algebra, however, are not considered. On the basis of these definitions and the proposal in [Cod79] to consider sure and maybe versions of operations, a simple evaluation method for relational algebra expression can be developed.

For every operation, a sure and a maybe version are defined: The sure version includes only tuples in the result fulfilling for sure the condition of the operation; the maybe version includes in addition every tuple which may not be excluded from the result (notice the difference to the definitions in [Cod79] where the maybe–result only includes these additional tuples). As an example consider the two versions of difference: Let R, S be relations on the same set of attributes; $R -_s S =_{def} \{t \mid t \in R \wedge \neg(\exists t' \in S)(t' \Delta t)\}, R -_m S =_{def} R - \{t \mid t \in S \wedge t \text{ total}\}$ where "$-$" is the ordinary set difference.

For the evaluation of an arbitrary expression e, sure and maybe versions of operations are chosen depending on the "level of negation" which is determined by the difference operations in e. For all operations on the outermost level of e, sure versions are used; for a difference operation, the operations at the outermost level of the subexpression representing the subtrahend are evaluated applying their maybe version. This proceeding is iterated switching from maybe to sure and vice versa.

Example: For $R - ((S - R[A = 2]) \cup T)$ we get $R -_s ((S -_m R[A = 2]_s) \cup_m T)$. We call this method *switch–evaluation*. It can be shown that applying switch–evaluation to relational algebra expressions and three–valued logic to relational calculus queries leads to the same results for algebra expressions and calculus queries being equivalent in case of complete databases ([Kln97]). This holds for completion as well as for close extension semantics. Thus the well–known equivalence theorem can be lifted to incomplete databases. Answers are always subsets of the sure uni–answer.

In order to approximate sure max–answers without loss of efficiency, extended definitions of algebra operations and multivalued–logics for calculus queries can be introduced ([Kln99b]). For all kinds of answers, approximations can be improved without loss of efficiency by exploiting identities of unknown values established with evaluation.

4.3   Literature Review

Many proposals for handling the *unknown*–interpretation of missing values present definitions of algebraic operators without giving declarative semantics for queries. Here we discuss some of them with respect to sure information in answers.

The approach of [Cod79] does not guarantee sure information in answers because null values are handled in set operations like ordinary values. Let $R = S = \{(1, unk)\}$; then $R \cap S$ is defined as $\{(1, unk)\}$. The combination of maybe–results for the evaluation of expressions may lead to results which are not sound in any possible world. Consider the expression $(R - R[B = 2]) \cap (R - R[B <> 2])$. We get R as result; under completion semantics the result is empty in every possible world.

In [GZ88], different types of null values are considered among them *unknown* with completion semantics and *open* with close subsumption semantics. Symbolic equality is used in set operations such that the preceding remarks hold as well. For *open* nulls, undecidability follows from undecidability in case of close extension semantics.

The weakness of "truth–preserving extension" based on representation systems (see Section 3.1) shows up in definitions of extended operators given in [LeLo99] for null values with completion semantics. Consider the definition of extended difference given there: $R -_e S =_{def} \{t \mid t \in R \wedge \neg(\exists u \in S)(u \leq t \vee t \leq u)\}$. Let $R = \{(1, unk)\}, S = \{(unk, 2)\}$. Then $R -_e S = \{(1, unk)\}$ since neither $\{(unk, 2)\} \leq \{(1, unk)\}$ nor $\{(1, unk)\} \leq \{(unk, 2)\}$. In order to restrict the result to sure information, compatibility has to be excluded for u and t (cf. our definition of the sure version of difference). Similar comments hold for extensions proposed in connection with the nested relation data model (e.g. [RKS89]).

A completely different approach to the problem of inefficiency in connection with the evaluation of queries against incomplete databases has been taken in [Lak89]. Query evaluation is based on intuitionistic logic and it is claimed that the missing of the tautology problem in this logic allows to compute complete answers efficiently. Of course, answers have a special meaning in this approach and it can be shown that they do not always represent sure information in the model–theoretic sense ([Kln97]).

It has been indicated an advantage of the proof–theoretic approach over to the model–theoretic approach that in case of incompleteness the introduction of new logics can be avoided ([Rei86]). However, a closer look at the evaluation methods proposed in [Rei86] and [Var86] shows that both give the same result as an appropriate interpretation of calculus formulas using three–valued logic for databases with variables as null values ([Kln98]). This means that peculiarities connected with three–valued logic (e.g.: the equivalence of $A = B \wedge B = C$ and $A = B \wedge A = C$ does not always hold) are present as well.

5   No Information Nulls and SQL

5.1   Semantics

Consider a relation type PERSON with 'Status' and 'Birthname' among its attributes. The attribute 'Birthname' shall not be applicable when a person is unmarried. Assume that for some person we do not know her or his family status. In consequence, the corresponding tuple in the PERSON relation has a null value in the attribute 'Status' interpreted as "value existent but unknown". If the family status of a person is unknown nothing is known about her or his birthname: it could be existent or not. Hence our tuple has a null value in the attribute 'Birthname' and for this null value a *no information* interpretation is appropriate representing the two cases "value existent but unknown" and "value does not exist" together.

[Zan84] proposes to base the definition of operators for relations with such *no information* nulls on equivalence classes of relations on extensible attribute sets. [Kel86] shows that there is no way to define relational algebra operators on equivalence classes of this kind so that the definitions are in accordance with ordinary set theory. The definitions of relational operators in [Zan84] do also not guarantee sure information in results. As an example take the definition of difference (missing values are represented by *null* in this section):
$R \tilde{-} S =_{def} \hat{\{t \mid t \in R \wedge (\forall s \in S)(\neg(s \geq t)\}}$. Here $\hat{\{..\}}$ denotes the equivalence class for $\{..\}$. For $R = \{(1,2)\}$ and $S = \{(1, null)\}$ we get $R \tilde{-} S = \hat{\{(1,2)\}}$.

The *no information* interpretation combines incomplete knowledge (*unknown*) and complete knowledge (*value does not exist*). As a consequence, possible world semantics for databases with *no information* nulls has to take care of a representation problem arising in connection with the closed world assumption.

Example 4. Consider the following relations with information on the names of invited guests of a party and their telephone number in GUESTS and the names of some guests together with the name of an accompanying person in COMPANY. For Frank we know that he comes alone whereas for Petty we know nothing with respect to her company.

GUESTS

Gname	Tel#
Tim	346537
Petty	660890
Frank	730286

COMPANY

Gname	Cname
Tim	Anna
Petty	null

When we consider all possible worlds then in one of these worlds the information is modeled that Petty is not accompanied. This means that in this possible world the null value is not replaced by a data value but it is interpreted as *value not existent*. Hence the information on the company of Frank is represented differently from the information on the company of Petty. This may lead to inconsistencies with respect to query evaluation. Consider the following SQL–query:

```
SELECT * FROM GUESTS x WHERE EXISTS (SELECT * FROM COMPANY y
                            WHERE y.Gname = x.Gname)
```

The SQL–answer to this query is {(Tim,346537), (Petty,660890)}. It does not represent sure information if we think of a formulation asking for guests accompanied by someone. A first idea is to specify *POSS*–semantics depending on a given query and to remove partial tuples in possible worlds if there are conflicts with the CWA. The next example demonstrates that this proceeding does not work in general without additional schema information.

Example 5. Consider the following relations modeling information on employees and project managers as indicated by the entity/relationship–diagram.

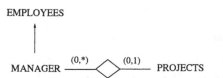

EMPLOYEES	
Name	Tel#
Ben	375
Jane	897
Peter	993

PROJMAN

Name	Proj#
Ben	P3
Jane	null

Look at the following query having the same structure as the query above:

```
SELECT * FROM EMPLOYEES x WHERE EXISTS (SELECT * FROM PROJMAN y
                            WHERE y.Name = x.Name)
```

Now the intention of the query could be to get the data of all employees being managers without requiring that they actually lead projects. Hence in the possible world corresponding to that one mentioned in the preceding example no tuple has to be discarded for solving conflicts with the CWA. This means that from the structure of the relations and queries alone we cannot derive a satisfying definition of sure information answers. Obviously, it is necessary to know more about *concepts* modeled in schemas and addressed in queries. In the GUESTS/COMPANY–example, a single concept "guest comes in company of a person" is modeled, in the EMPLOYEES/PROJMAN–example two concepts are modeled: "an employee *e* is a manager" and "a manager *e* controls a project *p*". The standard relational data model does not allow to address such concepts. An appropriate proposal to extend the model by specifying sets of attributes as units of interest can already be found in [Mai83] (see also [Sci79] and [IL82]). There the term *object* is used for characterizing allowable schemes for subtuples in relations with tuples padded out by placeholders. We refine this proposal by distinguishing mandatory and optional attributes in order to be able to differentiate values which are necessary for a combination of values to "make sense" from values which may be considered as supplementary information.

Definition 5. *A concept for a relation type is a subset of its set of attributes with elements classified either as mandatory or as optional.*

Example 6. Consider the relation type PARTY(Gname, Tel#, Cname, Contribution) for modeling information about names of guests, their telephone number, the name of the person in their company, and their contribution to the buffet. The following are reasonable concepts for this type: {(Gname,m), (Tel#,o)}, {(Gname,m), (Cname,m)}, and {(Gname,m), (Contribution,m)}.
For the relation type PROJMAN(Name, Proj#) from above we could choose: {(Name,m)} and {(Name,m), (Proj#,m)}.

Given a non–empty set of concepts for a relation type we can now define the set of tuples "making sense".

Definition 6. *Let (RT, α) be a relation type with set of concepts C. A partial tuple t on α is* allowed *with respect to $C \Leftrightarrow_{def}$*

 − *t is not totally undefined.*
 − *If t is defined in an attribute A then there is at least one concept in C containing A such that t is defined in all mandatory attributes of this concept.*

A relation R on α is allowed *with respect to $C \Leftrightarrow_{def}$ every tuple of R is allowed with respect to C.*

These definitions permit to move semantics at the level of syntax.

Definition 7. *Let (RT, α) be a relation type with set of concepts C and let R be an incomplete relation on α. A relation R' is called* possible relation *to R with respect to $C \Leftrightarrow_{def} R'$ can be obtained from an augmentation (close subsumption) R'' of R by discarding all tuples which are not allowed.*

Augmentations are appropriate if null values are not allowed in key attributes; otherwise, close subsumptions have to be considered. Since possible worlds may contain values with *nonexistent* interpretation, it has to be fixed how these values are treated in query evaluation. Taking the view that it makes no sense to compare "nothing" with a data value ([Vas79]) the following approach seems to be appropriate: Consider relational calculus as a reference language; make use of three–valued logic as in Section 4 with *nonexistent* entries in attributes treated like *unknown* values.
If both operands in a comparison expression are *nonexistent* values, a Boolean value could be assigned to the expression. There are, however, also arguments to determine that a nonexistent value does not satisfy any relational expression ([Zan84]). Furthermore, for such a refinement the addressing of concepts in queries has to be taken into account which is beyond the scope of this paper.

5.2   Peculiarities in SQL

In the following, we restrict our attention to *SELECT...FROM...WHERE*–clauses in SQL. Problems in connection with the duplicate removal rule in set operations applied in SQL have already been addressed in Section 4. Furthermore, we assume that the IS NULL predicate is not used in query formulations. Since this predicate refers directly to occurrences of the null value it is of a syntactical nature and must be considered separately. In reformulations of SQL queries, however, we make use of this predicate.

The SQL standard specifies the *null value* as "a special value, or mark, that is used to indicate the absence of any data value" ([ISO92]), a characterization fitting well the *no information* interpretation. Any comparison expression involving the null value returns *unknown* as result. For the evaluation of WHERE–clauses three–valued logic is applied with tables for the connectives as given in subsection 4.2. In the result of a FROM–clause only tuples are retained for which the WHERE–clause evaluates to *true*. This switch from three–valued to Boolean logic may lead to answers not representing sure information in the same way as the approaches discussed in Section 4.

Example 7. ("Maybe and maybe may be not maybe"): Let $R = \{(1, null)\}$ be a relation on $\{A, B\}$.

```
SELECT x.A FROM R x WHERE NOT EXISTS (SELECT * FROM  R y WHERE y.B
= 2)
                    AND NOT EXISTS (SELECT * FROM  R z WHERE z.B <> 2)
```

The SQL–answer to this query is $\{(1)\}$ since in both subqueries the WHERE–clauses evaluate to *unknown*. Obviously, the answer is empty if *null* is replaced by any ordinary value of the domain of B.

SQL–semantics allows a programmer to interpret occurrences of the null value by choosing an appropriate query formulation, i.e. to get results fitting the *unknown* or the *nonexistent* interpretation.

Example 8. Let tuples in the following relation represent information on orders with a unique number, the date of their entry, and the identification number of the employee working on it.

ORDER

Ordno	Date	Works_on
1	03-nov	null
3	03-nov	E2
4	04-nov	null

Consider the query 'Get the data of each order o such that there is no employee working on o and on further orders with the same date of entry as o.'
Assuming *nonexistent* interpretation for nulls ("no employee is working on the order") the appropriate answer is the relation ORDER. We get it by choosing the following formulation:

```
SELECT * FROM ORDER x WHERE NOT EXISTS (SELECT * FROM ORDER y
WHERE
  y.Date = x.Date AND y.Ordno <> x.Ordno AND y.Works_on = x.Works_on)
```

Assuming *unknown* interpretation, a sure information answer should consist only of the third tuple. The following formulation is appropriate:

```
SELECT * FROM ORDER x WHERE Works_on <> ALL (SELECT Works_on FROM
         ORDER y WHERE y.Date = x.Date AND y.Ordno <> x.Ordno)
```

This formulation is suited also if the *no information* interpretation is assumed.

In general, equivalences valid in Kleene's three–valued logic do not hold in SQL. Applying the equivalence of $A \wedge \neg B \Rightarrow \neg C$ and $C \wedge \neg B \Rightarrow \neg A$, for example, to the second formulation above we get a formulation equivalent to the first formulation:

```
SELECT * FROM ORDER x WHERE x.Date <> ALL (SELECT y.Date FROM
              ORDER y WHERE y.Works_on = x.Works_on AND y.Ordno <> x.Ordno)
```

By making use of the IS NULL predicate of SQL, it is always possible to choose a query formulation having a subset of the sure uni–answer as result for any of the above interpretations. How this can be done systematically by query transformation will be described in the following.

5.3   Modification of SQL Queries

In Section 4, we proposed the *switch–algebra* for the *unknown*–interpretation. Based on this idea, transformation rules can be developed for SQL such that transformed queries have sure information answers if concepts are taken into account appropriately.

Let an arbitrary WHERE–clause w be given. Because of the monotonicity property of three–valued logic, the evaluation of the WHERE–clause generates sure information provided that all Boolean values used for the evaluation represent sure information. For a negated subformula $NOT(w')$ of w this means that w' has to evaluate to *true* if it cannot be excluded that there is a possibility to replace occurrences of the null value such that w' has the value *true*. The possible loss of the monotonicity property by the transition from three–valued logic to Boolean logic in predicates has to be avoided by appropriate modification of the predicates. In analogy to the switch–algebra, *sure* and *maybe* versions of predicates are needed. *unknown* as result of a predicate or of the WHERE–clause of a predicate has to be turned into a *false* or *true* depending on the kind of result needed. The switch from *sure* to *maybe* and vice versa is caused by negation and universal quantification.
Concepts are taken into account by deleting at the "sure level" all tuples not allowed in the corresponding tables. An appropriate conditional expression can be formulated with the help of the IS NULL predicate. At the "maybe level" no tuple has to be deleted.

As an example, let us consider the sure version of the *quantified comparison predicate* with ALL as quantifier (for the maybe version and other predicates see [Kln94]). Its general form is $v \Theta ALL(S)$ where v is a row value expression, S stands for a suitable subquery, and Θ is a comparison operator. To get a sure result, the maybe result is needed for S. There are two critical cases to consider:

1) v evaluates to *null*.
In this case, the SQL answer is *true* if S evaluates to $\emptyset$ and *unknown*, otherwise. *unknown* can be changed to *false* by adding AND (v IS NOT NULL OR NOT EXISTS(S)) to the predicate.

2) *null* occurs in the result of the evaluation of S (e.g.: $5 \leq ALL\{7,8,null\}$).
false instead of the possible result *unknown* has to be guaranteed. For simplicity,

let S = SELECT B FROM ... WHERE *cond*. Add AND NOT EXISTS(S') to the predicate where S' = SELECT B FROM ...WHERE *cond* AND B IS NULL.

The application of this rule to the third formulation in Example 8 results in the following SQL query:

```
SELECT * FROM ORDER x WHERE x.Date <> ALL (SELECT y.Date FROM ORDER y
            WHERE (y.Works_on = x.Works_on OR y.Works_on IS NULL
                   OR x.Works_on IS NULL)  AND y.Ordno <> x.Ordno)
-           AND x.Date IS NOT NULL
-           AND NOT EXISTS (SELECT y.Date FROM ORDER y WHERE
-               y.Ordno <> x.Ordno AND y.Date IS NULL)
```

With $\{(\text{Ordno},m),(\text{Date},m)\}$ and $\{(\text{Ordno},m),(\text{Works\_on},m)\}$ as concepts, the condition AND (y.Ordno IS NOT NULL AND y.Date IS NOT NULL OR y.ORDNO IS NOT NULL AND y.Works_on IS NOT NULL) has to be added to the WHERE–clauses of both subqueries. The conditions in the lines marked by "-" are redundant and can be removed if occurrences of the null value in the attribute 'Date' are not allowed. Simplifications of this kind allow to reduce the size of the generated formulation in many cases.

Instead of answers representing sure information, "maybe" answers can be produced by starting with maybe versions of predicates (including simple comparison predicates) at the outer level. These "maybe" answers, however, may contain information not implied by any possible world because of contradictions which are not detected.

In many cases, users will not associate the *no information* interpretation to null values in attributes but the *unknown* or *nonexistent* interpretation. The 'Birthday' attribute in a PERSON relation type is a typical example for a situation where the *unknown* interpretation is natural. For the 'Works_on' attribute of the ORDER relation type in Example 8 a *nonexistent* interpretation could be in accordance with the processing of order data in a company. The question arises how to cope with such views in our approach. Since the set of possible worlds for the *unknown* as well as for the *nonexistent* interpretation is a subset of the set of possible worlds for the *no information* interpretation, the proposed method for transforming queries guarantees sure information answers for both interpretations. The approximation of sure uni–answers can be improved by not taking into account concepts in case of attributes for which the *unknown* interpretation is assumed. What has to be done is to consider occurrences of the null value in these attributes as ordinary values when determining allowed tuples.

6   Conclusions

There is much controversy about the way incomplete information should be represented and processed by relational database systems. We suggest that systems should provide interfaces avoiding the confrontation of the user with the intrinsic complexity of incompleteness. Different kinds of answers with well–defined non–procedural semantics should be requestable at different levels of completeness allowing users to explore incomplete information at costs they are willing to spend (cf. [KW85]). To have answers with different semantics and increasing levels of completeness at hand, may prevent users from wrong conclusions.

The use of three–valued logic has been criticized for producing incomplete results ([Grt77]). However, if this logic is carefully applied to avoid unallowed identification of missing information, it provides a good foundation for efficient computation of answers representing sure information. Evaluation methods based on this logic or their corresponding extended operators in relational algebra can be refined with moderate increase of costs such that better approximations to sure information answers can be achieved. As we have seen, some proposals in the literature fail to guarantee sure information in answers so that the interpretation of answers becomes problematic.

Sure information is not guaranteed in SQL answers because of the problems arising in connection with the *no information* interpretation of the null value and because three–valued logic is not applied consequently. "Maybe" tuples, for example, are discarded when the result of a subquery is generated independent of the context the subquery is used in. As a consequence, to get sure answers, subqueries have to be evaluated in dependence upon their context or they have to include "maybe" tuples marked uniquely. Additionally, concepts have to be introduced and considered appropriately. For the first approach, we have shown how to supplement SQL queries in order to take concepts and the context into account. For the second approach, SQL semantics has to be modified.

In this paper, we considered standard languages for the formulation of queries since this is the current reality we have to cope with. Another approach is to look for new languages better suited for the formulation of requests against databases with incomplete information ([Lip81], [Lib98]). This could help to get a common framework for dealing with the many facets incompleteness shows.

References

[AHV95] S. Abiteboul, R. Hull, and V. Vianu: Foundations of Databases. Addison–Wesley, Reading, 1995

[AKG91] S. Abiteboul, P. Kanellakis, and G. Grahne: On the representation and querying of sets of possible worlds. Theor. Comp. Science 78, pp. 159–187, 1991

[Bis83] J. Biskup: A foundation of Codd's relational maybe-operations. ACM Trans. on Database Systems 8 (4), pp. 608–636, 1983

[Cod72a] E.F. Codd: Further normalization of the database relational model. In: Data Base Systems. Current Computer Science Symposium 6 (R. Rustin, ed.), Prentice Hall, Englewood Cliffs, pp. 33–64, 1972

[Cod72b] E.F. Codd: Relational completeness of data base sublanguages. In: Data Base Systems. Current Computer Science Symposium 6 (R. Rustin, ed.), Prentice Hall, Englewood Cliffs, pp. 65–98, 1972

[Cod79] E.F. Codd: Extending the database relational model to capture more meaning. ACM Trans. on Database Systems 4 (4), pp. 397–434, 1979

[Cod86] E.F. Codd: Missing information (applicable and inapplicable) in relational databases. ACM SIGMOD RECORD 15 (4), pp. 53–77, 1986

[Dat86] C.J. Date: Null values in database management. In: Relational Database: Selected Writings, Addison–Wesley, Reading, 1986

[Dat90]   C.J. Date: Relational Database Writings 1985–1989. Addison-Wesley, Reading, 1990

[DD98]   C.J. Date and H. Darwen: Foundation for Object/Relational Databases. Addison Wesley, Reading, 1998

[Ges90]   G.H. Gessert: Four valued logic for relational database systems. ACM SIGMOD Record 19 (1), pp. 29–35, 1990

[GM90]   O.N. Garcia and M. Moussavi: A six–valued logic for representing incomplete knowledge. Proc. IEEE 20th Int. Symp. on Multiple–valued Logic (G. Epstein, ed.), Charlotte, pp. 110–114, 1990

[Gra91]   G. Grahne: The Problem of Incomplete Information in Relational Databases. LNCS 554, Springer–Verlag, Berlin, 1991

[Grt77]   J. Grant: Null values in relational data base. Inform. Proc. Letters 6 (5), pp. 156–157, 1977

[GZ88]   G. Gottlob and R. Zicari: Closed world databases opened through null values. Proc. 14th VLDB Conf., Los Angeles, pp. 50–61, 1988

[IL82]   T. Imielinski and W. Lipski: A systematic approach to relational database theory. Proc. ACM SIGMOD Conf., Orlando, pp. 8–14, 1982

[IL84]   T. Imielinski and W. Lipski: Incomplete information in relational databases. Journal of the ACM 31 (4), pp. 761–791, 1984

[ISO92]   International Standard ISO/IEC 9075, third edition. Genève, 1992

[Kel86]   A.M. Keller: Set–theoretic problems of null completion in relational databases. Inf. Proc. Letters 22, pp. 261–265, 1986

[Kle38]   S.C. Kleene: On a notation of ordinal numbers. The Journal of Symbolic Logic 3, pp. 150–155, 1938

[Kln94]   H.-J. Klein: How to modify SQL queries in order to guarantee sure answers. ACM SIGMOD Record 23 (3), pp. 14–20, 1994

[Kln97]   H.-J. Klein: Sure and possible answers to queries for relational databases with partial relations. Technical Report 9802, Inst. f. Informatik u. Prakt. Mathem., Univ. Kiel, 194 pages, 1997 (in German)

[Kln98]   H.-J. Klein: Model theoretic and proof theoretic view of relational databases with null values: a comparison, in: M.H. Scholl et al. (Eds.), 10. GI-Workshop "Grundlagen von Datenbanken", Konstanzer Schriften in Mathematik u. Informatik, Nr. 63, Univ. Konstanz, pp. 57–61, 1998

[Kln99a] H.-J. Klein: On the use of marked nulls for the evaluation of queries against incomplete relational databases. In: Fundamentals of Informations Systems (T. Polle, T. Ripke, and K.-D. Schewe, eds), Kluwer Academic Publ., Boston, pp. 82–99, 1999

[Kln99b] H.-J. Klein: Efficient algorithms for approximating answers to queries against incomplete relational databases. In: Proc. KRDB'99 (E. Franconi and M. Kifer, eds.), Linköping, pp. 26–30, 1999

[KW85]   A.M Keller and M.W. Winslett Wilkins: On the use of an extended relational model to handle changing information and nulls. IEEE Trans. on Soft. Eng., SE–11 (7), pp. 620–633, 1985

[Lak89]   V.S. Lakshmanan: Query evaluation with null values: how complex is completeness?. Proc. 9th Conf. on Found. of Software Technologies and Theor. Comp. Science, Bangalore, LNCS 405, Springer–Verlag, Berlin, pp. 204–222, 1989

[LeLi86] N. Lerat and W. Lipski: Nonapplicable nulls. Theoretical Computer Science 46, pp. 67–82, 1986

[LeLo99] M. Levene and G. Loizou: A Guided Tour of Relational Databases and Beyond. Springer–Verlag, London, 1999

[Lib98]   L. Libkin: A semantics based approach to design of query languages for partial information. In: Semantics in Databases (B. Thalheim, L. Libkin, eds.), LNCS 1358, pp. 170–208, 1998

[Lip79]   W. Lipski: On semantic issues connected with incomplete information databases. ACM Trans. on Database Systems 4 (3), pp. 262–296, 1979

[Lip81]   W. Lipski: On databases with incomplete information. J. of the ACM 18 (1), pp. 41–70, 1981

[LS90]   K.C. Liu and R. Sunderraman: Indefinite and maybe information in relational databases. ACM Trans. on Database Systems 15 (1), pp. 1–39, 1990

[Mai83]   D. Maier: The Theory of Relational Databases. Computer Science Press, Rockville, 1983

[MS97]   A. Motro and P. Smets (eds.): Uncertainty Management in Information Systems. Kluwer Academic Publ., Boston, 1997

[MUV84]   D. Maier, J.D. Ullman, and M.Y. Vardi: On the foundations of the universal relational model. ACM Trans. on Database Systems 9 (2), pp. 283–308, 1984

[OO93]   A. Ola and G. Ozsoyoglu: Incomplete relational database models based on intervals. IEEE Trans. on Knowledge and Data Eng. 5 (2), pp. 294–308, 1993

[Rei78]   R. Reiter: On closed world databases. In: Logic and Databases (H. Gallaire and J. Minker, eds.), Plenum Press, New York, pp. 55–76, 1978

[Rei86]   R. Reiter: A sound and sometimes complete query evaluation algorithm for relational databases with null values. J. of the ACM 33 (2), pp. 349–370, 1986

[RKS89]   M.A. Roth, H.F. Korth, and A. Silberschatz: Null values in nested relational databases. Acta Informatica 26, pp 615–642, 1989

[Sci79]   E. Sciore: Improving semantic specification in a relational database. Proc. ACM SIGMOD Conf., Boston, pp. 170–178, 1979

[Var86]   M.Y. Vardi: Querying logical databases. Journal of Comp. and System Sciences 33, pp. 142–160, 1986

[Vas79]   Y. Vassiliou: Null values in data base management: a denotational approach. Proc. ACM SIGMOD Conf., Boston, pp. 162–169, 1979

[Yue91]   K. Yue: A more general model for handling missing information in relational databases using 3–value logic. ACM SIGMOD RECORD 20 (3), pp. 43–49, 1991

[Zan84]   C. Zaniolo: Database relations with null values. J. of Comp. and System Sciences 28, pp. 142–166, 1984

[ZP97]   E. Zimányi and A. Pirotte: Imperfect information in relational databases. In: [MS97], pp. 35–87, 1997

Consistency Enforcement in Databases

Sebastian Link

Information Science Research Centre, Massey University,
Information Systems, Private Bag 11222, Palmerston North, New Zealand
s.link@massey.ac.nz

Abstract. Consistency enforcement aims at systematically modifying a database program such that the result is consistent with respect to a specified set of integrity constraints. This modification may be done at compile-time or at run-time. The commonly known run-time approach uses rule triggering systems (RTSs). It has been shown that these systems cannot solve the problem in general.

As an alternative *greatest consistent specializations* (GCSs) have been studied. This approach requires the modified program specification to be a maximal consistent diminution of the original one with respect to some partial order. The chosen order is operational specialization. On this basis it is possible to derive a *commutativity* result and a *compositionality* result. The first one enables step-by-step enforcement for sets of constraints. The second one reduces the problem to providing the GCSs just for basic operations, whereas for complex programs the GCS can be easily determined. The approach turns out to be well-founded since the GCS for such complex programs is effectively computable if we require loops to be bounded.

Despite its theoretical merits the GCS approach is still too coarse. This leads to the problem of modifying the chosen specialization order and to relax the requirement that the result should be unique. One idea is to exploit the fact that operational specialization is equivalent to the preservation of a set of transition invariants. In this case a reasonable order arises from a slight modification of this set, in which case we talk of a *maximal consistent effect preserver* (MCE). However, a strict theory of MCEs is still outstanding.

1   Introduction

In database design a lot of attention has to be paid to integrity constraints. Static integrity constraints restrict the set of legal database states, whereas dynamic integrity constraints restrict the set of state sequences. In this article we only deal with static integrity constraints. The problem in database programming then is to guarantee *consistency*. With respect to static integrity constraints consistency means that a program starting in a legal database state should also terminate in a legal database state. Consistency has also been taken as one of the key properties of transactions (ACID principle).

The easiest approach to guarantee consistency leads to *consistency checking* at run-time. Whenever a transaction is to be committed, all static integrity

L. Bertossi et al. (Eds.): Semantics in Databases, LNCS 2582, pp. 139–159, 2003.

constraints are checked on the final database state. If they are satisfied, the transaction commits, otherwise it will be aborted. Then the crucial problem is to minimize the number of tuples (or 'database objects' in general).

Choosing a compile-time approach instead leads to *consistency verification* in which case theorem provers or proof assistants are heavily used. Both checking and verification do not allow changes to the program. These are either consistent and hence accepted or not. There is also no feedback to the designers on how to change programs in order to achieve consistency.

The alternative to these approaches is *consistency enforcement*. Generally speaking, enforcement aims at systematically modifying a program such that the result is consistent. This has some tradition in the field of deductive databases [3, 4,5,6,10,12,19,20] and also appears as one of the goals of active databases [2,7,14, 17]. The former group of articles investigates additional insertions, deletions and sometimes also updates that guarantee consistency; the latter group of articles emphasizes the use of triggers.

It has been shown that the trigger approach is not able to solve the consistency enforcement problem in general. A severe drawback is that the resulting consistent state may have invalidated all the effects that were intended by the original program. In Section 2 we briefly review these problems.

In contrast, the work in [15,16,18] approaches consistency enforcement at compile-time, hence starts from arbitrary program specifications instead of sets of insertions and deletions. In addition, the work tries to achieve a solid theoretical foundation starting to characterize the goals of consistency enforcement.

1.1   The Problem of Consistency Enforcement

It does not make much sense to allow arbitrary changes to program specifications in consistency enforcement. Therefore, the primary guideline underlying our work is that there should be a theoretically justified characterization of the desired result.

- Obviously, the resulting program specification should be consistent and it should be terminating. Depending on how general the original program specification may be we can weaken the termination requirement such that termination is only required, if the original program specification terminates. These two requirements are also claimed in the trigger approach [2,7].
- It is also desirable that the result should be unique. We shall see that this requirement may be debatable, as there exist equally justified solutions. In the GCS approach in [16] this is reflected by non-deterministic choices; in other approaches such as MCE [15] these options are reflected by chosen enforcement policies. The uniqueness requirement in the trigger approach only refers to the confluence of the trigger rules which is a purely technical issue not questioning the adequacy of the rules as such.
- In order to abolish the arbitrariness in program rewriting we should require that the resulting program specification is close to the original one. Intuitively, we would like to claim that consistency enforcement should 'preserve

the effects' of the original program. In order to formalize this we may think of a partial order (or at least a preorder) $\sqsubseteq$ on programs such that $T \sqsubseteq S$ means that T preserves the effects of S. The problem then is to choose the right order and to find a new consistent program that is smaller than the original one, but still close to it.

- It is desirable that the new program can be effectively and efficiently computed from the original one and the constraints. In addition, the resulting program has to be effective itself. In particular, all conditions to be checked on the database must be decidable.
- As we normally consider more than one integrity constraint, it is desirable to be able to enforce consistency step by step. If we require uniqueness this includes commutativity, i.e., the order of the constraints in the process of stepwise consistency enforcement should be irrelevant.
- At compile-time we must consider complex program specifications, whereas at run-time it is sufficient to consider the effect of a transaction on the database, i.e., a set of insertions and deletions. Therefore, it is desirable to obtain compositionality, i.e., to be able to reduce the problem to consistency for basic operations such as inserts, deletes and updates.

1.2 The Theory of Greatest Consistent Specializations

The approach in [15,16,18] requires the modified program specification to be a maximal consistent diminution of the original one with respect to some partial order $\sqsubseteq$. If we are given a program specification S and an invariant $\mathcal{I}$, we are looking for a new program specification $S_{\mathcal{I}}$ that is consistent with respect to $\mathcal{I}$, satisfies $S_{\mathcal{I}} \sqsubseteq S$ and is maximal with these properties.

As stated above the intention behind the partial order is the preservation of effects, i.e., state changes due to S should be preserved as state changes by T, whenever $T \sqsubseteq S$ holds. The GCS choice for the partial order on semantic equivalence classes of program specifications is operational specialization. The starting point is to consider the state variables that are affected by the given program specification S. With respect to these state variables, a specialization $T \sqsubseteq S$ only allows executions that already appear with S. However, state variables not affected by S may be handled arbitrarily. Operational specialization can be formalized quite easily exploiting predicate transformer semantics in the style of Dijkstra and Nelson [13].

Specialization leads to unique *greatest consistent specializations* (GCSs) $S_{\mathcal{I}}$. These were investigated in detail in [16]. Besides existence and uniqueness (up to semantic equivalence) there are two important results on GCSs. The first one is a commutativity result, which states that the GCS $S_{\mathcal{I}_1 \wedge \cdots \wedge \mathcal{I}_n}$ can be build sequentially taking any order of the invariants $\mathcal{I}_i$. This in turn justifies to concentrate on one invariant.

The second result concerns compositionality. Using guarded commands as in [13] we may build complex program specifications from simple ones. The simplest specifications are given by assignments and a few constants such as *skip*. In general it is not sufficient to replace the very simple specifications in S

by their GCSs. The result will not be the GCS $S_{\mathcal{I}}$. However, under some mild technical restrictions we obtain at least a generalization of the GCS, and the GCS itself results from adding a precondition.

In Section3 we briefly review the achievements of the GCS approach.

1.3   Effectivity Issues

In order to shift the GCS approach from a purely theoretical framework to an applicable theory we have to investigate computability of GCSs and decidability of preconditions that must be built. For these purposes it is preferable to obtain a tight connection with classical recursion theory [1]. This has been done in [8, 9].

We start in Section 4 with a brief review of arithmetic logic. Then we show the existence of predicate transformers with respect to this logic. In particular, relational program semantics becomes equivalent to predicate transformer semantics provided we guarantee the property of universal conjunctivity and the pairing condition. Moreover, recursion theory can be extended to the arithmetic case, at least, if we are restricted to certain WHILE-loops.

With this background it has been shown that the GCS approach carries over to arithmetic logic. Many of the proofs in [16] only require slight changes. Computability cannot be guaranteed in general, since the building of least fixpoints requires to test for semantic equivalence, which is undecidable. For the case of FOR-loops, however, GCSs are computable. Furthermore, we show that effective computation is possible, i.e., preconditions will remain decidable.

Therefore, the GCS approach is a theoretically well-founded approach to consistency enforcement.

1.4   Tailored Approaches to Consistency Enforcement

Despite its theoretical strength the GCS approach is debatable from a pragmatic point of view. The changes to state variables not affected by the original program specification allow too much freedom, whereas the approach may be too restrictive for the other state variables. In particular, in databases these state variables take set values, and the insertion of a value into a set should not disable the insertion of another value into the same set.

So, we either have to give up some of the theoretical strengths and allow more pragmatic approaches to the problem or develop an even better theory. In the first case we would nevertheless prefer to stay as close to the GCS approach as possible. If we restrict ourselves to deterministic branches of GCSs, the approach also enables a run-time approach. In this case only sequences of assignments have to be considered which allows to ignore the technical and computability restrictions.

Another idea to weaken the effect preservation order considers the set of transition invariants that are satisfied by the original program specification. These are called δ-*constraints*. As operational specialization is equivalent to the preservation of all δ-constraints of S involving only the state variables affected by S, the simple idea is to preserve a different set of δ-constraints. This leads to the definition of a *maximal consistent effect preserver* (MCE). Informally, an MCE arises as the GCS of a slightly extended program specification, but there is no more need to consider the set of affected state variables.

The choice of the set of δ-constraints to be preserved also introduces some pragmatics. We may refer to each of the possible choices as giving us a policy for consistency enforcement. We outline the MCE idea in Section 5. We show that reasonable examples are covered by the approach. So far, however, a complete theory of MCEs is still outstanding.

2   The Failure of the RTS Approach

The use of rule triggering systems (RTSs) has been suggested from the very beginning as an approach to consistency enforcement [2,7]. However, there has never been a satisfactory theory behind this. The work in [17] investigates the limitations of the approach.

– For each set of constraints we can always find a sequence of insertions and deletions that is *non-repairable*. Even worse, checking repairability is equivalent to the implication problem for the admitted class of constraints.
– Even if we only consider repairable program executions, the RTS may contain critical cycles, i.e., the triggers may undo some (or all) of the effects of the program, but nevertheless change the final state.

For the theoretical details we refer to [14,17]. Here we only illustrate the second problem with a little example [14].

Example 2.1. Let us define a schema with some simple functional and inclusion constraints. For simplicity we omit all types. The relation schemata are

WIRE = { wire_id, connection, wire_type, voltage, power } ,
TUBE = { tube_id, connection, tube_type } and
CONNECTION = { connection, from, to }

These are used to express that there are tubes between two locations and wires in these tubes. In addition consider the following constraints:

$FD_1 \equiv$ WIRE : wire_id $\rightarrow$ connection, wire_type, voltage, power
$FD_2 \equiv$ TUBE : tube_id $\rightarrow$ connection, tube_type
$FD_3 \equiv$ CONNECTION : connection $\rightarrow$ from, to

$ID_1 \equiv$ WIRE[connection] $\subseteq$ TUBE[connection]
$ID_2 \equiv$ TUBE[connection] $\subseteq$ CONNECTION[connection]

The first three functional dependencies express that the values of wire_id, tube_id and connection are unique in relations over WIRE, TUBE and CONNECTION, respectively. The latter inclusion constraints express that there is no wire nor tube without a corresponding tuple in a relation over CONNECTION.

Then the following relations define an instance of the schema:

WIRE

wire_id	connection	wire_type	voltage	power
4711	HH-HB	Koax	12	600
4814	HH-H	Tel	12	600

TUBE

tube_id	connection	tube_type
8314	HH-H	GX44
8511	HH-HB	GX44
023	HB-H	T33

CONNECTION

connection	from	to
HH-H	Hamburg	Hannover
HH-HB	Hamburg	Bremen
HB-H	Bremen	Hannover

It is easy to see that this instance satisfies the constraints above.

Now consider the operation $insert_{WIRE}(t)$. This may lead to a violation of constraint ID_1, in which case we must add a tuple to TUBE. Hence it can be replaced by

$insert_{WIRE}(t)$;
IF   connection$(t) \notin$ TUBE[connection] THEN
        $insert_{TUBE}(?, $connection$(t), ?)$
ENDIF

Here the question marks stand for arbitrarily chosen values of the corresponding data type. Similarly, the operation $delete_{TUBE}(t)$ may also violate ID_1. Therefore, we may replace $delete_{TUBE}(t)$ by

$delete_{TUBE}(t)$;
IF   connection$(t) \in$ WIRE[connection] $-$ TUBE[connection] THEN
FOR ALL t' WITH connection$(t') = $ connection(t) DO
        $delete_{WIRE}(t')$
ENDFOR
ENDIF

In order to enforce FD_2 we may then replace $insert_{TUBE}(t)$ by

IF $\forall t' \in$ TUBE . tube_id$(t) \neq$ tube_id(t') THEN
        $insert_{WIRE}(t)$
ENDIF

Let us now add the exclusion constraint $ED \equiv$ WIRE[wire_id] $\|$ TUBE[tube_id]. In order to enforce this constraint insertions into one of WIRE or TUBE should be followed by deletions in the other. The resulting transactions are

$insert_{WIRE}(t)$;
FOR ALL $t' \in$ TUBE WITH tube_id$(t') = $ wire_id(t) DO
        $delete_{TUBE}(t')$
ENDFOR

and

$delete_{TUBE}(t)$;
FOR ALL $t' \in$ WIRE WITH wire_id$(t') = $ tube_id(t) DO
        $delete_{WIRE}(t')$
ENDFOR

If we now take together FD_2, ID_1 and ED we must be very careful. E.g., if we execute $insert_{WIRE}$(8511,HH-HB,Koax,12,600) on the instance above, we may first delete the tuple (8511,HH-HB,GX44) in TUBE in order to enforce ED and then the two tuples (4711,HH-HB,Koax,12,600) and (8511,HH-HB,Koax,12,600) in WIRE in order to enforce ID_1. The resulting instance would be (omitting CONNECTION):

WIRE

wire_id	connection	wire_type	voltage	power
4814	HH-H	Tel	12	600

TUBE

tube_id	connection	tube_type
8314	HH-H	GX44
023	HB-H	T33

Thus, the "effect" of the original operation, i.e. insertion of a tuple into WIRE, is completely destroyed. The new effect is a deletion in WIRE and TUBE.

There is a necessary and sufficient condition on the set of constraints that guarantees the absence of critical cycles, but checking this condition is **NP**-hard [17]. There is also a weaker condition which is only sufficient [14,17]. Fortunately, this condition is at least satisfied for constraints that are implicitly contained in simple conceptual database schemata.

3   The Theory of Greatest Consistent Specializations

In order to develop a theory of consistency enforcement we adapt the state-based approach to formal specifications. This means that we support explicitly the concepts of state and state transition. Program specifications will be taken in the form of extended guarded commands as defined by Nelson [13]. Semantics will be expressed by predicate transformers that are defined over infinitary logic [11] as in [16].

3.1   Program Specifications

The original theory [16] of Consistency Enforcement has been developed with repect to the many-sorted, infinitary, first-order logic $\mathcal{L}^{\omega}_{\infty\omega}$. Therefore, we assume a fixed interpretion structure (D, ω), where $D = \bigcup_{T:type} T$ is the semantic domain and ω assigns type-compatible functions $\omega(f) : T_1 \times \ldots \times T_n \to T$ and $\omega(p) : T_1 \times \ldots \times T_n \to \{true, false\}$ to n-ary function symbols f and n-ary predicate symbols p, respectively [11]. We extend ω in the usual way to the terms and formulae of $\mathcal{L}^{\omega}_{\omega\infty}$. This logic restricts the set of formulae on those with finitely-many variables. In addition, we assume the *domain closure property*, ie., for each $d \in D$ there is some closed term t in $\mathcal{L}^{\omega}_{\infty\omega}$ with $\omega(t) = d$. This property can be easily achieved by extending the set of constants.

Definition 3.1. A *state space* is a finite set X of variables of $\mathcal{L}^{\omega}_{\infty\omega}$ such that for each $x \in X$ there is an associated type $\sharp x$. A *state* on X is a type-compatible variable assignment $x \mapsto \sigma(x) \in \sharp x$ for each $x \in X$. We use Σ to denote the set of all states on X.

Next we define program specifications on a given state space.

Definition 3.2. Let X be a state space. The language of *guarded commands* contains *skip, fail, loop*, simultaneous assignment $x_{i_1} := t_{i_1} \| \ldots \| x_{i_k} := t_{i_k} \in \mathcal{S}(X)$ for state variables $x_{i_j} \in X$ and terms t_{i_j} (of corresponding type $\sharp x_{i_j}$), sequential composition $S_1; S_2$, choice $S_1 \square S_2$, restricted choice $S_1 \boxtimes S_2$, guard $\mathcal{P} \to S$ with an X-formula $\mathcal{P}$, unbounded choice $@y(:: \sharp y) \bullet S$ with a variable y, and least fixed points $\mu S.f(S)$, with a program variable S and a guarded command expression $f(S)$, in which S may occur at the place of a basic command, i.e., *skip, fail, loop* or an assignment.

In order to define the semantics of guarded commands on the basis of the introduced logics, we associate with S two *predicate transformers* $wlp(S)$ and $wp(S)$—i.e., functions from (equivalence classes) of formulae to (equivalence classes) of formulae—with the standard informal meaning:

- $wlp(S)(\varphi)$ characterizes those initial states σ such that each terminating execution of S starting in σ results in a state τ satisfying φ.
- $wp(S)(\varphi)$ characterizes those initial states σ such that each execution of S starting in σ terminates and results in a state τ satisfying φ.

The notation $wlp(S)(\varphi)$ and $wp(S)(\varphi)$ corresponds to the usual *weakest (liberal) precondition* of S with respect to the postcondition φ. In order to save space we shall often use the notation $w(l)p(S)(\varphi)$ to refer to both predicate transformers at a time. If this occurs in an equivalence, then omitting everything in parentheses gives the wp-part, whereas omitting just the parentheses results in the wlp-part.

Now consider the following definition of predicate transformers for our language of guarded commands (for a justification see [13,16])

$$w(l)p(skip)(\varphi) \Leftrightarrow \varphi$$
$$w(l)p(fail)(\varphi) \Leftrightarrow true$$
$$w(l)p(loop)(\varphi) \Leftrightarrow false(\lor true)$$
$$w(l)p(x_{i_1} := t_{i_1} \| \ldots \| x_{i_k} := t_{i_k})(\varphi) \Leftrightarrow \{x_{i_1}/t_{i_1}, \ldots, x_{i_k}/t_{i_k}\}.\varphi$$
$$w(l)p(S_1; S_2)(\varphi) \Leftrightarrow w(l)p(S_1)(w(l)p(S_2)(\varphi))$$
$$w(l)p(S_1 \square S_2)(\varphi) \Leftrightarrow w(l)p(S_1)(\varphi) \land w(l)p(S_2)(\varphi)$$
$$w(l)p(S_1 \boxtimes S_2)(\varphi) \Leftrightarrow w(l)p(S_1)(\varphi) \land (wp(S_1)^*(true) \lor w(l)p(S_2)(\varphi))$$
$$w(l)p(@x_j \bullet S)(\varphi) \Leftrightarrow \forall x_j.w(l)p(S)(\varphi)$$
$$w(l)p(\mathcal{P} \to S)(\varphi) \Leftrightarrow \mathcal{P} \Rightarrow w(l)p(S)(\varphi)$$
$$w(l)p(\mu S.f(S))(\varphi) \Leftrightarrow \bigvee_{\alpha} \left(\bigwedge_{\alpha} \right) w(l)p(f^{\alpha}(loop))(\varphi)$$

We say that S is an X-*command* for some state space X iff $w(l)p(S)(\varphi) \Leftrightarrow \varphi$ hold for each Y-formulae φ, where $X \cap Y = \emptyset$, and X is minimal with this property.

3.2 Consistency and GCSs

Static constraints are formulae that can be evaluated in states. This enables us to distinguish between legal states, i.e., those satisfying the constraints, and non-legal states.

Definition 3.3. A *static invariant* on a state space X (short: an X-constraint or X-formula) is a formula $\mathcal{I}$ of $\mathcal{L}^{\omega}_{\infty\omega}$ with free variables in X ($fr(\mathcal{I}) \subseteq X$).

We would like to derive a proof obligation that characterizes when a program specification S is consistent with respect to a static constraint $\mathcal{I}$ (see [16]).

Definition 3.4. Let S be a program specification and $\mathcal{I}$ a static constraint, both on X. Then S is *consistent* with respect to $\mathcal{I}$ (short: $\mathcal{I}$-consistent) iff $\mathcal{I} \Rightarrow wlp(S)(\mathcal{I})$ holds.

Operational specialization aims at reducing existing executions and at the same time extending the state space and allowing arbitrary additional changes on new state variables [9,16].

Definition 3.5. Let S and T be program specifications on X and Y with $X \subseteq Y$. Then S is *specialized* by T ($T \sqsubseteq S$) iff $w(l)p(S)(\varphi) \Rightarrow w(l)p(T)(\varphi)$ holds for all state formulae φ on X.

We remark, that for the wp-part it is enough to require that $wp(S)(true) \Rightarrow wp(T)(true)$ holds. Operational specialization defines a partial order $\sqsubseteq$ on semantic equivalence classes of program specifications with a minimum *fail*. The proof obligation for consistency and the definition of operational specialization are sufficient to define the central notion of the approach.

Definition 3.6. Let S be a Y-command and $\mathcal{I}$ a constraint on X with $Y \subseteq X$. The *greatest consistent specialization* (GCS) of S with respect to $\mathcal{I}$ is an X-command $S_{\mathcal{I}}$ with $S_{\mathcal{I}} \sqsubseteq S$, such that $S_{\mathcal{I}}$ is consistent with respect to $\mathcal{I}$ and each consistent specialization $T \sqsubseteq S$ satisfies $T \sqsubseteq S_{\mathcal{I}}$.

The first important result concerns the *commutativity*, i.e., GCSs with respect to conjunctions can be built successively using any order of the constraints.

Proposition 3.1. *For two constraints $\mathcal{I}$ and $\mathcal{J}$ we always obtain that $\mathcal{I} \wedge \mathcal{J} \to S_{\mathcal{I} \wedge \mathcal{J}}$ and $\mathcal{I} \wedge \mathcal{J} \to (S_{\mathcal{I}})_{\mathcal{J}}$ are semantically equivalent.*

It would be nice, if building the GCS for a complex program specification S simply required the basic operations in S to be replaced by their GCSs. Let $S'_{\mathcal{I}}$ denote the result of such a naïve syntactic replacement. However, in general $S'_{\mathcal{I}}$ is not the GCS $S_{\mathcal{I}}$. It may not even be a specialization of S, or it may be a consistent specialization, but not the greatest one.

There exists a technical condition which implies at least $S_{\mathcal{I}} \sqsubseteq S'_{\mathcal{I}}$ holds. The corresponding result was called the *upper bound theorem* in [16].

We need the notion of a *deterministic branch* S^+ of a command S, which requires $S^+ \sqsubseteq S$, $wp(S)^*(true) \Leftrightarrow wp(S^+)^*(true)$ and $wlp(S^+)^*(\varphi) \Rightarrow wp(S^+)(\varphi)$ to hold for all φ. Furthermore, we need the notion of a δ-constraint for an X-command S. This is a constraint $\mathcal{J}$ on $X \cup X'$ with a disjoint copy X' of X, for which $\{x'/x\}.wlp(S')(\mathcal{J})$ holds, where S' results from S by renaming all x_i to x'_i. We write φ_σ to denote the characterizing formula of a state σ.

Definition 3.7. Let $S = S_1; S_2$ be a Y-command such that S_i is a Y_i-command for $Y_i \subseteq Y$ ($i = 1, 2$). Let $\mathcal{I}$ be some X-constraint with $Y \subseteq X$. Let $X - Y_1 = \{y_1, \ldots, y_m\}$, $Y_1 = \{x_1, \ldots, x_l\}$ and assume that $\{x'_1, \ldots, x'_l\}$ is a disjoint copy of Y_1 disjoint also from X. Then S is in $\mathcal{I}$-*reduced form* iff for each deterministic branch S_1^+ of S_1 the following two conditions – with $\boldsymbol{x} = (x_1, \ldots, x_l)$, $\boldsymbol{x}' = (x'_1, \ldots, x'_l)$ – hold:

- For all states σ with $\models_\sigma \neg\mathcal{I}$ we have, if $\varphi_\sigma \Rightarrow \{\boldsymbol{x}/\boldsymbol{x}'\}.(\forall y_1 \ldots y_m.\mathcal{I})$ is a δ-constraint for S_1^+, then it is also a δ-constraint for S_1^+ ; S_2.
- For all states σ with $\models_\sigma \mathcal{I}$ we have, if $\varphi_\sigma \Rightarrow \{\boldsymbol{x}/\boldsymbol{x}'\}.(\forall y_1 \ldots y_m.\neg\mathcal{I})$ is a δ-constraint for S_1^+, then it is also a δ-constraint for S_1^+ ; S_2.

It is straightforward to extend this definition to arbitrary commands other than sequences. The upper bound theorem already has a flavour of compositionality, but it does not yet give the GCS. The idea of the main theorem on GCSs is to cut out from $S'_{\mathcal{I}}$ those executions that are not allowed to occur in a specialization of S. This leads to the following theorem [16].

Theorem 3.1. *Let $\mathcal{I}$ be an invariant on X and let S be some $\mathcal{I}$-reduced Y-command with $Y \subseteq X$. Let $S'_{\mathcal{I}}$ result from S as follows:*

- *Each restricted choice $S_1 \boxtimes S_2$ occurring within S will be replaced by*

$$S_1 \square wlp(S_1)(false) \rightarrow S_2 .$$

- *Then each basic command will be replaced by their GCSs with respect to $\mathcal{I}$.*

Let Z be a disjoint copy of the state space Y. With the formulae

$$\mathcal{P}(S, \mathcal{I}, \boldsymbol{x}') \equiv \{z/y\}.wlp(S''_{\mathcal{I}}; z = x' \rightarrow skip)(wlp(S)^*(z = y)) \quad ,$$

where $S''_{\mathcal{I}}$ results from $S'_{\mathcal{I}}$ by renaming the Y to Z, the GCS $S_{\mathcal{I}}$ is semantically equivalent to

$$@\boldsymbol{x}' \bullet \mathcal{P}(S, \mathcal{I}, \boldsymbol{x}') \rightarrow S'_{\mathcal{I}}; y = x' \rightarrow skip \quad .$$

Note that if we consider deterministic branches as a pragmatic approach suggested in [16], then the unbounded choice in Theorem 3.1 disappears. We omit further details.

The characterization of GCSs according to Theorem 3.1 makes it possible to reduce consistency enforcement to a simple syntactical replacement (the forming of $S'_{\mathcal{I}}$) and to an investigation of a guard, namely $\mathcal{P}(S, \mathcal{I}, \boldsymbol{x}')$.

3.3   GCS Branches

Due to the definition of the specialization preorder, GCSs in general are highly non-deterministic, even if the original program specification was not. From a pragmatic point of view, however, it will be enough to have a deterministic result. Therefore, it is a natural idea to consider *deterministic GCS branches* or at least *quasi-deterministic GCS branches*. Here quasi-determinism means determinism up to the selection of values [16].

Let us now look at an example how to construct GCS branches on the basis of Theorem 3.1.

Example 3.1.   Consider the state space $X = \{x_1, x_2\}$. Although types have been left implicit so far, let us assume that values for both state variables are sets of pairs. We consider the following three X-constraints:

$$
\begin{aligned}
\mathcal{I}_1 &\equiv \pi_1(x_1) \subseteq \pi_1(x_2) \\
\mathcal{I}_2 &\equiv \forall x, y.x \in x_2 \land y \in x_2 \land \pi_2(x) = \pi_2(y) \Rightarrow \pi_1(x) = \pi_1(y) \qquad \text{and} \\
\mathcal{I}_3 &\equiv \pi_2(x_1) \cap \pi_2(x_2) = \emptyset
\end{aligned}
$$

with the projection functions π_i onto the i'th components. Then we consider a program specification S on $\{x_1\}$ defined by the simple assignment $x_1 := x_1 \cup \{(a, b)\}$ with some constants a and b.

Step 1.   First consider the constraint $\mathcal{I}_1$. Since S is just an assignment, it is $\mathcal{I}_1$-reduced. We then replace S by a quasi-deterministic branch of its GCS with respect to $\mathcal{I}_1$ and obtain

$$ x_1 := x_1 \cup \{(a, b)\} \,;\, (a \notin \pi_1(x_2) \to @c \bullet x_2 := x_2 \cup \{(a, c)\} \boxtimes skip) \quad, $$

which is an X-operation. Let this be our new S_1.

Step 2.   Now consider the constraint $\mathcal{I}_2$. It can be shown that S_1 is $\mathcal{I}_2$-reduced. We have to remove the restricted choice and then replace the assignment to x_2 by the deterministic GCS branch $c \notin \pi_2(x_2) \to x_2 := x_2 \cup \{(a, c)\}$ with respect to $\mathcal{I}_2$. For the resulting operation $S'_{\mathcal{I}}$ we compute $P(S_1, \mathcal{I}_2) \Leftrightarrow true$. After some rearrangements we obtain the following GCS branch of S with respect to $\mathcal{I}_1 \land \mathcal{I}_2$:

$$ x_1 := x_1 \cup \{(a, b)\} \,;\, ((a \notin \pi_1(x_2) \to @c \bullet c \notin \pi_2(x_2) \to x_2 := x_2 \cup \{(a, c)\}) $$
$$ \square\, a \in \pi_1(x_2) \to skip) $$

Let this be our new specification S_2.

Step 3.   Now regard the constraint $\mathcal{I}_3$. Again we can show $\mathcal{I}_3$-reducedness, but dispense with the formal proof. We replace the assigment to x_1 in S_2 by the deterministic GCS branch

$$ x_1 := x_1 \cup \{(a, b)\} \,;\, x_2 := x_2 - \{x \in x_2 \mid \pi_2(x) = b\} \quad. $$

Analogously, we replace the assignment to x_2 in S_2 by the deterministic GCS branch

$$x_2 := x_2 \cup \{(a, c)\} \; ; \; x_1 := x_1 - \{x \in x_1 \mid \pi_2(x) = c\} \; .$$

Then we compute

$$P(S_2, \mathcal{I}_3) \; \Leftrightarrow \; b \notin \pi_2(x_2) \wedge (a \notin \pi_1(x_2) \Rightarrow \forall c.(c \notin \pi_2(x_2) \Rightarrow c \notin \pi_2(x_1) \cup \{b\})) \; .$$

After some rearrangements the final result is

$$b \notin \pi_2(x_2) \; \rightarrow \; x_1 := x_1 \cup \{(a, b)\} \; ; \; ((a \notin \pi_1(x_2) \rightarrow @c \bullet$$
$$c \notin \pi_2(x_2) \wedge c \notin \pi_2(x_1) \rightarrow x_2 := x_2 \cup \{(a, c)\}) \square \, a \in \pi_1(x_2) \rightarrow skip \,) \; ,$$

which is a branch of the GCS of S with respect to $\mathcal{I}_1 \wedge \mathcal{I}_2 \wedge \mathcal{I}_3$.

3.4   A GCS Run-Time Version for Consistency Enforcement

Example 3.1 demonstrates the potential of the GCS approach. However, we have to face some drawbacks. Besides the general computability and decidability problems we also have to check $\mathcal{I}$-reducedness. It has been shown in [16] that this is only necessary for sequences $S = S_1; S_2$, where S is defined on the same space as S_1. If S is defined on a larger state space than S_1, there is nothing to show.

Furthermore, $\mathcal{I}$-reducedness is only violated in "pathological" cases, where the program specification has been written in such a way that S_2 simply undoes the effect of S_1. For example, this is the case for an insertion immediately following a deletion. From a pragmatic point of view it may be a good idea to ignore $\mathcal{I}$-reducedness. If this affects the result at all, then we will still obtain a consistent specialization, but not the greatest one. In other words, a few computations that should be present in the result for theoretical reasons, will be discarded. Whether this loss is tolerable or not may be subject to an empirical study.

Another idea to handle the problem of $\mathcal{I}$-reducedness is to shift the approach to run-time. Of course, this is only possible for executable program specifications S.

Before building a GCS branch of S, we execute S on a copy of the current state. We cannot use the state itself, because consistency enforcement may increase partiality. In other words, the execution may be disallowed completely. After the execution we precisely know which state variables were changed in which way. Thus, we may replace S by a sequence of assignments to different state variables. Of course, this sequence is state-dependent.

Due to the aforementioned result in [16] such a sequence is indeed $\mathcal{I}$-reduced. Furthermore, according to [8,9] there is no problem with computability or decidability in this case. Hence we may apply the compositionality result from Theorem 3.1 as done in Example 3.1 to obtain a new program specification S. This process has to be repeated until all static constraints have been processed. The final result can then be applied to the real state.

4   Arithmetic Theory of Consistency Enforcement

In this section we summarize the theory investigated in detail in [8,9]. The theory is based on first-order arithmetic logic [1, Ch.7], i.e., our logical language contains just the function symbols 0, s, $+$ and $*$ of arity 0, 1, 2 and 2. The informal meaning is as usual: the constant 0, the successor function, addition and multiplication. By convenience $+$ and $*$ are written as infix operators. The only predicate symbol is the equality symbol $=$. Variables in our language will be $x_1, x_2, x_3, \ldots$.

We use the notation $\mathbb{T}$ for the set of terms and $\mathbb{F}$ for the set of formulae. In addition, let V denote the set of variables. We allow all standard abbreviations including formulae *true* and *false*.

Semantically, we fix a structure with domain $\mathbb{N}$, the set of non-negative integers. Then 0, s, $+$, $*$ and $=$ are interpreted in the usual way. For an interpretation it is then sufficient to consider a function $\sigma : V \to \mathbb{N}$. By the coincidence theorem it is even sufficient to be given the values $\sigma(x_i)$ for the free variables x_i in a term or a formula. In particular, we may always write σ as a k-tuple, if the number of free variables is k.

Finally, a k-ary relation $R \subseteq \mathbb{N}^k$ is called *arithmetical* iff it can be represented by a formula $Q \in \mathbb{F}$ in arithmetic logic (with free variables $x_1, \ldots, x_k$), i.e. $(a_1, \ldots, a_k) \in R$ holds iff $\models_\sigma Q$ holds for the interpretation defined by $\sigma(x_i) = a_i$ $(i = 1, \ldots, k)$.

4.1   Predicate Transformers and Program Semantics

We define *state spaces, states, the set of all states, the language of guarded-commands* and *invariants* according to the theory from Section 3 within arithmetic logic.

Then any pair of formulae $(\Delta(S), \Sigma_0(S))$ with $2k$ and k free variables, respectively, may be considered as defining the *relational semantics* of a program specification S. For convenience assume the first k free variables in $\Delta(S)$ to coincide with the free variables of $\Sigma_0(S)$.

According to our notation we sometimes write $\Delta(S)(\boldsymbol{x}, \boldsymbol{y})$ and $\Sigma_0(S)(\boldsymbol{x})$. So $\Delta(S)$ can be interpreted by state pairs, whereas $\Sigma_0(S)$ allows an interpretation by states. We interpret (σ, τ) with $\models_{(\sigma,\tau)} \Delta(S)$ as an *execution* of S with start state σ and a final state τ. Similarly, a state σ satisfying $\Sigma_0(S)$ is considered as a start state for S, in which a non-terminating execution of S exists.

From our introduction of $\Delta(S)$ and $\Sigma_0(S)$ the following definition is straightforward.

Definition 4.1.   The *predicate transformers* associated with a program specification S on a state space X are defined as

$$wlp(S)(\varphi(\boldsymbol{x})) \Leftrightarrow \forall \boldsymbol{y}.\Delta(S)(\boldsymbol{x}, \boldsymbol{y}) \Rightarrow \varphi(\boldsymbol{y}) \qquad \text{and}$$
$$wp(S)(\varphi(\boldsymbol{x})) \Leftrightarrow (\forall \boldsymbol{y}.\Delta(S)(\boldsymbol{x}, \boldsymbol{y}) \Rightarrow \varphi(\boldsymbol{y})) \wedge \neg\Sigma_0(S)(\boldsymbol{x})$$

for arbitrary X-formulae φ.

The next step is to show that predicate transformers satisfying some nice conditions are sufficient for the definition of program specifications S. The conditions are the *pairing condition* and a slightly modified *universal conjunctivity property*. This gives the equivalence between the relational and the predicate transformer semantics. For a proof see [8,9].

We use the standard notation $w(l)p(S)^*(\varphi) \Leftrightarrow \neg w(l)p(S)(\neg\varphi)$ and refer to $wlp(S)^*$ and $wp(S)^*$ as the *dual predicate transformers*.

Proposition 4.1. *The predicate transformers $w(l)p(S)$ satisfy the following conditions:*

$$wp(S)(\varphi) \Leftrightarrow wlp(S)(\varphi) \wedge wp(S)(true) \qquad\qquad and$$
$$wlp(S)(\forall \boldsymbol{y}.Q(\boldsymbol{y}) \Rightarrow \varphi(\boldsymbol{x}, \boldsymbol{y})) \Leftrightarrow \forall \boldsymbol{y}.Q(\boldsymbol{y}) \Rightarrow wlp(S)(\varphi(\boldsymbol{x}, \boldsymbol{y})) \qquad .$$

Conversely, any pair of predicate transformers satisfying these two conditions defines $\Delta(S)(\boldsymbol{x}, \boldsymbol{y}) \Leftrightarrow wlp(S)^*(\boldsymbol{x} = \boldsymbol{y})$ and $\Sigma_0(\boldsymbol{x}) \Leftrightarrow wp(S)^*(false)$.

The Inversion Theorem (Proposition 4.1) allows to define semantics for the language of guarded-command via predicate transformers. Apart from the least fixed point $\mu S.f(S)$ the definition is exactly the same as for the original theory in Section 3.1.

4.2   Recursion

We are going to investigate recursive programs expressed as least fixpoints $\mu T.f(T)$ with respect to a suitable order $\preceq$ (see [9]). This order will be the standard Nelson-order [13].

Unfortunately, we are not able to carry over the very general recursion theory from [13]. We have to restrict ourselves to simple WHILE-loops, i.e. $f(T) = \mathcal{P} \rightarrow S; T \square \neg \mathcal{P} \rightarrow skip$, where the variable T does not occur within S. For convenience, we introduce command variables $T_1, T_2, \ldots$.

The idea of the Nelson-order is that whenever $S_1 \preceq S_2$ holds, then each terminating execution of S_1 is preserved within S_2, but a terminating execution in S_2 may be "approximated" in S_1 by a non-terminating execution. This leads to the following definition.

Definition 4.2. The *Nelson-order* is defined by

$$S_1 \preceq S_2 \Leftrightarrow (wlp(S_2)(\varphi) \Rightarrow wlp(S_1)(\varphi)) \wedge (wp(S_1)(\varphi) \Rightarrow wp(S_2)(\varphi))$$

for all φ.

Particularly, we are interested in chains $\{f^i(loop)\}_{i \in \mathbb{N}}$ with respect to $\preceq$. Therefore, we define a Gödel numbering g of guarded commands, which extends the Gödel numbering of terms and formulae from [1, p.327f.]. Let h denote this Gödel numbering for our logic.

It can be shown that with this Gödel numbering g we may express all formulae $w(l)p(f^i(loop))(\varphi)$ by two arithmetic predicate transformers. The proof ([16]) is rather technical.

Lemma 4.1. *Let $f(T) = \mathcal{P} \to S; T\square\neg\mathcal{P} \to skip$ such that T does not occur within S. Then for each $j \in \mathbb{N}$, there exist predicate transformers $\tau_l(j)$ and $\tau(j)$ on arithmetic predicates such that the following properties are satisfied:*

(i) for each arithmetic predicate $\varphi(\boldsymbol{x})$, the results of applying these predicate transformers are arithmetic predicates in i and x, say

$$\chi_j^1(i, \boldsymbol{x}) = \tau_l(j)(\varphi(\boldsymbol{x})) \qquad and \qquad \chi_j^2(i, \boldsymbol{x}) = \tau(j)(\varphi(\boldsymbol{x}))$$

(ii) for $j = h(\varphi)$ we obtain

$$\forall \boldsymbol{x}.\forall i.\left(\chi_j^1(i, \boldsymbol{x}) \Leftrightarrow wlp(f^i(loop))(\varphi(\boldsymbol{x}))\right) \qquad and$$
$$\forall \boldsymbol{x}.\forall i.\left(\chi_j^2(i, \boldsymbol{x}) \Leftrightarrow wp(f^i(loop))(\varphi(\boldsymbol{x}))\right)$$

with $\boldsymbol{x} = x_{i_1}, \dots, x_{i_k}$.

The arithmetic predicate transformers $\tau_l(j)$ and $\tau(j)$ from Lemma 4.1 enable us to define a *limit operator* $S = \lim_{k \in \mathbb{N}} f^k(loop)$ via

$$wlp(S)(\varphi(\boldsymbol{x})) \quad \Leftrightarrow \quad \forall k.\chi_{h(\varphi)}^1(k, \boldsymbol{x}) \qquad \text{and}$$
$$wp(S)(\varphi(\boldsymbol{x})) \quad \Leftrightarrow \quad \exists k.\chi_{h(\varphi)}^2(k, \boldsymbol{x}) \quad .$$

for $\chi_{h(\varphi)}^1(k, \boldsymbol{x}) = \tau_l(h(\varphi))(\varphi(\boldsymbol{x}))$ and $\chi_{h(\varphi)}^2(k, \boldsymbol{x}) = \tau(h(\varphi))(\varphi(\boldsymbol{x}))$. Is is technical, but straightforward to observe that the definition of $S = \lim_{i \in \mathbb{N}} f^i(loop)$ is sound.

Now, we are going to show how to obtain the semantics for WHILE-loops. It is easy to see that the function f on guarded commands is monotonic in the Nelson order [13]. Then an immediate consequence of the limit operator's definition is the existence of a least upper bound for the $\preceq$-chain $\{f^i(loop) \mid i \in \mathbb{N}\}$. This is simply $\lim_{i \in \mathbb{N}} f^i(loop)$.

Finally, we can show the existence of a least fixed point $\mu T.f(T)$ for simple WHILE-loops $f(T)$.

Proposition 4.2. *Let $f(T) = \mathcal{P} \to S; T\square\neg\mathcal{P} \to skip$. Then f has a least fixpoint with respect to $\preceq$, which is $\mu T.f(T) = \lim_{i \in \mathbb{N}} f^i(loop)$.*

4.3 Computability

Having defined the program semantics with respect to arithmetic logic we see that the whole theory of greatest consistent specializations (Section 3.2) can be carried over. For each definition and for every result within the original framework, we obtain an arithmetic version. Please refer to [8] or [9] for a detailed development of this theory.

Taking the general form of the GCS in the arithmetic version of Theorem 3.1 we may now ask, whether we can find an algorithm to compute the GCS. We may further ask, whether the result is effective. The answer to both questions is negative in general, but we will identify subcases, for which effective GCSs can be computed. First consider the computability problem. Taking our Gödel numberings h for terms and formulae and g for commands, we obtain the following immediate consequence.

Lemma 4.2. *For each $n \in \mathbb{N}$ it is decidable, whether n is the Gödel number of a term, a formula or a guarded command.*

Next we consider the upper bound $S'_{\mathcal{I}}$ that occurs in the GCS. Since this is only a syntactic transformation, we may now conclude that $(S,\mathcal{I}) \mapsto S'_{\mathcal{I}}$ is computable. Hence it is sufficient to investigate the computability for the pre-condition $\mathcal{P}(S,\mathcal{I},x')$ for arbitrary x'.

These conditions involve the predicate transformers $wlp(S)$ and $wlp(S'_{\mathcal{I}})$. According to our definition of axiomatic semantics for commands, we know that building these predicate transformers is simple done by syntactic replacement operations. By exploiting our Gödel numbering h again, we conclude that for recursion-free S the mapping

$$(S,\mathcal{I},x') \mapsto \mathcal{P}(S,\mathcal{I},x')$$

– and hence $(S,\mathcal{I}) \mapsto S_{\mathcal{I}}$, too – is computable.

However, if S involves a loop, then $S'_{\mathcal{I}}$ also involves a loop. In order to determine $wlp(S)$ and $wlp(S'_{\mathcal{I}})$ we have to use the limit operator. For a loop $\mu T_j.f(T_j)$ this means to build $wlp(f^i(loop))$ for all $i \in \mathbb{N}$. This is only possible, if there is some $n \in \mathbb{N}$ such that $wlp(f^n(loop)) = wlp(f^m(loop))$ holds for all $m \geq n, m \in \mathbb{N}$. This means that we have a bounded loop (or equivalently a FOR-loop).

Proposition 4.3. *If recursive guarded commands are restricted to bounded loops, then GCSs are computable, i.e. the function $(S,\mathcal{I}) \mapsto S_{\mathcal{I}}$ is computable. In general, however, the GCS cannot be computed.*

4.4   Effective GCSs

Even, if the GCS $S_{\mathcal{I}}$ can be computed from a given command S and the invariant $\mathcal{I}$, the result still contains the preconditions $\mathcal{P}(S,\mathcal{I},x')$. If such a precondition is undecidable, then the GCSs will not be effective. We will demonstrate how effective GCSs can be computed.

Therefore, we consider the arithmetic version of the upper bound theorem from [9] which, in particular, leads to the following lemma.

Lemma 4.3. *Let T be a program specification on Y and $\mathcal{I}$ a static constraint on X with $Y \subseteq X$.*

(i)   If $T = P \rightarrow S$, then $T_{\mathcal{I}} = P \rightarrow S_{\mathcal{I}}$.
(ii)   If $T = S_1 \square S_2$, then $T_{\mathcal{I}} = (S_1)_{\mathcal{I}} \square (S_2)_{\mathcal{I}}$.
(iii)   If $T = @y \bullet S$, then $T_{\mathcal{I}} = @y \bullet S_{\mathcal{I}}$.

Proof. The upper bound theorem shows specialization in one direction. For the reverse specialization, one shows straightforwardly that $P \rightarrow S_{\mathcal{I}}$, $(S_1)_{\mathcal{I}} \square (S_2)_{\mathcal{I}}$ and $@y \bullet S_{\mathcal{I}}$ are $\mathcal{I}$-consistent specializations of $P \rightarrow S$, $S_1 \square S_2$ and $@y \bullet S$, respectively.

Note, that Lemma 4.3 does not hold for the case of sequences, even if they are δ-$\mathcal{I}$-reduced. Although the upper bound theorem gives us of course specialization in one direction, the reverse specialization does not hold in general. The reason why $(S_1)_\mathcal{I}; (S_2)_\mathcal{I}$ is not a specialization of $S_1; S_2$ is that $wlp(S_2)(\varphi)$ is not necessarily a state formula of the underlying S_1 state space.

The next lemma will give us a computation of effective GCSs for program specifications S that only use basic commands, choices, guards and sequences. We dispense with the case of restricted choices since we will not use them later.

Lemma 4.4. *Let S be a program specification on X built of basic commands, choices, guards with decidable preconditions and sequences. If φ is a decidable state formula on X, then $wlp(S)(\varphi)$ and $wlp(S)^*(\varphi)$ are decidable as well.*

Proof. The proof is a straightforward structural induction on guarded commands that makes use of closure properties for decidable arithmetical predicates.

It it well-known that every first-order predicate formula φ is equivalent to a formula $\mathcal{Q}_1 x_1 \ldots \mathcal{Q}_k x_k.\psi$ where $\mathcal{Q}_i \in \{\forall, \exists\}$ for $i = 1, \ldots, k$ and ψ is quantifier-free. This result carries immediately over to guarded commands with respect to the @-operator.

Lemma 4.5. *Each guarded command S, whose occurences of loops are all bounded, can be written in the form $@x_1 \bullet \ldots @x_n \bullet S'$ such that S' does not contain an unbounded choice operator @.*

Proof. The only interesting case is the one for bounded loops. Applying the predicate transformer wlp here results in a finite conjunction, whereas wp gives a finite disjunction.

Let us now consider a program specification S for which all occurences of loops are bounded and all preconditions are decidable. In a first step, we replace all occurences of the restricted choice operator $\boxtimes$ in the usual way. Then we apply Lemma 4.5 that provides us with a specification $T = @y_1 \bullet \ldots @y_n \bullet R$ that is semantically equivalent to S. Lemma 4.3 tells us not to worry about the occurences of unbounded-choice operators, i.e., $T_\mathcal{I} = @y_1 \bullet \ldots @y_n \bullet R_\mathcal{I}$. We apply the main theorem (Theorem 3.1) to compute $R_\mathcal{I}$ and conclude by Lemma 4.4 that all preconditions of the form $\mathcal{P}(S', \mathcal{I}, x')$ are decidable. Consequentely we obtain the following result.

Theorem 4.1. *Let S be a program specification such that every loop is bounded and all preconditions are decidable. Let $\mathcal{I}$ be a decidable static constraint. Then we can compute the GCS $S_\mathcal{I}$ in the form $S_\mathcal{I} = @y_1 \bullet \ldots @y_n \bullet T_\mathcal{I}$, where $T_\mathcal{I}$ has the form of Theorem 3.1 with all preconditions $\mathcal{P}(T', \mathcal{I}, x')$ being decidable.*

We conclude that GCS consistency enforcement is a well-founded approach and a big step forward to the systematic development of correct programs.

5   Towards a Tailored Theory of Consistency Enforcement

In the introduction we already explained that GCSs are not the only possible choice for formalizing consistency enforcement. If a program specification S affects state variables in Y and the constraint is defined on X with $Y \subseteq X$, then specialization may be too restrictive on Y and too liberal on $X - Y$. Maximal consistent effect preservers (MCEs) intend to overcome both these weaknesses of the GCS-approach.

5.1   A Motivating Example

The key concept is the one of a δ-constraint introduced in Section 3. This is a transition constraint $\mathcal{J}$ satisfied by the given program specification S. Thus, the δ-constraints of S express some kind of effect of S. Consequently, "preservation of effects" could be formalized by the preservation of δ-constraints. Let us first look at an example.

Example 5.1. Let us look back again at the enforcement strategies in Example 3.1 underlying the construction of a GCS branch. With respect to the inclusion constraint $\mathcal{I}_1$ the GCS branch was chosen in such a way that an insertion into x_1 was followed by an insertion into x_2, if necessary. Alternatively, we may like to replace $x_1 := x_1 \cup \{(a, b)\}$ by $a \notin \pi_1(x_2) \to x_1 := x_1 \cup \{(a, b)\} \boxtimes skip$. This would mean to restrict insertions by adding a precondition and to do nothing, if this condition is violated. Such a strategy is not possible with the GCS-approach.

Analogously, for the assignment $x_2 := x_2 \cup \{(a, c)\}$ and the constraint $\mathcal{I}_2$ from Example 3.1 on the state space $\{x_2\}$ we may like to replace it by $x_2 := \{(x, y) \in x_2 \mid y \neq c\} \cup \{(a, c)\}$. Taking these ideas together we would first replace S by S_1, which is

$$a \notin \pi_1(x_2) \to x_1 := x_1 \cup \{(a, b)\} \boxtimes skip$$

in order to enforce consistency with respect to $\mathcal{I}_1$. Then there is nothing to do to enforce $\mathcal{I}_2$. Finally, we replace

$$b \notin \pi_2(x_2) \to x_1 := x_1 \cup \{(a, b)\} \boxtimes skip$$

for the assignment $x_1 := x_1 \cup \{(a, b)\}$ to enforce the constraint $\mathcal{I}_3$. Thus, the final result would be

$$a \notin \pi_1(x_2) \wedge b \notin \pi_2(x_2) \to x_1 := x_1 \cup \{(a, b)\} \boxtimes skip \quad ,$$

which appears to be a reasonable alternative to the result obtained in Example 3.1.

5.2   Formalizing Effects by Transition Constraints

We may always associate a program specification

$$S(\mathcal{J}) = loop \square @x' \bullet \mathcal{J} \to x := x'$$

with each δ-constraint $\mathcal{J}$. In order to preserve $\mathcal{J}$ we must require to specialize $S(\mathcal{J})$. However, since $S(\mathcal{J})$ allows non-termination in all starting states, we need a stronger version. Consider

$$\bar{S}(\mathcal{J}) = @\boldsymbol{x}' \bullet \mathcal{J} \rightarrow \boldsymbol{x} := \boldsymbol{x}' \ \square \ \neg wp(S)(true) \rightarrow loop \quad .$$

which allows non-termination only in the same cases as S does. We should require to specialize $\bar{S}(\mathcal{J})$.

The basic idea of the MCE-approach is now to consider not all δ-constraints, but only some of them. Thus, we do no longer build the GCS of S with respect to $\mathcal{I}$, but the GCS of some $\bar{S}(\mathcal{J})$.

If some δ-constraints of S are omitted in $\mathcal{J}$, then $\bar{S}(\mathcal{J})$ will allow executions that do not occur in any specialization of S. In this way, we can capture reasonable changes to S. Taking any such δ-constraint is much too weak. $\bar{S}(\mathcal{J})$ should only add executions that are consistent with $\mathcal{I}$. This justifies to define δ-constraints that are compatible with a given static constraint $\mathcal{I}$ on X in the sense that building the GCS $\bar{S}(\mathcal{J})_{\mathcal{I}}$ does not increase partiality.

Definition 5.1. A δ-constraint $\mathcal{J}$ for a program specification S is *compatible* with a static constraint $\mathcal{I}$ iff $wp(\bar{S}(\mathcal{J})_{\mathcal{I}})(false) \Rightarrow wlp(S)(false)$ holds.

In general, the conjunction of all δ-constraints of S will not be compatible with $\mathcal{I}$. This suggests to consider the implication order on δ-constraints. We say that $\mathcal{J}_1$ is *stronger* than $\mathcal{J}_2$ iff $\mathcal{J}_1 \Rightarrow \mathcal{J}_2$ holds. Unfortunately, there is no smallest δ-constraint compatible with $\mathcal{I}$ and we cannot consider the "strongest" $\mathcal{I}$-compatible δ-constraint for S, but we may consider minimal elements in this order.

Definition 5.2. A δ-constraint $\mathcal{J}$ for S is *low* with respect to $\mathcal{I}$ iff it is $\mathcal{I}$-compatible and there is no strictly stronger $\mathcal{I}$-compatible δ-constraint.

Now we are prepared to define maximal consistent effect preservers. For these we choose a low δ-constraint $\mathcal{J}$ which formalizes "effects" of S to be preserved. Then we take a consistent program specification $S^{\mathcal{I},\mathcal{J}}$ that preserves this effect, but remains undefined, whereever S is undefined. Finally, we require $S^{\mathcal{I},\mathcal{J}}$ to be maximal with these properties with respect to the specialization order.

Definition 5.3. Let S be a program specification and $\mathcal{I}$ a static constraint on X. Let $\mathcal{J}$ be a low δ-constraint of S with respect to $\mathcal{I}$. A program specification $S^{\mathcal{I},\mathcal{J}}$ on X is called a *maximal consistent effect preserver* (MCE) of S with respect to $\mathcal{I}$ and $\mathcal{J}$ iff

 - $\mathcal{J}$ is a low δ-constraint for $S^{\mathcal{I},\mathcal{J}}$,
 - $wp(S)(false) \Rightarrow wp(S^{\mathcal{I},\mathcal{J}})(false)$ holds,
 - $S^{\mathcal{I},\mathcal{J}}$ is consistent with respect to $\mathcal{I}$ and
 - any other program specification T with these properties specializes $S^{\mathcal{I},\mathcal{J}}$.

Note that in this definition the state space on which S is defined is no longer important. It "vanishes" inside the chosen $\mathcal{J}$.

The last part of the definition employs the specialization order $\sqsubseteq$ again. This seems to be surprising for the first moment, but it turns out to be a natural definition as shown in the following proposition.

Proposition 5.1. *Let S be a program specification and $\mathcal{I}$ a static constraint on X. Let $\mathcal{J}$ be a low δ-constraint of S with respect to $\mathcal{I}$. Then $\neg wp(S)(false) \rightarrow \bar{S}(\mathcal{J})_{\mathcal{I}}$ is the MCE with respect to $\mathcal{I}$ and $\mathcal{J}$.*

The proposition suggests that there is also a strict theory for MCEs leading to commutativity and compositionality results. This theory, however, has not yet been developed, but we can check that the construction in Example 5.1 led indeed to an MCE.

6   Conclusion

In this article we discussed consistency in databases. We started showing that rule triggering systems (RTSs) cannot solve the problem, as the approach only looks at the constraints and does not pay enough attention to complex operations as the source of inconsistencies. We argue that any reasonable approach to consistency enforcement should 'prefer the effects' of such complex operations.

The central question is how to formalize effect preservation. We may think of a partial order on operations (more precisely: on semantic equivalence classes of operations) such that smaller operations preserve effects of larger operations. Then enforcement would mean to find a maximal consistent diminution of the original operation, which reduces the problem to find the most suitable order and to develop a theory for it.

For such a theory there are some indispensible desiderata concerning compositionality with respect to the operations and the constraints. This means that step-by-step enforcement for a set of constraints should be justified as well as a reduction to basic operations. These properties are guaranteed, if the order is taken to be operational specialization. In this case we even have uniqueness of the result and hence may talk of greatest consistent specializations (GCSs), and commutativity, i.e., the order of constraints in a step-by-step enforcement process does not matter. From a computability point of view, the number of cases, where GCSs can be effectively computed and the results are effective, is large enough for databases.

Nevertheless, this theory of greatest consistent specializations has some problems to be addressed. The order is too coarse to capture all intuitively justified possibilities for consistency enforcement and the result in general is highly non-deterministic. A pragmatic approach considers only deterministic branches of GCSs and shifts the GCS approach to run-time. This guarantees that there is no problem with computability at all, but of course does not remove the problem with the order being too coarse.

The idea for an enhanced enforcement theory starts from the observation that specialization is equivalent to the preservation of all those transition constraints that only refer to the variables read or changed by the operation. Such constraints are called δ-constraints. The idea is then to ask for a different set of δ-constraints to be preserved. It has been shown that this means to build the GCS of a slightly modified operation. However, a strict theory for this approach is still outstanding.

References

1. J. Bell, M. Machover. *A Course in Mathematical Logic.* North-Holland 1977.
2. S. Ceri, P. Fraternali, S. Paraboschi, L. Tanca: Automatic Generation of Production Rules for Integrity Maintenance. *ACM TODS* 19(3), 1994, 367–422.
3. I. A. Chen, R. Hull, D. McLeod. An Execution Model for Limited Ambiguity Rules and its Applications to Derived Data Update. *ACM ToDS* 20, 1995, 365–413.
4. L. Console, M. L. Sapino, D. Theseider. The Role of Abduction in Database View Updating. *Journal of Intelligent Information Systems* 4, 1995, 261–280.
5. H. Decker. One Abductive Logic Programming Procedure for two Kinds of Update. *Proc. DYNAMICS'97*, 1997.
6. M. Dekhtyar, A. Dikovsky, S. Dudakov, N. Spyratos. Maximal Expansions of Database Updates. In K.-D. Schewe, B. Thalheim (Eds.). *Foundations of Information and Knowledge Systems*, 72–87. Springer LNCS 1762, 2000.
7. M. Gertz. Specifying Reactive Integrity Control for Active Databases. *Proc. RIDE '94*, 1994, 62–70.
8. S. Link. *Eine Theorie der Konsistenzerzwingung auf der Basis arithmetischer Logik.* M.Sc. Thesis (in German). TU Clausthal 2000.
9. S. Link, K.-D. Schewe. An Arithmetic Theory of Consistency Enforcement. *Acta Cybernetica.* vol. 15 . 2002 . 379–416 .
10. J. Lobo, G. Trajcevski. Minimal and Consistent Evolution in Knowledge Bases. *Journal of Applied Non-Classical Logics* 7, 1997, 117–146.
11. M. Makkai. Admissible Sets and Infinitary Logic. In J. Barwise (Ed). *Handbook of Mathematical Logic.* North Holland, Studies in Logic and Foundations of Mathematics. vol. 90: 233–281. 1977.
12. E. Mayol, E. Teniente. Dealing with Modification Requests During View Updating and Integrity Constraint Maintenance. In K.-D. Schewe, B. Thalheim (Eds.). *Foundations of Information and Knowledge Systems*, 192–212. Springer LNCS 1762, 2000.
13. G. Nelson. A Generalization of Dijkstra's Calculus. *ACM TOPLAS.* vol. 11 (4): 517–561. 1989.
14. K.-D. Schewe. Consistency Enforcement in Entity-Relationship and Object-Oriented Models. *Data and Knowledge Engineering* 28, 1998, 121–140.
15. K.-D. Schewe. Fundamentals of Consistency Enforcement. In H. Jaakkola, H. Kangassalo, E. Kawaguchi (eds.). *Information Modelling and Knowledge Bases X:* 275–291. IOS Press 1999.
16. K.-D. Schewe, B. Thalheim. Towards a Theory of Consistency Enforcement. *Acta Informatica.* vol. 36: 97–141. 1999.
17. K.-D. Schewe, B. Thalheim. Limitations of Rule Triggering Systems for Integrity Maintenance in the Context of Transition Specifications. *Acta Cybernetica.* vol. 13: 277–304. 1998.
18. K.-D. Schewe, B. Thalheim, J. Schmidt, I. Wetzel. Integrity Enforcement in Object Oriented Databases. In U. Lipeck, B. Thalheim (eds.). *Modelling Database Dynamics:* 174–195. Workshops in Computing. Springer 1993.
19. E. Teniente, A. Olivé. Updating Knowledge Bases while Maintaining their Consistency. *The VLDB Journal* 4, 1995, 193–241.
20. B. Wüthrich. On Updates and Inconsistency Repairing in Knowledge Bases. *Proc. ICDE'93*, 1993, 608–615.

Automata- and Logic-Based Pattern Languages for Tree-Structured Data

Frank Neven[1]* and Thomas Schwentick[2]

[1] University of Limburg
[2] Philipps-Universität Marburg

Abstract. This paper surveys work of the authors on pattern languages for tree-structured data with XML as the main application in mind. The main focus is on formalisms from formal language theory and logic. In particular, it considers attribute grammars, query automata, tree-walking automata, extensions of first-order logic, and monadic second-order logic. It investigates expressiveness as well as the complexity of query evaluation and some optimization problems. Finally, formalisms that allow comparison of attribute values are considered.

Keywords: attribute grammars, automata, formal languages, logic, query evaluation, XML

1   Introduction

There is a powerful evolving technology around XML. A lot of proposals for query languages, pattern languages and the like are emerging. Some of these are complemented with working implementations. However, as such, there is no real standard for an XML query language. Although there are some requirements for an XML query language [31] and the latest XQuery working draft receives a lot of attention [6], it is not clear what the desired expressive power of such a query language should be. Furthermore, by and large, the expressive power and complexity of the existing languages is not well understood. This paper surveys research done by the authors on the expressiveness of query constructs for tree-structured data (that is, XML). The main focus is on logic and formalisms from formal language theory.

Let us look back at the history of query languages for relational databases. There a standard for expressive capabilities emerged from the tight connection between some query mechanisms and logic. More precisely, the well-known equivalence between relational algebra, core SQL and, first-order logic. First-order logic is such a natural and robust notion that it is no surprise that there are so many equivalent formalisms of the same expressive power. But the connection did not only help to find a standard. Also database theory profited a lot from it, e.g., it helped to understand what kinds of queries can be expressed or not expressed and to clarify semantical issues. The compositional nature of

---

* Research Assistant of the Fund for Scientific Research, Flanders.

L. Bertossi et al. (Eds.): Semantics in Databases, LNCS 2582, pp. 160–178, 2003.

first-order logic further supported the development of a theory of views. In a sense, first-order logic can be seen as a link between the operational nature of relational algebra and the declarative, user-oriented nature of SQL.

Coming back to XML the natural question arises: Is there a logic that could play a similar benchmark role for query languages for XML as first-order logic did for relational databases? One immediate difference is that XML data is primarily tree-structured. Obviously, this difference in representation affects the way queries are asked. However, when searching for a natural logic on trees there is one obvious candidate that was well investigated already 30 years ago: *monadic second-order logic* (MSO). That is, the extension of first-order logic (FO) where quantification over sets of nodes is allowed. There is a large body of research on that logic on trees. [30] The expressive power of MSO logic exactly matches the recognition power of several kinds of tree automata (e.g., bottom-up automata that combine information about the tree in one pass from the leaves to the root) and other formalisms like attributed grammars. It gives rise to a very robust class: the class of *regular tree languages* which is almost as robust as the class of regular string languages.

Further support for the consideration of MSO logic comes from an observation about existing query languages for XML. Most of them allow a kind of pattern matching along paths in an XML document; some of them by means of regular expressions. But there is a matching logic with the expressive power of regular expressions. By Büchi's Theorem it is MSO logic. Of course, there are also various kinds of string automata that capture the same level of expressive ability. By the way, there is an analogy to the case of relational databases that there is an operational way of specification by automata, a logical way of specification by MSO logic and a user-oriented declarative specification by regular expressions.

The mentioned connections strongly indicate that besides logic also automata might be helpful in the design and understanding of query formalisms for XML data. Further support in that direction is given by the very successful application of automata theoretic methods in the realm of verification. There also tree-structured data is very important and there is a tight interplay between logics (here: temporal logics) and automata.

Altogether we deal in this article with four main topics.

- The use of logical formulas as query mechanisms for tree-structured data, especially formulas of MSO and fragments of it;
- The use of parallel automata models (like bottom-up automata and attribute grammars). What is the impact of the knowledge about regular tree languages in the context of XML? There are two mismatches one has to deal with (a third one will be mentioned shortly): (1) For XML we are mainly interested in querying tree-structured documents rather than checking properties of a document; (2) For XML we have to deal with trees of unbounded degree (i.e., unranked trees) in contrast to the bounded degree trees in classical regular tree languages.
- The use of sequential automata models. Although it seems obvious that sequential automata models on trees have limited power compared with par-

allel automata this has not been proved yet. As these automata are related to XSLT, the understanding of the exact expressive power of such automata is a relevant topic.

- The fourth issue is orthogonal to the first three and it is related to the third mismatch. XML documents might contain arbitrary text, numbers, references etc., whereas the logics and automata mentioned so far work on trees with a fixed finite set of labels. Actually, whether one allows arbitrary data values or not is a main dividing line for theoretical work on tree-structured data. Whereas subqueries like `name = ''Johnson''` can still be handled in the finite alphabet framework, propositions like `value at vertex` x = `value at vertex` y cannot, in general. Whereas it is straightforward to equip MSO with the ability to deal with data values there are several possibilities how to extend automata models in that direction. We describe some results on such models.

It should be noted, that queries that do not compare values of different nodes are very common and are also frequent as subqueries. Therefore it is reasonable to try to understand them as good as possible. Nevertheless, the general formal model allows not only a finite set of labels but also attributes that might take values from an infinite domain. But we restrict access to these data values as we only allow equality and inequality comparisons of data values.

The main goals of the work surveyed here are as follows.

- Understand the expressive power of the various formalisms;
- Find out their evaluation complexity; Maybe find new formalisms with a better evaluation complexity;
- In particular, find formalisms that have a good combined evaluation complexity, i.e., formalisms for which evaluation is efficient even if the query itself is considered as part of the input;
- Investigate decidability issues, e.g., related to satisfiability and containment of queries in the various formalisms.

The rest of the paper is structured as follows. In Section 2 we introduce the necessary definitions. In Section 3, we consider pattern languages equivalent to MSO. In Section 4, we look at sequential formalisms. In Section 5, we study pattern languages based on fragment of MSO. In Section 6, we investigate formalisms with the ability to compare attribute values. Finally, we present some conclusions in Section 7.

2   Preliminaries

In this section we introduce the basic formalisms used throughout the paper, including the notion of (attributed) trees as abstraction of XML documents and the logical framework.

2.1   Trees and XML

We start with the necessary definitions regarding trees. Let Σ be a finite alphabet. The Σ-symbols will correspond to the element names of the XML document. To use trees as adequate abstractions of actual XML documents, we extend them with attributes from a finite set A that take values from an infinite (recursively enumerable) domain $\mathbf{D} = \{d_1, d_2, \dots\}$. In the sequel we assume some fixed Σ and $\mathbf{D}$.

A *tree domain* τ *over* $\mathbb{N}$ is a subset of $\mathbb{N}^*$, such that if $v \cdot i \in \tau$, where $v \in \mathbb{N}^*$ and $i \in \mathbb{N}$, then $v \in \tau$. Here, $\mathbb{N}$ denotes the set of natural numbers without zero. If $i > 1$ then also $v \cdot (i-1) \in \tau$. The empty sequence, denoted by ε, represents the root. We call the elements of τ *vertices*. A vertex w is a *child* of a vertex v (and v the *parent* of w) if $vi = w$, for some i.

In this article, we only consider *finite* tree domains.

Definition 1. An *attributed* (Σ, A)*-tree* is a triple $t = (\mathrm{dom}(t), \mathrm{lab}_t, (\lambda_t^a)_{a \in A})$, where $\mathrm{dom}(t)$ is a tree domain over $\mathbb{N}$, $\mathrm{lab}_t : \mathrm{dom}(t) \to \Sigma$ is a function, and for every $a \in A$, $\lambda_t^a : \mathrm{dom}(t) \to \mathbf{D}$ is a partial function.

When Σ and A are clear from the context or not important, we sometimes say tree rather than (Σ, A)-tree. Note that in the definition of a tree there is no a priori bound on the number of children that a node may have.

We describe next the representation of XML documents as (Σ, A)-trees. We represent XML elements by means of the finite set Σ of labels and attribute values by the functions λ_t^a. Maximal sequences of character data are represented by nodes (inevitable leaves) that are labeled by a special element TEXT of Σ that is not used otherwise. The actual data is represented by an attribute PC from $\mathbf{D}$ also not used otherwise (see example).

Example 1. The XML document

```
<beer name="Grimbergen Trippel">
  <alc> 9 </alc>
    <description>
      <color> blonde </color>
      <sort> trappist </sort>
    </description>
</beer>
```

is faithfully modeled by the tree

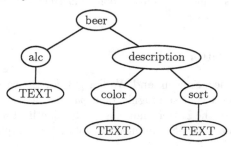

where $\lambda_t^{\text{name}}(\varepsilon) =$ "Grimbergen Trippel", $\lambda_t^{\text{PC}}(11) =$ "9", $\lambda_t^{\text{PC}}(211) =$ "blonde", and $\lambda_t^{\text{PC}}(221) =$ "trappist". Here, the domain $\mathbf{D}$ is assumed to contain the subset {"Grimbergen Trippel", "9", "blonde", "trappist"}.

The XML specification allows also mixed content elements like

```
<example>
  This is not a very
  <stress> meaningful </stress>
  sentence.
</example>
```

which can be modelled by the tree

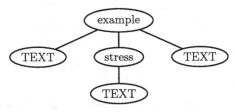

with $\lambda_t^{\text{PC}}(1) =$ "This is not a very", $\lambda_t^{\text{PC}}(21) =$ "meaningful" and $\lambda_t^{\text{PC}}(3) =$ "sentence".

□

As we already pointed out in the introduction, most of the query mechanisms considered in this article do not allow the comparison of data values of different nodes. It turns out that in these cases one can model XML documents adequately by trees without data values, i.e., by $(\Sigma, \emptyset)$-trees. We refer to such trees simply as Σ-*trees*. The reason for this is as follows. For the purposes of this article, there is no need to fix a uniform representation of XML documents into trees but rather it is sufficient to consider representations *relative to the given query*. E.g., if the query asks for trappist beers and more than 9 % alcohol, we can represent XML documents by trees over an alphabet that has a label *trappist & 9 or less % alcohol* as well as a label for *trappist and more than 9% alcohol* and so on. In this manner and in the absence of data value comparisons between nodes, only a finite amount of information (depending on the query) is needed from the potentially infinite set of data values. This finite amount of information can be represented by Σ.

Hence, in sections 3 and 5 we only consider Σ-trees whereas in section 6 we consider general (Σ, A)-trees.

2.2   Queries and Patterns

As argued by Fernandez, Siméon, and Wadler [13] for the case of XML, queries on tree-structured data consist roughly of a *pattern* clause and a *constructor* clause. The purpose of the pattern language is to identify the different parts of the document that have to be combined to obtain the output document. The constructing part, on the other hand, indicates how the selected parts should be assembled. Such queries can, for instance, be written as

$$\text{WHERE } \varphi(\bar{x}), \text{ CONSTRUCT result}(t),$$

where φ is a pattern selecting vertices and t is a tree containing at leaves special constructs like $\text{yield}(x)$, $\text{lab}(x)$, $\text{subtree}(x)$ indicating that at this position the yield, the label, or the subtree rooted at the matched vertex for x should be plugged in. Note that the WHERE clause maps trees to an intermediate result of relational nature which is then transformed into a tree by the CONSTRUCT clause.

Pattern languages, therefore, form the basic building blocks of more general query languages transforming documents into other documents. Clearly, the choice of the pattern language can affect tremendously the expressive power of the overall query language.

In this article we focus on pattern languages. In the rest of the article, the terms *query* and *pattern* will both refer to mappings from trees to relations corresponding to pattern clauses.

Many of the pattern mechanisms considered in this article can not produce arbitrary relational results but are limited to unary results, i.e., results that are sets of vertices. This is justified as the identification of relevant parts of a tree is the most common purpose of pattern clauses (see e.g., the example queries of the W3C [31]).

2.3 Logic

We view trees as logical structures (in the sense of mathematical logic [10]) over the binary relation symbols E and $<$, and the unary relation symbols $(O_\sigma)_{\sigma \in \Sigma}$. We denote this vocabulary by τ_Σ. The domain of t, viewed as a structure, equals the set of nodes of t, i.e., $\text{dom}(t)$. Further, E is the edge relation and equals the set of pairs $(v, v \cdot i)$ where $v, v \cdot i \in \text{dom}(t)$. The relation $<$ specifies the ordering of the children of a node, and equals the set of pairs $(v \cdot i, v \cdot j)$, where $i < j$ and $v \cdot j \in \text{dom}(t)$. For each σ, O_σ is the set of nodes that are labeled with a σ.

We consider first-order (FO) and monadic second-order logic (MSO) over these structures. MSO is FO extended with quantification over set variables. We refer the unfamiliar reader to, e.g., the books by Ebbinghaus and Flum [10], or the chapter by Thomas [30].

To be able to deal with attribute values (in section 6), we allow atomic formulas of the forms $a(x) = d$ where a is an attribute and $d \in \text{dom}$ and of the form $a(x) = b(y)$. The former formula holds for u in t iff $\lambda_t^a(u) = d$, the latter holds for u and v in t iff $\lambda_t^a(u) = \lambda_t^b(v)$.

2.4 Complexity Issues

Besides comparing the expressive power of the different formalisms we are typically interested in the following kinds of computational problems related to the classes $\mathcal{C}$ of queries under consideration (where $\mathcal{C}$ is a set of *representations* of queries, e.g., automata or formulas).

NON-EMPTINESS Given a query $Q \in \mathcal{C}$, is there a tree t such that $Q(t) \neq \emptyset$?
CONTAINMENT Given queries $Q_1, Q_2 \in \mathcal{C}$, is $Q_1(t) \subseteq Q_2(t)$, for all trees t?
EQUIVALENCE Given queries $Q_1, Q_2 \in \mathcal{C}$, is $Q_1(t) = Q_2(t)$, for all trees t?
QUERY EVALUATION Given a query $Q \in \mathcal{C}$ and a tree t, compute $Q(t)$?

For NON-EMPTINESS and CONTAINMENT the first question is whether they are decidable, for a given class $\mathcal{C}$. If so, for most cases we consider the complexity is EXPTIME. Note that NON-EMPTINESS is also called SATISFIA-BILITY in the case of Boolean queries. The QUERY EVALUATION problem comes in two flavors. Either the query is considered fixed or it is considered as part of the input (so the complexity is a function in the length of the representation of Q and t). We refer to the former as the *data complexity* of QUERY EVALUATION and to the letter as the *combined complexity* of QUERY EVALUATION.

The linear time results and the complexity results in Section 5 assume Random Access Machines with a unit cost measure as computational model. Further they require a suitable representation of the input trees.

3   Pattern Languages Equivalent to MSO

As demonstrated in Section 2, XML documents can be faithfully represented by attributed trees. Trees have been studied in depth in the area of formal language theory for over 30 years and many formalisms have been proposed. We re-consider some of these formalisms from the viewpoint of pattern languages. More precisely we re-examine attribute grammars, tree automata, and tree transducers.

As mentioned in the introduction, there are two essential differences between trees studied in formal language theory and attributed trees representing XML documents as considered in this paper: (i) trees are unranked, that is, there is no fixed bound on the number of children of a node; and (ii) XML trees have attributes instantiated by elements coming from an infinite domain. In this section we do only consider pattern languages which disallow comparisons of attribute values of different vertices. Hence, as explained in Section 2 we consider only Σ-trees.

3.1   Tree Automata

The basic computational model in tree language theory is the tree automaton. To warm-up, let us first consider finite state machines (FSMs) on strings. FSMs process input strings from left to right by first assigning the initial state to the first letter of the input string; a state for an inner position i is then determined via the transition function based on the label of $i - 1$ and the state at $i - 1$. The input string is accepted when a final state is reached at the end of the string.

On trees there are two quite different types of generalizations of this mechanism. They can be characterized as *sequential* and *parallel* tree automata, respectively. Sequential tree automata are similar to string automata as they have only *one head* which moves around the tree. In order to inspect the whole tree they have to visit some vertices several times. Sequential automata will be discussed in Section 4. On the other hand, most parallel tree automata keep the *visit positions only once* paradigm of string automata. Bottom-up tree automata, for instance, process trees in one pass from the leaves to the root. That is, first initial states are assigned to leaf nodes (depending on the symbol they carry). Second, the state at an inner node is determined via the transition function based on the label of the inner node and on the states assumed at their children. Finally, the automaton accepts if a final state is reached. In this subsection we study parallel tree automata as query mechanisms for trees.

How do we define the transition function for a bottom-up tree automaton? If the arity of the input trees is bounded by n, a transition function is simply a function $\delta : \bigcup_{i=0}^{n} Q^n \times \Sigma \to Q$. However, when the arity is unbounded, we can not adapt this approach as it would imply having to define transitions for an infinite number of cases. Brüggemann-Klein, Murata, and Wood [4] proposed a solution based on regular sets of states which we explain next.

Formally, a (bottom-up) tree-automaton is a tuple $T = (Q, \Sigma, \delta, q_0, F)$ where Q is a set of states, Σ is the alphabet of labels, q_0 is the initial state and F is the set of accepting states, as in the definition of finite state machines, Finally, $\delta : Q \times \Sigma \to 2^{Q^*}$ is a transition function mapping every pair (q, σ) to a regular language over Q. T accepts a tree t if there is a labeling function $\gamma : \mathrm{dom}(t) \to Q$ such that (*i*) for every vertex v, with children $v_1 \cdots v_n$, $\gamma(v1) \cdots \gamma(vn) \in \delta(\gamma(v), \mathrm{lab}_t(v))$; [1] and (*ii*) $\gamma(\varepsilon) \in F$, that is, a final state is reached. A tree language, that is, a set of trees, is *regular* if it is accepted by a tree-automaton.

Using regular string languages to specify properties of the sequence of children in unranked trees lies at the core of many of the following formalisms.

3.2   Query Automata

This section is based on [23]. A *query automaton* (QA) is a two-way (i.e., up and down) deterministic finite tree automaton with a distinguished set of selecting states. At each step of a computation, there is a partial mapping s from $\mathrm{dom}(t)$ to Q. Intuitively, the set C of vertices v, for which $s(v)$ is defined contains those vertices of the tree that are currently visited by a head of the automaton. This set C has to contain exactly one vertex of each path from the root to a leaf. A computation step consists of either a *down transition* or an *up transition*. I.e., either in C, a vertex is replaced by its children, or, if all children of a certain vertex v are in C, they might be replaced by v itself. In both cases, the new vertices in C are assigned states depending on the transition function.

A vertex is selected by a QA if it is visited at least once in a selecting state. Hence, a QA can compute unary queries in a natural way: the result of a QA on a tree consists of all those nodes that are selected during the computation of the QA on that tree. We stress that QAs are quite different from the tree acceptors studied in formal language theory [14]. Although two-way tree *acceptors* are

---

[1] Note, that in particular, when v is a leaf, ε should be in $\delta(\gamma(v), \mathrm{lab}_t(v))$.

equivalent to one-way acceptors [20] it is straightforward to see that (two-way) QAs are not equivalent to bottom-up query automata. Indeed, a bottom-up QA, for example, cannot compute the query "select all leaves if the root is labeled with σ", simply because it cannot know the label of the root when it starts at the leaves.

Theorem 1. *On ranked trees, QAs express exactly the MSO definable unary patterns.*

For QAs on unranked trees the picture looks quite different. QAs over unranked trees cannot even express all FO definable unary patterns. The basic reason is that in the unranked case very little information can be passed from one sibling to another. To resolve this, QAs are equipped with *stay transitions* where a two-way string-automaton reads the string formed by the states at the children of a certain node, and then outputs for each child a new state. Automata are then restricted to make only a constant number of stay transitions for the children of each node. These automata are called *strong* QAs.

Theorem 2. *On unranked trees, strong QAs compute exactly all MSO-definable unary patterns.*

W.r.t. standard decision problems we obtain the following.

Theorem 3. *NON-EMPTINESS, CONTAINMENT, and EQUIVALENCE of strong QAs are complete for EXPTIME.*

Although the run time of a QA might be quadratic in the size of the tree query evaluation can be done more efficiently.

Theorem 4. *The data complexity of QUERY EVALUATION for QAs is linear time. The combined complexity of QUERY EVALUATION for QAs is PTIME.*

3.3 Attribute Grammars

Attribute grammars (AGs), as defined by Knuth [16], constitute a deeply studied general computational model for trees [9]. In brief, an AG consists of a context-free grammar and a set of rules defining annotations (called attributes) of nodes of derivation trees. As defined by Knuth, the domain of attributes can be anything and the semantic rules can be any recursive function. Further, attributes can be defined in a top-down (inherited) or a bottom-up (synthesized) way.

Towards pattern languages for tree-structured data, Neven and Van den Bussche [27] considered Boolean-valued AGs (BAGs) where attributes are restricted to be Boolean-valued and rules are propositional formulas over attributes. As an example consider the following BAG working on a list of paragraphs

$$L \to Lp \qquad odd(0) := \neg odd(1)$$
$$L \to p \qquad odd(0) := \text{true}.$$

Here, the attribute *odd* will be true for all L-labeled nodes on an odd position when starting counting from the bottom. In brief, *odd* is true for the last L-labeled node; for an inner node *odd* is the negation of the *odd*-value of the next L. In the semantic rules the numbers 0 and 1 refer to the left-hand side and the first non-terminal of the right-hand side of a production, respectively. By designating *odd* as the result attribute, on an input tree the BAG retrieves all nodes for which *odd* is true. Although BAGs constitute a seemingly simple formalism, it can be shown that BAGs express precisely all MSO definable unary patterns [27,3].

XML documents, however, are usually described by *extended* context-free grammars (DTDs). These are grammars with regular expressions on the right-hand sides of productions. The above context-free grammar could be specified, for instance, like List $\rightarrow L^*$, $L \rightarrow p$. One problem that arises when defining AGs to work on such grammars is that semantic rules should be able to depend on an unbounded number of attributes (as the production List $\rightarrow L^*$ does not put any restriction on the number of L's). The latter problem can be resolved by using regular languages over attribute values as semantic rules.[2] Consider the rule

$$\text{List} \rightarrow L^* \qquad \text{odd}(1) := L^* \# L(LL)^*$$

and the tree consisting of a root with label List and four children with label L. To determine whether *odd* is true for, say, the second child, we consider the string $L\#LLL^*$ where we insert the marker $\#$ in front of the second position. As this string matches the regular expression $L^*\#L(LL)*$, *odd* will be true for the second child. In general, semantic rules can also depend on the values of other attributes, as opposed to labels of nodes; we refer the interested reader to [22]. Again unary patterns can be expressed by designating some attribute as the result attribute.

The next theorem provides evidence for the robustness of the definition of AGs on unranked trees.

Theorem 5. *AGs compute exactly all MSO-definable unary patterns on unranked trees.*

Theorem 6. *NON-EMPTINESS, CONTAINMENT, and EQUIVALENCE of AGs on unranked trees are complete for* EXPTIME.

The latter result can be used to improve optimization of Region Algebra expressions [8]. Indeed, by exhibiting a linear time translation from Region Algebra expressions to attribute grammars, one obtains that testing equivalence of Region Algebra is in EXPTIME. This should be contrasted to the hyper-exponential upper bound of Milo and Consens [8].

Corollary 1. *EQUIVALENCE of Region Algebra expressions is in* EXPTIME.

The data complexity of QUERY EVALUATION for AGs is linear time. This is clear because of the equivalence with Query Automata. But there is also a straightforward evaluation strategy which gives this time bound directly, if the evaluation of single rules can be done in constant time. The combined complexity is PTIME.

---

[2] Actually, some more problems arise. See [22].

3.4 Related Work

We briefly discuss some related formalisms for unranked trees. Neumann and Seidl define a μ-calculus for specifying unary patterns [28] and a push-down tree automaton model for evaluating them. These expressions can be evaluated by a one-pass traversal of the tree. Murata defines a pattern language for expressing unary patterns based on tree-regular expressions [21]. These correspond exactly to MSO. Finally, we mention that tree-transducers (over unranked trees) as a formal model for XSLT have been considered in [17]. We come back to XSLT in Section 6.

4   Pattern Languages Based on Sequential Automata

The automata model discussed in the previous section is a parallel one. For instance, a down transition assigns states to all children of the current vertex which in turn are processed rather independently. So the control of the automaton is at several nodes of the input tree simultaneously, rather than at just one. In contrast, the finite control of a tree-walking automaton (TWA) is always at one node of the input tree. The computation starts at the root in the initial state. Based on the label of the current node and its location (that is, root, first child, last child, or leaf) the automaton changes state and steps to one of the neighboring nodes (that is, parent, first child, last child, left sibling, right sibling). An automaton can express selections in the same way as query automata: all nodes that are visited in a selecting state are selected.

It is an open problem whether TWAs can express all MSO definable selection patterns. This is related to the question whether TWAs can define all regular tree languages. To be precise, a TWA accepts a tree when it selects the root. Engelfriet and Hoogeboom conjecture that TWAs are strictly weaker than tree automata [11,12] which would imply that TWA cannot define all MSO selection patterns. In [25] it is shown that over ranked trees, TWAs accept precisely the set of trees definable in a fragment of deterministic unary transitive closure logic. Further, the conjecture is proved for a restriction of TWAs: TWAs that visit each subtree only once (and a mild generalization thereof) cannot express the set of all regular tree languages.

In research on tree-structured data, tree-walking automata are used for various purposes and appeared in various forms. Milo, Suciu, and Vianu [19], for instance, use a transducer model based on tree-walking automata as a formal model for an XML transformer encompassing most current XML transformation languages. Another occurrence of tree-walking automata is embodied in the actual XML transformation language XSLT [7] proposed by the W3C. In formal language theoretic terms, this query language can be best described as a *tree-walking tree transducer* [2].

As a third example we mention the caterpillar expressions of Brüggeman-Klein and Wood. [5] These are regular expressions over moving instructions {up, left, down, right, isLeaf, isFirst, isLast, isRoot} and Σ-symbols. For instance, the expression $a\,\text{isLeaf}\,(\text{up}^* \, b\,\text{down}^* c)$, selects all a-labeled leaves that have a

b-labeled ancestor who in turn has a c-labeled descendant. Again, a caterpillar accepts a tree when it selects the root. Brüggeman-Klein and Wood leave it as an open question whether caterpillar expressions can define all regular (unranked) tree languages.

Hence, results on the expressiveness of tree-walking automata could give insight in the expressiveness of actual XML transformation languages.

5   Pattern Languages Based on Fragments of MSO

In this section we consider fragments of MSO as query mechanisms for tree-structured data. We still keep the restriction that the comparison of data values other than comparisons with fixed constants are not allowed. Hence, as discussed in Section 2 we deal with Σ-trees without data values. Section 6 treats the case where data values are present. This section is based on the work reported in [24,29]. First we introduce an intermediate logic (FOREG) between FO and MSO logic which attempts to capture the expressive power of languages with regular path expressions in a very general way. The basic idea is to allow regular expressions over formulas that are evaluated along a vertical path or along the children of a node. Then we turn our attention to fragments of MSO, FOREG and FO respectively, for which the QUERY EVALUATION problem has low combined complexity. These fragments are obtained by restricting quantification of variables in the spirit of guarded logics [1] on one hand and by allowing vertical path expressions on the other hand.

5.1   Regular Expressions

As mentioned before, the logic FOREG is an extension of FO by regular path expressions. Further, it is designed to capture also arbitrary nesting of such expressions. It uses the following two kinds of path formulas.

- If P is a regular expression (in the usual sense) over formulas with free variables r and s then $\varphi = [P]_{r,s}^{\downarrow}(x, y)$ is a *vertical path formula*.
- If P is a regular expression over formulas with free variable r then $[P]_{r}^{\rightarrow}(x)$ is a *horizontal path formula*.

We refer to path formulas also by the term *path expressions*. A simple example of a horizontal path formula is $[(O_a(r))^* O_b(r)]_{r}^{\rightarrow}(x)$.

The semantics of such formulas is defined as follows.

- Let $\varphi = [P]_{r,s}^{\downarrow}(x, y)$ be a vertical path formula. Let t be a tree and let v, w be vertices of t. Then, $t \models \varphi[v, w]$, iff w is in the subtree rooted at v and there is a labeling of the edges on the path from v to w with formulas, such that (1) each edge (u, u') is labeled with a formula $\theta(r, s)$ such that $t \models \theta[u, u']$, and (2) the sequence of labels along the path from v to w matches P.
- Let $\psi = [P]_{r}^{\rightarrow}(x)$ be a horizontal path formula. Then $t \models \psi[v]$, iff there is a labeling of the children of v with formulas, such that (1) each child w of v is labeled with a formula $\theta(r)$ such that $t \models \theta[w]$, and (2) the sequence of labels is matched by P.

Example 2. (a) The example formula $[(O_a(r))^* O_b(r)]_r^{\rightarrow}(x)$ holds at a vertex v, iff the rightmost child of v is labeled with a b and the remaining children are labelled with an a.

(b) For a more involved example consider first the formula $\varphi(x) = [(\text{true true})^*]_r^{\rightarrow}(x)$. Here, true stands for a formula like $r = r$ which always holds. Hence, φ holds at a vertex v if the number of children of v is even. Then the formula $\psi(x, y) = [(\varphi(r)(\neg\varphi(r)))^*]_{r,s}^{\downarrow}(x, y)$ holds for vertices v and w, if w is below v, the path from v to w is of even length and the vertices above w on this path have alternatingly even and odd degree. An example is given in Figure 1.

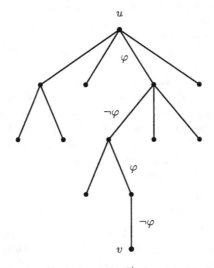

Fig. 1. The formula $\psi(x, y) = [(\varphi(r)(\neg\varphi(r)))^*]_{r,s}^{\downarrow}(x, y)$ holds for the pair (u, v) as the path can be labeled by the sequence $\varphi\neg\varphi\varphi\neg\varphi$.

The logic FOREG is the extension of FO by vertical and horizontal path formulas.[3] More formally,

(1) Every FO formula is an FOREG formula.
(2) If P is a regular expression over FOREG formulas then $\varphi = [P]_{r,s}^{\downarrow}(x, y)$ is an FOREG formula.
(3) If P is a regular expression over FOREG formulas then $[P]_r^{\rightarrow}(x)$ is an FOREG formula.

---

[3] It should be noted that this logic was called FOREG* in [24]. There, the analogue logic in which path expressions are only formed over atomic formulas was called FOREG.

Of course, FOREG can express all FO queries and every FOREG query can be expressed in MSO. It was shown in [24] that these inclusions are strict, i.e., there are MSO queries that can not be expressed in FOREG and there are FOREG queries that can not be expressed in FO.

5.2 Guarded Fragments

There is a significant difference between the data complexity and the combined complexity of the QUERY EVALUATION problem for MSO formulas. Whereas the data complexity is linear time the combined complexity is non-elementary (i.e., there is *no fixed tower of exponentials* that captures the complexity) [18]. In particular, the translation of an MSO formula into an equivalent tree automaton is non-elementary. Even for first-order formulas the combined complexity of QUERY EVALUATION is PSPACE-complete.

In this section we consider syntactic restrictions GFO, GFOREG and GMSO of the logics FO, FOREG and MSO, respectively, that have a much better combined complexity, linear time in the tree size and exponential time in the formula size. But they still can express all queries of the respective full logic. Clearly, this means, e.g., that there are MSO formulas that are much more succinct than their counterpart in the restricted fragment. But we believe that the restricted formulas allow the formulation of natural queries in a transparent way.

For the moment, we restrict attention to unary queries. In particular, all formulas considered have one free variable. Whether a vertex v is selected by a query might depend on properties of the subtree rooted at v, on properties of the path from the root to v and on properties of the remainder of the tree. It is reasonable to assume that in many cases properties of the subtree at v and of the path to v will be more important than properties of the rest of the tree.

In this spirit the fragments we define express properties of vertices in a tree in a modular way, i.e., by Boolean combinations of formulas $\varphi(x)$ that only speak about the subtree rooted at x and formulas $[P]_{r,s}^{\downarrow}(\text{root}, x)$ speaking about the path from the root to x and, if necessary, about the rest of the tree. In the latter kind of formulas, P is a regular expression over formulas $\psi(r, s)$ each of which is restricted to the subtree rooted at r. Therefore such *subtree-restricted quantification* is the basic ingredient of our logic.

We first define the syntax of GFOREG formulas. GFO formulas are then obtained as the restriction of GFOREG where only star-free regular expressions are allowed.

First of all, we make use of an additional partial order. For vertices u, v of a tree t it holds $u \preccurlyeq v$ if v is a vertex of the subtree of t that is rooted at u.

Besides the usual kind of variables $x, y, x', x_1, x_2, \ldots$ (to which we refer as *quantifier variables* in the following) we use a second kind of variables, called *expression variables*. They are only used in path expressions and are denoted by symbols like r and s.

The syntax of GFOREG formulas is defined as follows.

(i) Every atomic formula is a GFOREG formula.

(ii) If y is a quantifier variable and φ is a GFOREG formula with free quantifier variables from $\{x, y\}$ then $\exists y(x \preccurlyeq y \wedge \varphi)$ is a GFO formula.

(iii) If P is a regular expression over GFOREG formulas without free quantifier variables then $[P]_{r,s}^{\downarrow}(x, y)$ is a GFOREG formula.

(iv) If P is a regular expression over GFOREG formulas without free quantifier variables then $[P]_r^{\rightarrow}(x)$ is a GFOREG formula.

(v) Any Boolean combination of GFOREG formulas is a GFOREG formula.

GMSO formulas have the additional ability to quantify over sets but quantification is restricted to the current subtree. I.e., if φ is a GMSO formula, free$(\varphi) \subseteq \{x, X\}$ then $\exists X(x \preccurlyeq X \wedge \varphi)$ is a GMSO formula.

A closer inspection of the definitions above shows that the fragments are essentially two-variable fragments of their respective full logics (in the case of MSO with one additional free set variable). In so far and in the general way of thinking about trees they are quite similar in spirit to temporal logics on trees. Actually, a design goal for the development of a usable query language based on these fragments could be to get rid of the explicit use of variables in a similar way as variables are omitted in temporal logics.

As already mentioned, we get the following results about the expressive power of the defined fragments.

Theorem 7. *For each FO (FOREG, MSO) formula $\varphi(x)$ there is a Boolean combination of GFO (GFOREG, GMSO) formulas $\psi(x)$ and formulas $[P]_{r,s}^{\downarrow}(root, x)$ where P consists of GFO (GFOREG, GMSO) formulas which is equivalent to φ on Σ-trees.*

The evaluation complexity is as follows.

Theorem 8. *1. There is an algorithm which computes on input (t, φ), where t is a tree and φ is a GFOREG formula, the set of all vertices v of t such that $t \models \varphi(v)$ in time $O(|t|2^{|\varphi|})$.*

2. There is an algorithm which computes on input (t, φ), where t is a tree and φ is a GMSO formula, the set of all vertices v of t such that $t \models \varphi(v)$ in time $O(|t|2^{|\varphi|^2})$.

3. There is an algorithm which computes on input (t, φ), where t is a tree and φ is a GMSO formula in which no vertical path formulas occur within the scope of any set quantification, the set of all vertices v of t such that $t \models \varphi(v)$ in time $O(|t|2^{|\varphi|})$.

In principal, the logical approach makes it very easy to go from unary queries to queries of arbitrary arity by simply passing from formulas with one free variable to formulas with more free variables. This can not be done directly in the case of the fragments considered here as the restricted use of variables is essential for them. Nevertheless, it was shown in [29] that queries of arbitrary arity can be expressed in a very similar way with one additional kind of path expressions which talk about the children of a vertex v between those two children that contain vertices u and w in their subtree, respectively. The results about expressive power go through. In the first two statements on evaluation complexity of Theorem 8 one has to replace the factor $|t|$ by $|t|^k$ in the case of k-ary queries. Furthermore, on input (t, φ) one can compute in time $O(|t|^2 2^{|\varphi|})$ a data structure which allows to check in time $O(|\overline{v}|)$, whether $t \models \varphi(\overline{v})$ holds, for a given tuple $\overline{v}$.

6   Formalisms with Comparisons between Data Values

In this section we allow comparisons between attribute values as described in Section 2.3. In the above we showed that many formalisms for Σ-trees are equivalent to MSO. Obtaining such a correspondence for trees with data values is much harder if comparisons of values of different vertices are allowed. One of the reasons is that the data complexity of QUERY EVALUATION increases quite a bit. More precisely, it is shown in [26] that for every $i \in \mathbb{N}$, there are MSO formulas φ_i and ψ_i such that the QUERY EVALUATION on (Σ, A)-trees for φ_i and ψ_i is hard for Σ_i^P and Π_i^P, respectively. The prime reason is that attributed trees can encode graphs in a way such that even FO formulas can decode them. So when defining computational devices over attributed trees one should define powerful (complex) formalisms to capture MSO or settle for less. We choose for the latter.

This section is based on [26]. It should be noted that the results in that paper are formulated for automata on *strings* over data values. But the results we mention here easily carry over to tree automata. Therefore, our exposition moves back and forth between string automata and tree automata.

To keep our operational models simple and manageable, we only consider formalisms which are included in the regular languages when restricted to ordinary trees. It is useful to observe that for attributed trees it is no longer sufficient to equip automata with states alone. Indeed, automata should at least be able to check equality of attributes. There are two main ways to do this:

- store a finite set of positions and allow equality tests between the symbols on these positions;
- store a finite set of symbols only and allow equality tests with these symbols.

The first approach, however, leads to multi-head automata, immediately going beyond regular languages. Therefore, we instead equip tree-walking automata with a finite set of *pebbles* whose use is restricted by a stack discipline. That is, all pebbles have a number and pebble i can only be lifted when pebble $i + 1$ is not placed. The automaton can test equality by comparing the attributes of the pebbled symbols. We refer to these automata as pebble automata. In the second approach, we follow[4] Kaminski and Francez [15] and extend finite tree-walking automata with a finite number of *registers* that can store attribute values. When walking on a tree, an automaton compares an attribute value with values in the registers; based on this comparison it can decide to store the current symbol in some register. We refer to these automata as register automata. Both models can express unary queries by means of a selection function.

There is a great mismatch in expressive power between register and pebble automata. Indeed, register automata do not fit nicely into the framework of logically defined queries or languages.

---

[4] Their model is based on strings but easily translates to trees.

Theorem 9. $-$ *There are Boolean FO queries not expressible by any register automaton.*
$-$ *There are Boolean queries expressible by register automata that are not definable in MSO.*

On the one hand, register automata cannot even define all properties in FO. The basic idea is as follows. Consider strings of the form $u\#v$ where $u, v \in \Sigma^*$. Here, $\#$ functions as a delimiter. The only way an automaton can compare the data values of the left side with the ones from the right side is by storing values in the registers. Based on communication complexity theoretic arguments one can show that this does not suffice to express all of FO. On the other hand, register automata are quite strong as they can express properties not even expressible in MSO. To see this, consider strings of the same form $u\#v$ as before. Define N_u and N_v as the set of symbols occurring in attributes in u and v, respectively. It can be shown that there is a register automaton that accepts $u\#v$ iff $|N_u| = |N_v|$ while there is no such MSO sentence.

The formal model of XSLT of [2] is based on tree-walking transducers with registers. By applying Theorem 9 we get that XSLT programs without nested calls cannot define all of FO.

Pebble automata behave much better, their expressiveness lies in between FO and MSO. Indeed, pebbles provide a mechanism to instantiate variables occurring in formulas. A brute force approach to check, for instance, the formula $\exists x \exists y \delta(x, y)$, where $\delta(x, y) \equiv E(x, y) \wedge a(x) = b(y)$, is the following. Assign a pebble to x and one to y, say i_x and i_y. Put them subsequently on all possible combinations of vertices and test whether $\delta(i_x, i_y)$ holds. The latter only involves local checks. The automaton accepts when it finds such two vertices. Further, the enforced stack discipline makes sure the model behaves in a regular way. Indeed, it can be shown that pebble automata can be defined in MSO.

Theorem 10. $-$ *All unary FO queries are expressible by pebble automata.*
$-$ *Every query defined by a pebble automaton is also definable in MSO.*

The evaluation complexity is as follows.

Theorem 11. *The data complexity of QUERY EVALUATION for register and pebble automata is in* PTIME.

7   Discussion

We surveyed work investigating well-known formalisms from formal languages and logic as a pattern languages for tree-structured data. The main focus was on expressiveness, evaluation complexity, and decision problems relevant to optimization. Although attribute grammars as well as query automata are quite expressive they are quite complicated formalisms and do not seem to be the basis for an easy-to-use pattern language. Tree-walking automata are much more intuitive. Therefore, we need to understand better their expressiveness. However, the problem whether they capture MSO hac been open for a while now and therefore appears to be difficult. The restricted logics we considered might be useful in the design of a pattern language but this requires further work. We merely touched upon the issue of comparison of data values. Undoubtedly it deserves a lot more investigation.

References

1. H. Andréka, J. van Benthem, and I. Németi. Modal languages and bounded fragments of predicate logic. *Journal of Philosophical Logic*, 27:217–274, 1998.
2. G. J. Bex, S. Maneth, and F. Neven. A formal model for an expressive fragment of XSLT. *1st International Conference on Computational Logic*, pages 1137–1151, *Lecture Notes in Artificial Intelligence*, volume 1861. Springer, 2000.
3. R. Bloem and J. Engelfriet. A comparison of tree transductions defined by monadic second order logic and by attribute grammars. *Journal of Computer and System Sciences*, 61(1):1–50, 2000.
4. A. Brüggemann-Klein, M. Murata, and D. Wood. Regular Tree and Regular Hedge Languages over Unranked Alphabets: Version 1, April 3, 2001. Technical report, HKUST-TCSC-2001-05, Hong Kong University of Science & Technology, 2001.
5. A. Brüggemann-Klein and D. Wood. Caterpillars: A Context Specification Technique. *Markup Languages*, 2(1):81–106, 2000.
6. D. Chamberlin, D. Florescu, J. Robie, J. Siméon, and M. Stefanescu. XQuery: a query language for XML. Latest version: http://www.w3.org/TR/xquery/.
7. J. Clark. XSL Transformations (XSLT) Version 1.0. Latest version: http://www.w3.org/TR/xslt.
8. M. Consens and T. Milo. Algebras for querying text regions: Expressive power and optimization. *Journal of Computer and System Sciences*, 3:272–288, 1998.
9. P. Deransart, M. Jourdan, and B. Lorho. *Attribute Grammars: Definition, Systems and Bibliography*, volume 323 of *Lecture Notes in Computer Science*. Springer, 1988.
10. H.-D. Ebbinghaus and J. Flum. *Finite Model Theory*. Springer, 1995.
11. J. Engelfriet and H. J. Hoogeboom. Tree-walking pebble automata. In J. Karhumäki, H. Maurer, G. Paun, and G.Rozenberg, editors, *Jewels are forever, contributions to Theoretical Computer Science in honor of Arto Salomaa*, pages 72–83. Springer-Verlag, 1999.
12. J. Engelfriet, H.J. Hoogeboom, and J.-P. van Best. Trips on trees. *Acta Cybernetica*, 14:51–64, 1999.
13. M. Fernandez, J. Siméon, and P. Wadler, editors. *XML Query languages: Experiences and Exemplars*, 1999.
    http://www-db.research.bell-labs.com/user/simeon/xquery.html.
14. F. Gécseg and M. Steinby. Tree languages. In G. Rozenberg and A. Salomaa, editors, *Handbook of Formal Languages*, volume 3, chapter 1. Springer, 1997.
15. M. Kaminski and N. Francez. Finite-memory automata. *Theoretical Computer Science*, 134(2):329–363, 1994.
16. D.E. Knuth. Semantics of context-free languages. *Mathematical Systems Theory*, 2(2):127–145, 1968. See also *Mathematical Systems Theory*, 5(2):95–96, 1971.
17. S. Maneth and F. Neven. Structured Document Transformations Based on XSL. In R. Conner, A. Mendelzon, editors, *Research Issues in Structured and Semistructured Database Programming*, volume 1949 of *Lecture Notes in Computer Science*, pages 80–98. Springer 2000.

18. A. R. Meyer. Weak monadic second-order theory of successor is not elementary recursive. In R. Parikh, editors, *Logic Colloquim*, volume 453 of Lecture Notes in Mathematics, pages 132–154. Springer, 1975.

19. T. Milo, D. Suciu, and V. Vianu. Type checking for XML transformers. In *Proceedings of the Nineteenth ACM Symposium on Principles of Database Systems*, pages 11–22. ACM Press, 2000.

20. E. Moriya. On two-way tree automata. *Information Processing Letters*, 50:117–121, 1994.

21. M. Murata. Extended Path Expressions for XML. To appear in *Proceedings of the Twentieth ACM Symposium on Principles of Database Systems*. ACM Press, 2001.

22. F. Neven. Extensions of attribute grammars for structured document queries. In R. Conner, A. Mendelzon, editor, *Research Issues in Structured and Semistructured Database Programming*, volume 1949 of *Lecture Notes in Computer Science*, pages 99–116. Springer 2000.

23. F. Neven and T. Schwentick. Query automata. To appear in *Theoretical Computer Science*. Extended abstrast in *Proceedings of the Eighteenth ACM Symposium on Principles of Database Systems*, pages 205–214. ACM Press, 1999.

24. F. Neven and T. Schwentick. Expressive and efficient pattern languages for tree-structured data. In *Proc. 19th Symposium on Principles of Database Systems (PODS 2000), Dallas*, pages 145–156, 2000.

25. F. Neven and T. Schwentick. On the power of tree-walking automata. To appear in *Information and Computation*. Extended abstract in *27th International Colloquium on Automata, Languages and Programming*, pages 547–560, *Lecture Notes in Computer Science*, volume 1853. Springer, 2000.

26. F. Neven, T. Schwentick, and V. Vianu. Towards regular languages over infinte alphabets. Submitted.

27. F. Neven and J. Van den Bussche. Expressiveness of structured document query languages based on attribute grammars. To appear in the *Journal of the ACM*. Extended abstract appeared in *Proceedings of the Seventeenth ACM Symposium on Principles of Database Systems*, pages 11–17. ACM Press, 1998.

28. A. Neumann and H. Seidl. Locating matches of tree patterns in forests. In V. Arvind and R. Ramanujam, editors, *Foundations of Software Technology and Theoretical Computer Science*, pages 134–145, *Lecture Notes in Computer Science*, volume 1530. Springer, 1998.

29. T. Schwentick. On Diving in Trees. In M. Nielsen and B. Rovan, editors, *Mathematical Foundations of Computer Science (MFCS 2000)*, pages 660–669, *Lecture Notes in Computer Science*, volume 1893. Springer, 2000

30. W. Thomas. Languages, automata, and logic. In G. Rozenberg and A. Salomaa, editors, *Handbook of Formal Languages*, volume 3, chapter 7. Springer, 1997.

31. World Wide Web Consortium. XML Query Requirements. Latest version: http://www.w3.org/TR/xmlquery-reg.

Cardinality Constraints in Disjunctive Deductive Databases

Dietmar Seipel[1] and Ulrich Geske[2]

[1] University of Würzburg, Department of Computer Science
Am Hubland, D – 97074 Würzburg, Germany
seipel@informatik.uni-wuerzburg.de
[2] Fraunhofer FIRST Berlin
Kekuléstraße 7, D – 12489 Berlin, Germany
geske@first.fhg.de

Abstract. We investigate *cardinality constraints* of the form $M \hookrightarrow_\Theta K$, where M is a set and Θ is one of the comparison operators "=", "≤", or "≥"; such a constraint states that "exactly", "at most", or "at least", respectively, K elements out of the set M have to be chosen.

We show how a set $\mathcal{C}$ of constraints can be represented by means of a positive–disjunctive deductive database $\mathcal{P}_\mathcal{C}$, such that the models of $\mathcal{P}_\mathcal{C}$ correspond to the solutions of $\mathcal{C}$. This allows for embedding cardinality constraints into applications dealing with *incomplete knowledge*.

We also present a *sound calculus* represented by a definite logic program $\mathcal{P}_{cc}$, which allows for directly reasoning with sets of exactly–cardinality constraints (i.e., where Θ is "="). Reasoning with $\mathcal{P}_{cc}$ is very efficient, and it can be used for performance reasons before $\mathcal{P}_\mathcal{C}$ is evaluated. For obtaining *completeness*, however, $\mathcal{P}_\mathcal{C}$ is necessary, since we show the theoretical result that a sound and complete calculus for exactly–cardinality constraints does not exist.

Keywords: disjunctive logic programming, constraint logic programming, cardinality constraints, hyperresolution

1 Introduction

Practical applications of deductive databases [1] often require the use of extended features like *disjunctive information, aggregation operators, default negation, computation of cheapest models*; several extensions of deductive databases have been proposed for these reasons. Applications dealing with *cardinality constraints* may for instance arise in assignment problems, such as course planning, or in games. E.g. there may be rules like "every student has to take 3 of the courses in a given set M". Also cardinality constraints for *binary relationship types* in *entity–relationship models* can be expressed easily using our concept of cardinality constraints.

Constraint logic programming (CLP) [6] offers very efficient methods for solving problems involving cardinality constraints. But if cardinality constraints have

L. Bertossi et al. (Eds.): Semantics in Databases, LNCS 2582, pp. 179–199, 2003.

to be mixed with other formalisms for representing incomplete knowldege, then it becomes necessary to embed them into a richer framework, e.g., disjunctive logic programming. It might for instance be necessary to represent generic incomplete information such as "every student has to finish two small labs or one big lab, except if he enrolled himself before a certain deadline".

Various *sound and complete calculi* are known from literature in relational databases, e.g. the calculus given by the well–known *Armstrong axioms* for deriving implied functional dependencies. In deductive databases, reasoning with the well–known consequence operator $\mathcal{T}_{\mathcal{P}}$ of van Emden and Kowalski (cf. [4]) is sound and complete for deriving all atomary consequences of a definite logic program $\mathcal{P}$; it is not necessary to derive other formulas, such as disjunctions of literals, as intermediate results. Similarly, we would like to *limit the search space* in a special calculus for cardinality constraints. We will show that cardinality constraints can be derived based on the disjunctive consequence operator $\mathcal{T}_{\mathcal{P}}^s$ of Minker and Rajasekar (cf. [7]), which generalizes $\mathcal{T}_{\mathcal{P}}$. But we will also show that, unfortunately, there exists no sound and complete calculus that is dealing with cardinality constraints as intermediate results only.

The rest of the paper is organized as follows. In Section 2 we define cardinality constraints and we present the theoretical background of disjunctive deductive databases. In Section 3 we are dealing with calculi for reasoning with cardinality constraints. Some experimental results obtained for reasoning with cardinality constraints based on disjunctive deductive databases and on CLP, respectively, are reported in Section 4. Finally, Section 5 presents some conclusions.

2   Basic Definitions and Notations

In this section we will define the concepts of disjunctive deductive databases and of cardinality constraints, and we will show how cardinality constraints can be expressed by disjunctive deductive databases.

2.1   Cardinality Constraints

In the following we will define the concept of cardinality constraints. We expand on the results of an earlier paper [12].

Syntax. Given a universe $\mathcal{At}$ of atoms, e.g. the set of atoms of a first order language $\mathcal{L}$, a subset $M \subseteq \mathcal{At}$, and natural numbers $K, K_1, K_2 \in \langle\, 0, |M| \,\rangle$.[1]

- A *cardinality constraint* is given by $M \hookrightarrow_\Theta K$, where $\Theta \in \{=, \leq, \geq\}$ is a comparison relation on $\mathbb{N}_0$.
- $M \hookrightarrow_= K$ is called an *exactly–cardinality constraint* (i.e. $\Theta = $ "$=$"), and it is called *trivial*, if $M = \emptyset$ and $K = 0$.

---

[1] By $\mathbb{N}_+$ we denote the set $\{\,1, 2, 3, \dots\,\}$ of positive natural numbers, whereas $\mathbb{N}_0$ denotes the set $\{\,0, 1, 2, \dots\,\}$ of all natural numbers. $\langle\, n, m \,\rangle$ denotes the interval $\{\,n, n+1, \dots, m\,\}$ of natural numbers.

- An *interval–cardinality constraint* is given by $M \hookrightarrow_\in \langle K_1, K_2 \rangle$, where $\langle K_1, K_2 \rangle$ denotes an interval of natural numbers.

E.g., $\{a, b, c\} \hookrightarrow_= 2$ is an exactly–cardinality constraint.

Semantics. An interpretation is a set $I \subseteq H_B = gnd\,(At)$ of ground atoms. I satisfies a cardinality constraint $c = M \hookrightarrow_\Theta K$, for short $I \models c$, if $|M\sigma \cap I| \,\Theta\, K$, for all ground substitutions σ of M. I.e., c imposes a condition on the cardinality of the selected subset $M\sigma \cap I$ of $M\sigma$. We further define $M \hookrightarrow_\in \langle K_1, K_2 \rangle$ by

$$I \models M \hookrightarrow_\in \langle K_1, K_2 \rangle \text{ if } I \models M \hookrightarrow_\geq K_1 \text{ and } I \models M \hookrightarrow_\leq K_2.$$

Then it holds for instance $M \hookrightarrow_= K \iff M \hookrightarrow_\in \langle K, K \rangle$. E.g., $I = \{a, c\} \models \{a, b, c\} \hookrightarrow_= 2$, since $|\{a, b, c\} \cap I| = 2$. For $H_B = \{p(a), p(b), q(a), q(b)\}$ and $c = \{p(X), q(X)\} \hookrightarrow_= 1$ we get $I = \{p(a), q(b)\} \models c$, since $|\{p(a), q(a)\} \cap I| = |\{p(b), q(b)\} \cap I| = 1$, but $I = \{p(a), q(a)\} \not\models c$, since $|\{p(a), q(a)\} \cap I| = 2$.

Given a set $\mathcal{C}$ of cardinality constraints. I is a *model* of $\mathcal{C}$, for short $I \models \mathcal{C}$, if I satisfies all constraints $c \in M$. $\mathcal{C}$ implies a single cardinality constraint c, $\mathcal{C} \models c$, if for all models I of $\mathcal{C}$ it holds $I \models c$.

A set $\mathcal{C}$ of exactly–cardinality constraints *covers* $c = M \hookrightarrow_= K$, for short $\mathcal{C} \triangleright c$, if there exist constraints $M_i \hookrightarrow_= K_i \in \mathcal{C}$, $i \in \langle 1, k \rangle$, such that the sets M_i are pairwise disjoint, $M = \cup_{i=1}^{k} M_i$, and $K = \sum_{i=1}^{k} K_i$. It can be shown that $\mathcal{C} \triangleright c$ always implies $\mathcal{C} \models c$, and that $\mathcal{C} \triangleright c$ if c is trivial.

Computational Complexity. The computational complexity of the *satisfiability problem* for sets $\mathcal{C}$ of exactly–cardinality constraints over ground atoms is in $\mathcal{NP}$ (since for a given interpretation I of $\mathcal{C}$, the problem of testing whether $I \models \mathcal{C}$ is polynomial). As a consequence of this and a recent result of Kaye [3], the satisfiability problem is also $\mathcal{NP}$–complete.

The Mine Sweeper Example. The well–known game Mine Sweeper takes place on an $N \times N$ board, where mines are located on some of the fields. The goal of the player is to investigate the fields of the board, such that he never investigates a field with a mine. Each time a field is investigated, the system informs the player about the number of mines on the neighbour fields. Of course, the intial investigations cannot rely on any knowledge, but after a few investigations in many cases it becomes possible to infer that some neighbour fields do not contain a mine, or that other fields must contain a mine.

Consider the following 6×6–board, which was obtained by investigating the upper left corner with the coordinates $(1, 1)$. Since the neighbours of $(1, 1)$ do not contain any mines, they have been investigated as well, etc., until finally a border line with non–zero fields was reached, cf. Figure 1. This board gives rise to the set $\mathcal{C} = \{c_1, \ldots, c_6\}$ of cardinality constraints, which are derived by scanning the 6 fields that are marked by non–zero numbers from left to right (and looking at their neighbours). E.g., since the field $(3, 3)$ is marked by "2",

Initial board:

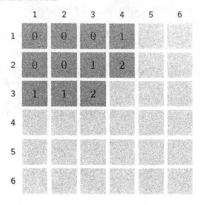

Fig. 1. Mine Sweeper Board after investigating the upper left corner $(1, 1)$

two of its non–investigated neighbours, namely $(2, 4)$, $(3, 4)$, $(4, 4)$, and $(4, 3)$, must contain a mine, which is modelled by c_3:

$$(1, 3) \mapsto c_1 = \{(1, 4), (2, 4)\} \hookrightarrow_= 1,$$
$$(2, 3) \mapsto c_2 = \{(1, 4), (2, 4), (3, 4)\} \hookrightarrow_= 1,$$
$$(3, 3) \mapsto c_3 = \{(2, 4), (3, 4), (4, 4), (4, 3)\} \hookrightarrow_= 2,$$
$$(3, 2) \mapsto c_4 = \{(4, 3)\} \hookrightarrow_= 1,$$
$$(4, 2) \mapsto c_5 = \{(4, 3), (5, 3), (5, 2), (5, 1)\} \hookrightarrow_= 2,$$
$$(4, 1) \mapsto c_6 = \{(5, 2), (5, 1)\} \hookrightarrow_= 1.$$

The Course Planning Example. A course plan for a student is given by a set $\mathcal{E}$ of atoms $e(Student, Course, Semester)$, which state that a student is enrolled in a course in a given semester. There could be constraints such as "every student S must be enrolled in at least K_{min} and at most K_{max} courses per semster":

$$C(S, T) = \{\, e(S, c_1, T), \ldots, e(S, c_n, T)\,\} \hookrightarrow_\in \langle\, K_{min}, K_{max}\,\rangle,$$

where $\chi = \{c_1, \ldots, c_n\}$ is the set of all possible courses. Furthermore, there may be groups $\psi = \{c_{i_1}, \ldots, c_{i_m}\} \subseteq \chi$ of courses, such that "every student S must be enrolled in at least K_{min} and at most K_{max} of the courses in ψ at some time":

$$C(S) = \{\, e(S, c_{i_1}, \_), \ldots, e(S, c_{i_m}, \_)\,\} \hookrightarrow_\in \langle\, K_{min}, K_{max}\,\rangle.$$

There may be a rule that "every student S can repeat a course at most once":

$$C(S, Co) = \{\, e(S, Co, t_1), \ldots, e(S, Co, t_k)\,\} \hookrightarrow_\leq 2,$$

where $\{t_1, \ldots, t_k\}$ is the set of all semesters. $C(S, Co)$ is equivalent to a denial rule $\leftarrow e(S, Co, T_1) \wedge e(S, Co, T_2) \wedge T_1 \neq T_2$ in disjunctive deductive databases.

2.2 Disjunctive Deductive Databases

Syntax. Given a first order language $\mathcal{L}$, a *disjunctive deductive database* $\mathcal{P}$ (cf. [5]) consists of logical inference rules of the form

$$r = A_1 \vee \ldots \vee A_k \leftarrow B_1 \wedge \ldots \wedge B_m \wedge not\, C_1 \wedge \ldots \wedge not\, C_n, \qquad (1)$$

where A_i, B_i, and C_i are atoms in the language $\mathcal{L}$, $k, m, n \in \mathbb{N}_0$, and *not* is the negation–by–default operator. A rule is called a *fact* if $m = n = 0$. The set of all *ground instances* of the rules and facts in $\mathcal{P}$ is denoted by $gnd\,(\mathcal{P})$. A rule (or database) is called *positive–disjunctive* if it does not contain default negation (i.e. $n = 0$). It is called *normal* if $k = 1$, and *definite* if $k = 1$ and $n = 0$. It is called a *denial rule* if $k = 0$. A rule r of the form (1) is denoted for short as $r = \alpha \leftarrow \beta \wedge not \cdot \gamma$, where $\alpha = A_1 \vee \ldots \vee A_k$, $\beta = B_1 \wedge \ldots \wedge B_m$, and $\gamma = C_1 \vee \ldots \vee C_n.$[2] α is called the *head*, β is called the *positive body*, and $not \cdot \gamma$ is called the *negative body* of r.

For positive–disjunctive deductive databases $\mathcal{P}$ the *dual version* (cf. [13]) can been defined: for a positive–disjunctive rule $r = A_1 \vee \ldots \vee A_k \leftarrow B_1 \wedge \ldots \wedge B_m$ the dual version

$$r^d = B_1 \vee \ldots \vee B_m \leftarrow A_1 \wedge \ldots \wedge A_k$$

is obtained by exchanging the head and the body atoms of r. The dual version of $\mathcal{P}$ is given by $\mathcal{P}^d = \{\, r^d \,|\, r \in \mathcal{P} \,\}$.

The Herbrand base $HB_\mathcal{P}$ of a disjunctive deductive database $\mathcal{P}$ contains all ground atoms over the language of $\mathcal{P}$. The *disjunctive Herbrand base* $DHB_\mathcal{P}$ of $\mathcal{P}$ consists of all ground disjunctions $A_1 \vee \ldots \vee A_k$, $k \in \mathbb{N}_0$, of atoms A_i, $i \in \langle\, 1, k \,\rangle$. The *negative Herbrand base* $NHB_\mathcal{P}$ of $\mathcal{P}$ consists of all ground disjunctions $\neg A_1 \vee \ldots \vee \neg A_k$, $k \in \mathbb{N}_0$, of negated atoms $\neg A_i$, $i \in \langle\, 1, k \,\rangle$. A negative disjunction can equivalently be represented by a denial rule $\leftarrow A_1 \wedge \ldots \wedge A_k$. A subset $S \subseteq DHB_\mathcal{P}$ ($S \subseteq NHB_\mathcal{P}$) is called a disjunctive (negative) Herbrand state.

Semantics. An *Herbrand interpretation* of $\mathcal{P}$ is given by a subset $I \subseteq HB_\mathcal{P}$. It defines a mapping $I\colon HB_\mathcal{P} \to \{\mathsf{true}, \mathsf{false}\}$ that assigns a truth value to each atom in $HB_\mathcal{P}$: $I(A) = \mathsf{true} \Leftrightarrow A \in I.$[3] I *satisfies* a ground rule $r = \alpha \leftarrow \beta \wedge not \cdot \gamma$, for short $I \models r$, if

$$I(\beta) \wedge \neg I(\gamma) = \mathsf{true} \;\Rightarrow\; I(\alpha) = \mathsf{true}.$$

I is an *Herbrand model* of $\mathcal{P}$ if I satisfies all ground instances $r \in gnd\,(\mathcal{P})$ of the elements of $\mathcal{P}$. I is a *minimal model* of $\mathcal{P}$ if I is an Herbrand model of $\mathcal{P}$ and there is no other Herbrand model J of $\mathcal{P}$ such that $J \subsetneq I$. The set of all *minimal models* of $\mathcal{P}$ is denoted by $\mathcal{MM}(\mathcal{P})$.

---

[2] Note that γ is a disjunction, and, according to De Morgan's law, $not \cdot \gamma$ is taken to be a conjunction.

[3] For $A_i \in HB_\mathcal{P}$, $i \in \langle\, 1, k \,\rangle$, and a connective $\otimes \in \{\vee, \wedge\}$ we define $I(\,A_1 \otimes \ldots \otimes A_k\,) = I(A_1) \otimes \ldots \otimes I(A_k)$. For $k = 0$, the empty disjunction (i.e. $\otimes = \vee$) evaluates to false, whereas the empty conjunction (i.e. $\otimes = \wedge$) evaluates to true.

For positive–disjunctive deductive databases $\mathcal{P}$, the semantics is given by $\mathcal{MM}(\mathcal{P})$, and a formula ϕ is a *consequence* of $\mathcal{P}$ under the minimal model semantics if ϕ holds in all minimal models of $\mathcal{P}$.

The Consequence Operator $\mathcal{T}_\mathcal{P}^s$. The set of all positive ground disjunctions that are logical consequences of $\mathcal{P}$ can be derived as the least fixpoint $lfp\,(\mathcal{T}_\mathcal{P}^s)$ of the *disjunctive consequence operator*

$$\mathcal{T}_\mathcal{P}^s : 2^{D_{HB_\mathcal{P}}} \to 2^{D_{HB_\mathcal{P}}}$$

of Minker and Rajasekar (cf. [7]):

$$\mathcal{T}_\mathcal{P}^s(S) = \{\, \alpha \vee \alpha_1 \vee \ldots \vee \alpha_m \in D_{HB_\mathcal{P}} \mid$$
$$\exists\, \alpha \leftarrow B_1 \wedge \ldots \wedge B_m \in gnd\,(\mathcal{P}) :\ \forall i \in \langle\, 1, m \,\rangle :\ B_i \vee \alpha_i \in S \,\}.$$

This operator $\mathcal{T}_\mathcal{P}^s$ is monotonic and continuous (cf. [4]), and thus it reaches its least fixpoint $lfp\,(\mathcal{T}_\mathcal{P}^s)$ in at most ω steps. $lfp\,(\mathcal{T}_\mathcal{P}^s)$ consists exactly of all positive ground disjunctions that are consequences of $\mathcal{P}$ under the minimal model semantics.

Example. For the positive–disjunctive database $\mathcal{P} = \{\, a \leftarrow a_1 \wedge a_2 \,\}$ and the disjunctive Herbrand state $S = \{\, a_1 \vee b_1, a_2 \vee b_2 \,\}$ we can derive $\mathcal{T}_\mathcal{P}^s(S) = \{\, a \vee b_1 \vee b_2 \,\}$. The set of minimal models of $\mathcal{P}$ is $\mathcal{MM}(\mathcal{P}) = \{\, \{a_1, a_2, a\}, \{a_1, b_2\}, \{a_2, b_1\} \,\}$.

2.3   Cardinality Constraints in Disjunctive Logic Programming

Due to the declarative, model–theoretic semantics of cardinality constraints, which was given in Section 2.1, it is possible to allow cardinality constraints in disjunctive rules of the form (1). E.g.,

$$p(X, Y) \vee q(Y, Z) \leftarrow (\{\, p(X, Y), p'(Y, Z) \,\} \hookrightarrow_\geq 1) \wedge q(Z, Y),$$

would be a disjunctive rule with cardinality constraints. This approach has also been taken independently by Niemelä and Simmons [8] for the subclass of normal logic programs[4], who have extended their system **Smodels** to cope with cardinality constraints.

The main goal in this section, however, is to show how cardinality constraints can be expressed in disjunctive logic programming (DLP) by using the following *transformations*:

$$\mathcal{C}(M \hookrightarrow_\leq K) = \{\, \leftarrow A_1 \wedge \ldots \wedge A_{K+1} \mid \{A_1, \ldots, A_{K+1}\} \subseteq M \,\},$$
$$\mathcal{C}(M \hookrightarrow_\geq K) = \{\, A_1 \vee \ldots \vee A_{|M|-K+1} \mid \{A_1, \ldots, A_{|M|-K+1}\} \subseteq M \,\},$$
$$\mathcal{C}(M \hookrightarrow_= K) = \mathcal{C}(M \hookrightarrow_\leq K) \cup \mathcal{C}(M \hookrightarrow_\geq K).$$

---

[4] i.e., all rules must have $k = 1$ head atoms

E.g., for $K = |M|$ it holds $\mathcal{C}(M \hookrightarrow_\le K) = \emptyset$, and for $K = 0$ it holds $\mathcal{C}(M \hookrightarrow_\ge K) = \emptyset$, since these cardinality constraints do not pose any restrictions on a model.

For a singleton set $M = \{A_1\}$ we get

$$\mathcal{C}(\{A_1\} \hookrightarrow_= 0) = \{\leftarrow A_1\},$$
$$\mathcal{C}(\{A_1\} \hookrightarrow_= 1) = \{A_1\},$$

and for a set $M = \{A_1, A_2\}$ we get

$$\mathcal{C}(M \hookrightarrow_= K) = \begin{cases} \{\leftarrow A_1, \leftarrow A_2\}, & \text{if } K = 0, \\ \{A_1 \vee A_2, \leftarrow A_1 \wedge A_2\}, & \text{if } K = 1, \\ \{A_1, A_2\}, & \text{if } K = 2. \end{cases}$$

In general, the sizes of the disjunctive deductive databases $\mathcal{C}(M \hookrightarrow_\Theta K)$ are:

$$|\mathcal{C}(M \hookrightarrow_\le K)| = \binom{|M|}{K+1},$$
$$|\mathcal{C}(M \hookrightarrow_\ge K)| = \binom{|M|}{|M|-K+1} = \binom{|M|}{K-1},$$
$$|\mathcal{C}(M \hookrightarrow_= K)| = \binom{|M|}{K+1} + \binom{|M|}{K-1}.$$

The databases $\mathcal{C}(M \hookrightarrow_\Theta K)$ are always positive–disjunctive, they consist of ground formulas only, and it holds:

$$\mathcal{C}(M \hookrightarrow_\le K) \subseteq N_{HB_M},$$
$$\mathcal{C}(M \hookrightarrow_\ge K) \subseteq D_{HB_M}.$$

The universe of $M \hookrightarrow_\Theta K$ coincides with the Herbrand universe of $\mathcal{C}(M \hookrightarrow_\Theta K)$, i.e. $HB = HB_{\mathcal{C}(M \hookrightarrow_\le K)}$, and also their models coincide:

Lemma 2.1. *For all $I \subseteq HB$ it holds:*

$$I \models M \hookrightarrow_\Theta K \Leftrightarrow I \models \mathcal{C}(M \hookrightarrow_\Theta K).$$

Proof. 1. $\Theta = $ "$\le$": Obviously, the set $\mathcal{C}(M \hookrightarrow_\le K)$ of denial rules is not satisfied by an interpretation I iff there exists some subset $\{A_1, \ldots, A_{K+1}\} \subseteq M$, such that the denial rule $\leftarrow A_1 \wedge \ldots \wedge A_{K+1}$ is not satisfied in I, i.e. such that $\{A_1, \ldots, A_{K+1}\} \subseteq I$. This precisely holds iff $|I \cap M| \ge K + 1$. Conversely, $I \models \mathcal{C}(M \hookrightarrow_\le K)$, iff $|I \cap M| \le K$, i.e. iff $I \models M \hookrightarrow_\le K$.

2. $\Theta = $ "$\ge$": Obviously, the set $\mathcal{C}(M \hookrightarrow_\ge K)$ of denial rules is not satisfied by an interpretation I iff there exists some subset $\{A_1, \ldots, A_{|M|-K+1}\} \subseteq M$, such that the fact $A_1 \vee \ldots \vee A_{|M|-K+1}$ is not satisfied in I, i.e., such that $\{A_1, \ldots, A_{|M|-K+1}\} \cap I = \emptyset$. This means that at least $|M|-K+1$ elements of M are not in I, i.e. $|M \setminus I| \ge |M|-K+1$, which means that $|I \cap M| \le K-1$. Conversely, $I \models \mathcal{C}(M \hookrightarrow_\le K)$, iff $|I \cap M| \ge K$, i.e. iff $I \models M \hookrightarrow_\ge K$.

3. $\Theta =$ "=": As a consequence of the previous two items, $I \models \mathcal{C}(M \hookrightarrow_= K)$, iff $I \models M \hookrightarrow_= K$. □

Another possible encoding of $M \hookrightarrow_= K$ would be to represent the alternative models $I = \{A_1, \ldots, A_K\} \subseteq H_B$ of $M \hookrightarrow_= K$ as conjunctive formulas

$$\phi_I = A_1 \wedge \ldots \wedge A_K \wedge \neg A_{K+1} \wedge \ldots \wedge \neg A_{|M|},$$

where $H_B \setminus I = \{A_{K+1}, \ldots, A_{|M|}\}$. The conjunctive normal form of the disjunction of the ϕ_I yields a positive–disjunctive deductive database that is equivalent to $\mathcal{C}(M \hookrightarrow_= K)$. But it can be shown that $\mathcal{C}(M \hookrightarrow_= K)$ is the most compact representation of $M \hookrightarrow_\Theta K$ in DLP.

Given a set $\mathcal{C}$ of cardinality constraints, we denote the positive–disjunctive deductive database representing $\mathcal{C}$ by $\mathcal{P}_{\mathcal{C}} = \cup_{c \in \mathcal{C}} \mathcal{C}(c)$. It can be shown that $\mathcal{P}_{\mathcal{C}}$ always is non–recursive. Moreover, as a consequence of Lemma 2.1, the least fixpoints of the disjunctive consequence operators $\mathcal{T}_{\mathcal{P}}^s$ for $\mathcal{P} = \mathcal{P}_{\mathcal{C}}$ and its dual version $\mathcal{P}^d = \mathcal{P}_{\mathcal{C}}^d$ can tell whether a singleton cardinality constraint $\{A\} \hookrightarrow_= K$, $K \in \{0, 1\}$, is implied by $\mathcal{C}$.

Lemma 2.2. *For a set $\mathcal{C}$ of cardinality constraints and an atom $A \in H_B$ it holds:*

$$(\mathcal{C} \models \{A\} \hookrightarrow_= 1) \Leftrightarrow A \in \mathit{lfp}\,(\mathcal{T}_{\mathcal{P}_{\mathcal{C}}}^s),$$
$$(\mathcal{C} \models \{A\} \hookrightarrow_= 0) \Leftrightarrow A \in \mathit{lfp}\,(\mathcal{T}_{\mathcal{P}_{\mathcal{C}}^d}^s).$$

In Section 4 we will especially be interested in singleton cardinality constraints. For arbitrary cardinality constraints, testing whether $\mathcal{C} \models M \hookrightarrow_= K$ could be done by checking whether the Herbrand state

$$\mathit{lfp}\,(\mathcal{T}_{\mathcal{P}_{\mathcal{C}}}^s) \cup (\mathit{lfp}\,(\mathcal{T}_{\mathcal{P}_{\mathcal{C}}^d}^s))^d$$

implies $M \hookrightarrow_= K$, which results in a check for subsumption. But computing the set of all implied cardinality constraints is very complex due to the size of the search space.

2.4   Cardinality Constraints in Constraint Logic Programming

Most CLP systems contain a special construct, a so–called *global constraint*, for facilitating the encoding of cardinality relations. For instance in the CHIP system the global constraint among/4 may be used for expressing cardinality constraints:

$$\mathsf{among}([K_1, K_2, L], [X_1, \ldots, X_n], [C_1, \ldots, C_n], [V_1, \ldots, V_m]).$$

It expresses conditions on values of sets and subsets of specified domain variables $X_1, \ldots, X_n$. The first argument states that at least K_1 and at most K_2 values from the list $[V_1, \ldots, V_m]$ of integers – which must be given in increasing order – occur in any list $[X_i + C_i, \ldots, X_{i+L-1} + C_{i+L-1}]$ with L consecutive indices. Thus, an interval–cardinality constraint $\{A_1, \ldots, A_L\} \hookrightarrow_= \langle K_1, K_2 \rangle$ with ground atoms A_i can be encoded by the global constraint

$$\mathsf{among}([K_1, K_2, L], [X_{A_1}, \ldots, X_{A_L}], [0, \ldots, 0], [1]),$$

where X_{A_i} is a $\{0, 1\}$–variable corresponding to the atom A_i.

In SICSTUS–PROLOG cardinality constraints can be expressed using *reified constraints*.

3   Reasoning with Cardinality Constraints

In this section we will present a definite logic program representing a *sound calculus* for reasoning with exactly–cardinality constraints, and we will show that there exists no definite logic program representing a sound and *complete* calculus.

3.1   A Sound Calculus for Cardinality Constraints

The following logic program $\mathcal{P}_{cc} = \{r_1, \ldots, r_7\}$ represents a calculus of some *sound* inference rules for exactly–cardinality constraints, where the predicate symbols are written in infix notation:

$$r_1 = (M_2 \setminus M_1 \hookrightarrow_= K_2 - K_1) \leftarrow$$
$$(M_1 \hookrightarrow_= K_1) \wedge (M_2 \hookrightarrow_= K_2) \wedge (M_1 \subseteq M_2),$$
$$r_2 = (M_2 \setminus M_1 \hookrightarrow_= K_2 - K_1) \leftarrow$$
$$(M_1 \hookrightarrow_= K_1) \wedge (M_2 \hookrightarrow_= K_2) \wedge (|M_2 \setminus M_1| = K_2 - K_1),$$
$$r_3 = (M_2 \setminus M_1 \hookrightarrow_= K_2 - |M_2 \cap M_1|) \leftarrow$$
$$(M_1 \hookrightarrow_= |M_1|) \wedge (M_2 \hookrightarrow_= K_2),$$
$$r_4 = (M_2 \cup M_1 \hookrightarrow_= |M_2 \cup M_1|) \leftarrow$$
$$(M_1 \hookrightarrow_= |M_1|) \wedge (M_2 \hookrightarrow_= |M_2|),$$
$$r_5 = (M_2 \setminus M_1 \hookrightarrow_= K_2) \leftarrow$$
$$(M_1 \hookrightarrow_= 0) \wedge (M_2 \hookrightarrow_= K_2),$$
$$r_6 = (M_2 \cup M_1 \hookrightarrow_= K_2) \leftarrow$$
$$(M_1 \hookrightarrow_= 0) \wedge (M_2 \hookrightarrow_= K_2),$$
$$r_7 = (\{A\} \hookrightarrow_= 0) \leftarrow$$
$$(M \hookrightarrow_= 0) \wedge (A \in M).$$

$\mathcal{P}_{cc}$ is considered as a definite logic program with function symbols and the predicate symbols "$\hookrightarrow_=$", "$\subseteq$", and "$\in$". An interpretation of $\mathcal{P}_{cc}$ is basically given by a set of ground cardinality constraints; each ground cardinality constraint is of the form $M \hookrightarrow_= K$, where $M \subseteq gnd(\mathcal{At})$ is a set of ground atoms, and $K \in \mathbb{N}_0$. Note that $\mathcal{P}_{cc}$ is not considered as a disjunctive logic program with cardinality constraints, since in that case an interpretation of $\mathcal{P}_{cc}$ would be given by a set of ground atoms $A \in gnd(\mathcal{At})$.

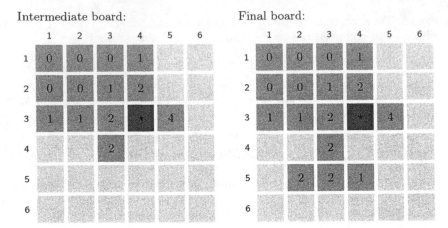

Fig. 2. Mine Sweeper Board after two further steps

Example. The calculus given by $\mathcal{P}_{cc}$ can be applied to the mine sweeper example of Section 2. Obviously, the cardinality constraint c_4 indicates that $(4,3)$ contains a mine. Further conclusions can be drawn using rule r_1. From c_1 and c_2 we can derive $c_7 = \{(3,4)\} \hookrightarrow_= 0$, since the set $\{(1,4),(2,4),(3,4)\}$ contains a single mine (cf. c_2) which must be located on one of the fields in the subset $\{(1,4),(2,4)\}$ (cf. c_1), i.e., $(3,4)$ cannot contain a mine. From c_5 and c_6 we can derive $c_8 = \{(4,3),(5,3)\} \hookrightarrow_= 1$, and from c_4 and c_8 we can derive $c_9 = \{(5,3)\} \hookrightarrow_= 0$, i.e., $(5,3)$ cannot contain a mine. Investigating the fields $(3,4)$ and $(5,3)$ and marking $(4,3)$ as a mine might result in the intermediate board shown in Figure 2. Now the only conclusions can be drawn from the cardinality constraints obtained by looking at the fields $(3,3)$ and $(3,4)$:

$$(3,3) \mapsto c_{10} = \{(2,4),(4,4)\} \hookrightarrow_= 1,$$
$$(3,4) \mapsto c_{11} = \{(2,4),(2,5),(3,5),(4,5),(4,4)\} \hookrightarrow_= 1.$$

Thus we can derive $c_{12} = \{(2,5),(3,5),(4,5)\} \hookrightarrow_= 0$, i.e., the fields $(2,5)$, $(3,5)$, and $(4,5)$ cannot contain a mine. Investigating these fields might result in the final board, cf. Figure 2. At this point it turns out that no further definite conclusions can be drawn as to whether any of the non–investigated fields contains a mine or not. It is only possible to derive for instance the indefinite conclusion that either $(2,4)$ or $(4,4)$ contains a mine.

It is easy to prove the *soundness* of $\mathcal{P}_{cc}$. Rule r_1 states the following: if K_1 elements are selected from a subset M_1 of M_2 and K_2 elements are selected from M_1, then $K_2 - K_1$ elements are selected from $M_2 \setminus M_1$. The rules $r_2, \ldots, r_7$ deal with *extremal* cardinality constraints $M \hookrightarrow_= K$, where K is either 0 or $|M|$, i.e., none or all of the elements of M should be selected. The rules $r_3, \ldots, r_7$ draw simple conclusions from such *extremal* constraints. Rule r_2 is a weaker form of a more general constraint that will be discussed below.

Of course there exist inference rules for cardinality constraints with other comparison relations $\Theta \in \{\leq, \geq\}$ as well. E.g. rule r_2 can be modified by shortening its body:

$$r_2' = (M_2 \setminus M_1 \hookrightarrow_\geq K_2 - K_1) \leftarrow (M_1 \hookrightarrow_= K_1) \wedge (M_2 \hookrightarrow_= K_2).$$

In the special case $|M_2 \setminus M_1| = K_2 - K_1$, the cardinality constraint $M_2 \setminus M_1 \hookrightarrow_\geq K_2 - K_1$ obviously implies $M_2 \setminus M_1 \hookrightarrow_= K_2 - K_1$. An even stronger inference rule would be

$$r_2'' = (M_2 \setminus M_1 \hookrightarrow_\geq K_2 - K_1) \leftarrow (M_1 \hookrightarrow_\geq K_1) \wedge (M_2 \hookrightarrow_\leq K_2).$$

In this paper, however, we will limit our attention to the restricted calculus for exactly–cardinality constraints given by $\mathcal{P}_{cc}$. The question arises, whether this calculus is also *complete* for inferring exactly–cardinality constraints, and – if not – whether $\mathcal{P}_{cc}$ can be extended by other rules to a larger definite logic program $\mathcal{P}_{cc}'$ that is sound and complete.

3.2   A Generic Example

For answering the questions above we will investigate the following generic set $\mathcal{C}_n$ of cardinality constraints.

$$\mathcal{C}_n = \{\, \{a_1, \ldots, a_n\} \hookrightarrow_= n - 1 \,\} \cup \{\, \{a_i, b_i, c\} \hookrightarrow_= 2 \mid 1 \leq i \leq n \,\},$$

where a_i, b_i, and c are pairwise different atoms.

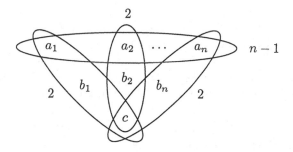

Lemma 3.1. *For the cardinality constraints*

$$cc_i = \{a_i, b_i\} \hookrightarrow_= 1, \ 1 \leq i \leq n,$$
$$cc_{n+1} = \{c\} \hookrightarrow_= 1,$$
$$cc_{n+2} = \{b_1, \ldots, b_n\} \hookrightarrow_= 1,$$

it holds

$$\mathcal{C}_n \models cc_i, \quad \text{for all } i \in \langle\, 1, n + 2 \,\rangle.$$

Proof. 1. The cardinality constraint $\{c\} \hookrightarrow_= 1$ can be derived by applying the disjunctive consequence operator $\mathcal{T}_P^s$ using the following subsets of $\mathcal{C}_n$:

$$\mathcal{C}(\{a_1,\ldots,a_n\} \hookrightarrow_\leq n-1) = \{\leftarrow a_1 \wedge \ldots \wedge a_n\},$$
$$\mathcal{C}(\{a_i,b_i,c\} \hookrightarrow_\geq 2) = \{a_i \vee b_i, a_i \vee c, b_i \vee c \,|\, i \in \langle 1,n \rangle\}.$$

Thus, for

$$\mathcal{P} = \{\leftarrow a_1 \wedge \ldots \wedge a_n\},$$
$$S = \{a_i \vee c \,|\, i \in \langle 1,n \rangle\}$$

we can conclude $\mathcal{T}_P^s(S) = \{c\}$, which implies $\{c\} \hookrightarrow_= 1$.

2. By applying the rule r_1 of $\mathcal{P}_{cc}$ to $\{c\} \hookrightarrow_= 1$ and $\{a_i,b_i,c\} \hookrightarrow_= 2$ we can derive $\{a_i,b_i\} \hookrightarrow_= 1$.

3. The constraint $\{b_1,\ldots,b_n\} \hookrightarrow_= 1$ obviously follows from the constraints $\{a_i,b_i\} \hookrightarrow_= 1$ and $\{a_1,\ldots,a_n\} \hookrightarrow_= n-1$. This can also be derived by applying the disjunctive consequence operator $\mathcal{T}_P^s$ for suitable $\mathcal{P}$ and S. From

$$\mathcal{P} = \{\leftarrow a_1 \wedge \ldots \wedge a_n\},$$
$$S = \{a_i \vee b_i \,|\, i \in \langle 1,n \rangle\},$$

we can derive $\mathcal{T}_P^s(S) = \{b_1 \vee \ldots \vee b_n\}$, which coincides with $\mathcal{C}(\{b_1,\ldots,b_n\} \hookrightarrow_\geq 1)$. We also use the following subsets of $\mathcal{C}_n$:

$$\mathcal{C}(\{a_1,\ldots,a_n\} \hookrightarrow_\geq n-1) = \{a_{i_1} \vee a_{i_2} \,|\, i_1,i_2 \in \langle 1,n \rangle, \, i_1 \neq i_2\},$$
$$\mathcal{C}(\{a_i,b_i\} \hookrightarrow_\leq 1) = \{\leftarrow a_i \wedge b_i \,|\, i \in \langle 1,n \rangle\}.$$

From the dualized rules in

$$\mathcal{P} = \{\leftarrow a_{i_1} \wedge a_{i_2} \,|\, i_1,i_2 \in \langle 1,n \rangle, \, i_1 \neq i_2\},$$
$$S = \{a_i \vee b_i \,|\, i \in \langle 1,n \rangle\},$$

we can derive $\mathcal{T}_P^s(S) = \{b_{i_1} \vee b_{i_2} \,|\, i_1,i_2 \in \langle 1,n \rangle, \, i_1 \neq i_2\}$, which is the dual version of $\mathcal{C}(\{b_1,\ldots,b_n\} \hookrightarrow_\leq 1) = \{\leftarrow b_{i_1} \wedge b_{i_2} \,|\, i_1,i_2 \in \langle 1,n \rangle, \, i_1 \neq i_2\}$. $\square$

3.3 The Incompleteness Result

Now we will answer the questions of whether the cardinality constraints cc_i, $i \in \langle 1,n+2 \rangle$, can be inferred from $\mathcal{C}_n$ using the calculus given by $\mathcal{P}_{cc}$ alone, i.e., without using the disjunctive consequence operator $\mathcal{T}_P^s$, and – more generally – whether there exists a definite logic program $\mathcal{P}'_{cc}$ (e.g., an extension of $\mathcal{P}_{cc}$) representing a sound and *complete* calculus for inferring exactly–cardinality constraints at all. Note that $\mathcal{P}'_{cc}$ should be finite and all of its rules can only contain a finite number of atoms of the form $M \hookrightarrow_= K$ in their bodies. Thus, there exists a rule $r_{max} \in \mathcal{P}'_{cc}$ with a maximal number n_{max} of atoms $M \hookrightarrow_= K$ in its body. In the following, however, we will show that for all strict subsets $C' \subsetneq C_n$ it holds

$$C' \not\models \{a_i, b_i\} \hookrightarrow_= 1, \ 1 \le i \le n,$$
$$C' \not\models \{c\} \hookrightarrow_= 1,$$
$$C' \not\models \{b_1, \ldots, b_n\} \hookrightarrow_= 1.$$

We can even prove – see appendix – a much stronger result, namely: there exists no non–trivial cardinality constraint $M \hookrightarrow_= K$ such that $C' \models M \hookrightarrow_= K$ and $\neg(C' \triangleright M \hookrightarrow_= K)$.

Theorem 3.1. *For all strict subsets $C' \subsetneq C_n$, where $n \ge 2$, and all non–trivial cardinality constraints $M \hookrightarrow_= K$ it holds:*

$$(C' \models M \hookrightarrow_= K) \Rightarrow (C' \triangleright M \hookrightarrow_= K).$$

Moreover it can be seen that $C' \triangleright M \hookrightarrow_= K$ implies that $M \hookrightarrow_= K \in C'$ for non–trivial cardinality constraints. As a consequence of Theorem 3.1, if $n > n_{max}$, then no sound calculus for reasoning with exactly–cardinality constraints that is given by a definite logic program $\mathcal{P}'_{cc}$ is able to derive the constraints cc_i, $i \in \langle 1, n+2 \rangle$, from the set C'.

These calculi can only derive exactly–cardinality constraints as intermediate results. Note, however, that other calculi which are deriving more general intermediate results could be able to obtain these conclusions. E.g., from the positive–disjunctive deductive database $\mathcal{P}_{C_n}$ the positive disjunctions and the denial rules in the sets $C(cc_i)$ can be derived due to Lemma 2.1.

3.4   A Mixed Calculus

The definite logic program $\mathcal{P}_{cc}$ is generic in that it can be used for reasoning with any set C of cardinality constraints. In the following we will show how the specific rules in the positive–disjunctive deductive database $\mathcal{P}_C$, which depends on C, can be transformed to rules that are reasoning about cardinality constraints as well.

The Program Transformation. For a disjunction $A_1 \vee \ldots \vee A_m$ and a conjunction $A_1 \wedge \ldots \wedge A_m$ of atoms A_i we define the formal notation

$$((A_1 \otimes \ldots \otimes A_m) \hookrightarrow_= K) = (\{A_1\} \hookrightarrow_= K) \otimes \ldots \otimes (\{A_m\} \hookrightarrow_= K),$$

where $\otimes \in \{\vee, \wedge\}$ and $K \in \{0, 1\}$. Thus, $(A_1 \vee \ldots \vee A_m) \hookrightarrow_= K$ denotes that at least one of the cardinality constraints $\{A_i\} \hookrightarrow_= K$ holds, and $(A_1 \wedge \ldots \wedge A_m) \hookrightarrow_= K$ denotes that all of them hold.

For a positive–disjunctive rule $r = \alpha \leftarrow \beta$ we can define another positive–disjunctive rule $r^{cc} = ((\alpha) \hookrightarrow_= 1) \leftarrow ((\beta) \hookrightarrow_= 1)$, i.e., for $r = A_1 \vee \ldots \vee A_k \leftarrow B_1 \wedge \ldots \wedge B_m$ we get

$$r^{cc} = (\{A_1\} \hookrightarrow_= 1) \vee \ldots \vee (\{A_k\} \hookrightarrow_= 1) \leftarrow$$
$$(\{B_1\} \hookrightarrow_= 1) \wedge \ldots \wedge (\{B_m\} \hookrightarrow_= 1).$$

The rules r and r^{cc} are basically equivalent, since for an interpretation I of $\mathcal{C}$ and an atom A it holds $I \models A \Leftrightarrow I \models (\{A\} \hookrightarrow_= 1)$.

By applying this transformation to $\mathcal{P}_C$, we define the positive–disjunctive logic program $\mathcal{P}_{\mathcal{C}}^{cc} = \{ r^{cc} \,|\, r \in \mathcal{P}_C \}$.[5] It holds $HB_{\mathcal{P}_{\mathcal{C}}^{cc}} \subseteq HB_{\mathcal{P}_{cc}}$, and it can be shown that $\mathcal{P}_{\mathcal{C}}^{cc}$ is able to derive all consequences of $\mathcal{C}$:

Lemma 3.2. *For a set C of exactly–cardinality constraints and an additional exactly–cardinality constraint c it holds* $\mathcal{C} \models c \Leftrightarrow \mathcal{P}_{\mathcal{C}}^{cc} \models c$.

Fixpoint Iteration. The specific positive–disjunctive logic program $\mathcal{P}_{\mathcal{C}}^{cc}$ can be mixed with the generic definite logic program $\mathcal{P}_{cc}$ to a program that allows for very efficiently deriving implied cardinality constraints. The positive–disjunctive logic program $\mathcal{P}_{\mathcal{C}}^{cc} \cup \mathcal{P}_{cc}$ operates on disjunctive Herbrand states.

During *fixpoint iteration*, whenever it is possible – of course – the definite rules from $\mathcal{P}_{cc}$ should be applied first as long as possible before disjunctive rules from $\mathcal{P}_{\mathcal{C}}^{cc}$ are applied. For sets such as the generic sets $\mathcal{C}_n$, $n \geq 2$ that were investigated in Section 3, however, $\mathcal{P}_{\mathcal{C}}^{cc}$ is able to draw conclusions, whereas – as a consequence of Theorem 3.1 – $\mathcal{P}_{cc}$ is not able to draw any conclusions.

Consider for example the sets $\mathcal{C}(\{a_1, \ldots, a_n\} \hookrightarrow_\leq n - 1)^{cc} = \{ r \}$ and $\mathcal{C}(\{a_i, b_i, c\} \hookrightarrow_\geq 2)^{cc} = \{ f_i, g_i, h_i \,|\, i \in \langle\, 1, n \,\rangle \}$, where

$$
\begin{aligned}
r &= \leftarrow (\{a_1\} \hookrightarrow_= 1) \wedge \ldots \wedge (\{a_n\} \hookrightarrow_= 1), \\
f_i &= (\{a_i\} \hookrightarrow_= 1) \vee (\{b_i\} \hookrightarrow_= 1), \\
g_i &= (\{a_i\} \hookrightarrow_= 1) \vee (\{c\} \hookrightarrow_= 1), \\
h_i &= (\{b_i\} \hookrightarrow_= 1) \vee (\{c\} \hookrightarrow_= 1).
\end{aligned}
$$

From the denial rule r and the disjunctive facts g_i we can derive $\{c\} \hookrightarrow_= 1$ by hyperresolution with the disjunctive consequence operator $\mathcal{T}_{\mathcal{P}_{\mathcal{C}}^{cc}}^s$. At this point the definite rule r_2 from $\mathcal{P}_{cc}$ becomes applicable, and it derives $\{a_i, b_i\} \hookrightarrow_= 1$ from $\{a_i, b_i, c\} \hookrightarrow_= 2$ and $\{c\} \hookrightarrow_= 1$. Now, the cardinality constraints $\{a_i, b_i\} \hookrightarrow_= 1$ can be accumulated to $\{a_1, b_1, \ldots, a_n, b_n\} \hookrightarrow_= n$ by iteratively applying the definite rule for covering

$$
\begin{aligned}
r_8 = \,&(M_2 \cup M_1 \hookrightarrow_= K_1 + K_2) \leftarrow \\
&(M_1 \hookrightarrow_= K_1) \wedge (M_2 \hookrightarrow_= K_2) \wedge (M_1 \cap M_2 = \emptyset),
\end{aligned}
$$

which we can add to $\mathcal{P}_{cc}$. Finally, $\{b_1, \ldots, b_n\} \hookrightarrow_= 1$ can be derived by applying $r_1 \in \mathcal{P}_{cc}$ to $\{a_1, \ldots, a_n\} \hookrightarrow_= n - 1$ and $\{a_1, b_1, \ldots, a_n, b_n\} \hookrightarrow_= n$.

Moreover, the fixpoint iteration can be tuned by incorporating a sophisticated *subsumption* mechanism for disjunctions of cardinality constraints.

---

[5] where for all rule $r \in \mathcal{P}_C$ either the head of the body is empty, i.e., $k = 0$ or $m = 0$.

4   Experimental Results

We have investigated a collection of test sets $\mathcal{C}$ of exactly–cardinality constraints of various sizes. In the following we report on sets that were derived from intermediate constellations of the game Mine Sweeper: each cardinality constraint $M \hookrightarrow_= K$ expresses how many of the non–investigated neighbours of a field on an $N \times N$ chess board contains a mine. Thus, the cardinalities of the sets M can range from 0 to 8, since each field has 8 neighbours. Larger sets of cardinality constraints could be obtained by enlarging the size N of the board.

Mixing $\mathcal{P}_\mathcal{C}$ and $\mathcal{P}_{cc}$

We generated about 320 sets $\mathcal{C}$ of cardinality constraints (for boards of the size $N = 8$), and we inferred the set $T_\mathcal{C}$ of all elements in the universe that definitely have to be chosen as a consequence of $\mathcal{C}$, and also the set $F_\mathcal{C}$ of all elements that definitely cannot be chosen:

$$T_\mathcal{C} = \{ A \in H_B \,|\, \mathcal{C} \models \{A\} \hookrightarrow_= 1 \},$$
$$F_\mathcal{C} = \{ A \in H_B \,|\, \mathcal{C} \models \{A\} \hookrightarrow_= 0 \}.$$

For about 99.4% of the generated sets $\mathcal{C}$ the sets $T_\mathcal{C}$ and $F_\mathcal{C}$ could be derived extremely efficiently using the calculus given by the definite logic program $\mathcal{P}_{cc}$.

The computation time needed when working with $\mathcal{P}_{cc}$ was orders of magnitude smaller than the time needed for computing with the disjunctive consequence operator $\mathcal{T}_\mathcal{P}^s$ of the positive–disjunctive deductive database $\mathcal{P}_\mathcal{C}$ according to Lemma 2.2, and yet the same result was derived. This is due to the fact that if K gets close to $|M|/2$, then the size of $\mathcal{C}(M \hookrightarrow_= K)$ becomes very large, and the computation time grows drastically with the size of $\mathcal{P}_\mathcal{C}$. Only for 0.6% (i.e., 2 out of 320) of the test cases, $\mathcal{T}_\mathcal{P}^s$ was able to derive knowledge, when $\mathcal{P}_{cc}$ did not return any conclusions.

Thus, it is a good strategy first to apply $\mathcal{P}_{cc}$ and see if it returns any results, and then to apply disjunctive reasoning only in case that no results have been obtained.

Comparison of Different Methods for Reasoning with $\mathcal{P}_\mathcal{C}$

For all of our generated test sets $\mathcal{C}$ of cardinality constraints, we have applied two semantically equivalent types of methods of disjunctive reasoning to obtain conclusions (cf. [9]):

1. *state generation*, i.e. positive hyperresolution with the disjunctive consequence operator $\mathcal{T}_\mathcal{P}^s$, and
2. *model generation* based on an operator $\mathcal{T}_\mathcal{P}^{INT}$, which computes the set $\mathcal{MM}(\mathcal{P})$ of minimal models of $\mathcal{P}$ as its least fixpoint first, i.e., $lfp\,(\mathcal{T}_\mathcal{P}^{INT}) = \mathcal{MM}(\mathcal{P})$, and then derives the conclusions from $\mathcal{MM}(\mathcal{P})$.

The methods are implemented in the system DISLOG (cf. [10,11]) for efficient reasoning in disjunctive deductive databases. In a variant of Case 1, the disjunctive consequence operator T_P^s was speeded up by using the compact *tree data structures* for disjunctive and negative Herbrand states of DISLOG (cf. [10]).

It is well–known that in general neither T_P^s nor T_P^{INT} is consistently superior to the other, which is due to the fact that for some boolean formulas the conjunctive normal form is smaller than the disjunctive normal form, and sometimes it is vice versa. But, interestingly it turned out that for the databases $\mathcal{P}_C$ obtained from cardinality constraints, T_P^s was always much faster than T_P^{INT}. Especially for some dual databases $\mathcal{P}_C^d$ the running time of T_P^{INT} was extremely large.

Based on the minimal models in $\mathcal{MM}(\mathcal{P}_C)$ it is possible to estimate the *probability* that an atom A is selected by a model I of C as n_1/n_2, where $n_1 = |\{ I \in \mathcal{MM}(\mathcal{P}_C) \,|\, A \in I \}|$ and $n_2 = |\mathcal{MM}(\mathcal{P}_C)|$.

Running Times. For obtaining an intuition about the complexity of reasoning with the positive–disjunctive deductive databases $\mathcal{P}_C$ obtained from cardinality constraints, we have related the computation times to the weight of $\mathcal{P}_C$. For this pupose we have considered the *following weight functions*:

$$w_n(\mathcal{C}(M \hookrightarrow_= K)) = \binom{|M|}{K+1} \cdot (K+1)^n + \binom{|M|}{K-1} \cdot (|M| - K + 1)^n,$$

and $w_n(\mathcal{C}) = \sum_{c \in \mathcal{C}} w_n(\mathcal{C}(c))$ for a set $\mathcal{C}$ of cardinality constraints. These functions compute the sum of the powers l^n, where l ranges over the lengths of the disjunctive rules in the sets $\mathcal{C}(M \hookrightarrow_= K)$, cf. Section 2.3. The motivation for using the powers l^n is the observation that on average reasoning with one long formula ϕ of length l in $\mathcal{C}(M \hookrightarrow_= K)$ is much more complex that reasoning with two shorter formulas ϕ_1 and ϕ_2 of the length l_1 and l_2, respectively, where $l_1 + l_2 = l$. Thus, the weight of ϕ should be larger than the sum of the weights of ϕ_1 and ϕ_2. Indeed this holds for $n \geq 2$, since $(l_1 + l_2)^n > l_1^n + l_2^n$.

For the weight function w_2 ($n = 2$) it turned out that the average computation time $t(w_2(\mathcal{C}))$ in seconds depending on the weight of $\mathcal{C}$ was given by a polynomial of small degree, which was approximately

$$t(w) \approx 3 \cdot 10^{-5} \cdot w^{3/2},$$

cf. Figure 3, which has got a doubly logarithmic scale. On the x–axis we show the size $w_2(\mathcal{C})$ of the problems, and on the y–axis we show the running times.

Four curves are shown for state generation, namely two for T_P^s (t_1^T for T_C, t_1^F for F_C) and two for the variant of T_P^s based on the tree data structure (t_2^T for T_C, t_2^F for F_C). For smaller values of $w = w_2(\mathcal{C}) < 2 \cdot 10^2$ we get $t_2^T(w) < t_2^F(w) < t_1^T(w) < t_1^F(w)$, whereas for larger values $w = w_2(\mathcal{C}) > 2 \cdot 10^2$ we get $t_1^T(w) < t_2^T(w) < t_2^F(w) < t_1^F(w)$. For extremal values of $w_2(\mathcal{C})$, the deviation from the straight line in Figure 3 can be explained by the *overlap* in the databases $\mathcal{P}_C$, which varied correspondingly and which we have not considered in our formula for $w_2(\mathcal{C})$.

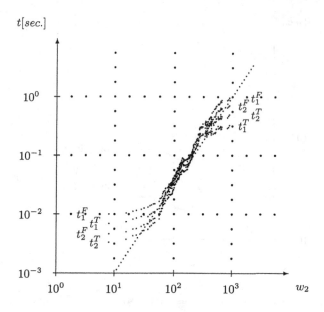

Fig. 3. Average Running Times for Reasoning with DLP

Comparison of $\mathcal{P}_{cc}$ and CLP

In Figure 4 the average running times for reasoning with LP (in SWI–PROLOG) and with CLP (in SICstus–PROLOG and in CHIP), respectively, are compared.

Our interest was to investigate, whether a problem that is hard to solve for one calculus is hard to solve for the other, too. It turned out that the computation time needed when working with $\mathcal{P}_{cc}$ (LP) was comparable to the time needed for computing with CLP; $\mathcal{P}_{cc}$ in SWI–PROLOG was faster for smaller examples, whereas CLP in CHIP was faster for bigger examples.

5   Conclusions

For reasoning with cardinality constraints the mixing of methods from disjunctive logic programming (DLP), which encode the constraints into disjunctive logic rules, with a calculus for directly dealing with the constraints proved itself to be very useful.

Interestingly, cardinality constraints turned out to be a domain where standard methods of reasoning in disjunctive deductive databases performed totally differently. It is well-known that in general neither state generation nor model generation is superior to the other. But on the restricted domain of disjunctive deductive databases derived from cardinality constraints it seems that state generation dominates model generation in most of the cases.

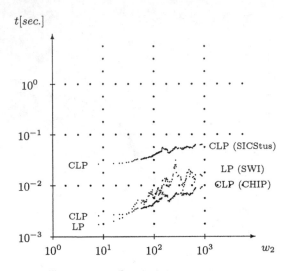

Fig. 4. Average Running Times for Reasoning with LP and CLP

A comparison with methods from constraint logic programming (CLP) showed that CLP is much faster than DLP. In our experiments CLP was comparable in performance to reasoning with our specially designed calculus for directly dealing with the constraints. But CLP is not applicable to problems where cardinality constraints and incomplete, disjunctive knowledge are envolved.

Acknowledgements. The authors would like to thank Joachim Baumeister for his useful comments on the paper and Johannes Lönne for giving an incentive to study the concept of cardinality constraints in the context of disjunctive deductive databases.

References

1. *S. Ceri, G. Gottlob, L. Tanca:* Logic Programming and Databases, Springer, 1990.
2. *P. Van Hentenryck, Y. Deville:* The Cardinality Operator: A new Logical Connective for Constraint Logic Programming, Proc. 8th Intl. Conference on Logic Programming 1991 (ICLP'91), MIT Press, 1991, pages 745–759.
3. *R. Kaye:* Minesweeper is NP–complete, Mathematical Intelligencer, 22(2), 2000, pages 9–15.
4. *J. W. Lloyd:* Foundations of Logic Programming, Second Edition, Springer, 1987.
5. *J. Lobo, J. Minker, A. Rajasekar:* Foundations of Disjunctive Logic Programming, MIT Press, 1992.
6. *K. Marriott, P. Stuckey:* Programming with Constraints – An Introduction, MIT Press, 1998.

7. *J. Minker, A. Rajasekar:* A Fixpoint Semantics for Disjunctive Logic Programs. *Journal of Logic Programming,* 9(1), 1990, pages 45–74.
8. *I. Niemelä, P. Simmons:* Extending the Smodels System with Cardinality and Weight Constraints, In Jack Minker (ed.): *Logic–Based Artificial Intelligence,* Kluwer, 2000, pages 491–522.
9. *D. Seipel, J. Minker, C. Ruiz:* Model Generation and State Generation for Disjunctive Logic Programs, *Journal of Logic Programming,* 32(1), 1997, pages 48–69.
10. *D. Seipel, H. Thöne:* DISLOG – A System for Reasoning in Disjunctive Deductive Databases, Proc. Intl. Workshop on the Deductive Approach to Information Systems and Databases 1994 (DAISD'94), pages 325–343.
11. *D. Seipel:* DISLOG – A Disjunctive Deductive Database Prototype, Proc. 12th Workshop on Logic Programming (WLP'97), 1997, pages 136–143. DISLOG is available at "http://www-info1.informatik.uni-wuerzburg.de/databases/DisLog".
12. *D. Seipel, U. Geske:* Cardinality Constraints in Disjunctive Deductive Databases, In *Workshop on Deductive Databases and Logic Programming (DDLP'2000)* at the International Conference on Applications of PROLOG (INAP'2000), 2000.
13. *A.H. Yahya:* Minimal Model Generation for Refined Answering of Generalized Queries in Disjunctive Deductive Databases. *Journal of Data and Knowledge Engineering,* 34(3), 2000, pages 219–249.

Appendix

Theorem 3.1 (Incompleteness)

For all strict subsets $\mathcal{C}' \subsetneq \mathcal{C}_n$, where $n \geq 2$, and all non–trivial cardinality constraints $M \hookrightarrow_= K$ it holds:

$$(\mathcal{C}' \models M \hookrightarrow_= K) \Rightarrow (\mathcal{C}' \rhd M \hookrightarrow_= K).$$

Proof. $\mathcal{C}_n$ has got precisely n models I_n, $i \in \langle\, 1, n\,\rangle$, given by

$$I_i = \{\, a_1, \ldots, a_{i-1}, b_i, a_{i+1}, \ldots, a_n, c\,\}.$$

Assume that $M \hookrightarrow_= K$ is a cardinality constraint, such that M is minimal w.r.t. set inclusion amoung all cardinality constraints satisfying

$$(\mathcal{C}' \models M \hookrightarrow_= K) \wedge \neg (\mathcal{C}' \rhd M \hookrightarrow_= K).$$

Obviously, the index sets

$$\Lambda_{a,b} = \{\, i \in \langle\, 1, n\,\rangle \mid a_i \in M,\, b_i \in M\,\},$$
$$\Lambda_{a,-} = \{\, i \in \langle\, 1, n\,\rangle \mid a_i \in M,\, b_i \notin M\,\},$$
$$\Lambda_{-,b} = \{\, i \in \langle\, 1, n\,\rangle \mid a_i \notin M,\, b_i \in M\,\},$$
$$\Lambda_{-,-} = \{\, i \in \langle\, 1, n\,\rangle \mid a_i \notin M,\, b_i \notin M\,\},$$

form a partition of $\langle\, 1, n\,\rangle$. Let $\lambda_{\mu,\nu} = |\Lambda_{\mu,\nu}|$. E.g., for $M = \{\, a_1, b_1, a_3, b_4, a_5\,\}$ we get $\Lambda_{a,b} = \{1\}$, $\Lambda_{a,-} = \{3,5\}$, $\Lambda_{-,b} = \{4\}$, $\Lambda_{-,-} = \{2\}$, and $\lambda_{a,b} = \lambda_{-,b} = \lambda_{-,-} = 1$, $\lambda_{a,-} = 2$.

Since $\mathcal{C}' \subsetneq \mathcal{C}_n$, there exists an element $C \in \mathcal{C}_n \setminus \mathcal{C}'$. In the following we will consider an arbitrary such element C.

1. If $C = \{a_k, b_k, c\} \hookrightarrow_= 2$, for some index $k \in \langle 1, n \rangle$, then C' has got the following $n + 2$ additional models, which are not models of C:
   - For $i = k$: $J_k^1 = I_k \setminus \{b_k\}$,

   $$J_k^2 = \{a_1, b_1, \ldots, a_{k-1}, b_{k-1}, a_{k+1}, b_{k+1}, \ldots, a_n, b_n\},$$

   and $J_k^3 = J_k^2 \cup \{b_k\}$.
   - For $i \neq k$: $J_i^1 = I_i \cup \{b_k\}$.

In the following three steps we will prove that $b_k \notin M$, $\lambda_{-,-} \neq n$, and $\lambda_{a,-} \neq n$.

If $b_k \in M$, then

$$|M \cap J_k^2| = |M \cap J_k^3| - 1,$$

i.e. $M \hookrightarrow_= K$ cannot hold both in J_k^2 and in J_k^3, i.e. $C \not\models M \hookrightarrow_= K$. Thus, it holds $b_k \notin M$.

If $\lambda_{-,-} = n$, then $M = \{c\}$, since $M \neq \emptyset$. Then for all $i \in \langle 1, n \rangle$ it holds

$$|M \cap I_i| = 1 \neq 0 = |M \cap J_k^1|,$$

i.e. $M \hookrightarrow_= K$ cannot hold both in I_i and in J_k^1, i.e. $C \not\models M \hookrightarrow_= K$. Thus, it holds $\lambda_{-,-} \neq n$.

If $\lambda_{a,-} = n$, then $\{a_1, \ldots, a_n\} \subseteq M$. Then M is not minimal in $M \hookrightarrow_= K$, since $M \hookrightarrow_= K$ can be simplified to $M' \hookrightarrow_= K - n + 1$, where $M' = M \setminus \{a_1, \ldots, a_n\} \subsetneq M$. Thus, it holds $\lambda_{a,-} \neq n$.

For the interpretations I_k and J_k^2 it holds

$$|M \cap I_k| = \lambda_{a,b} + \lambda_{a,-} - x + y,$$
$$|M \cap J_k^2| = 2 \cdot \lambda_{a,b} + \lambda_{a,-} - x + \lambda_{-,b},$$

where $x = |M \cap \{a_k\}|$ and $y = |M \cap \{c\}|$. Since $|M \cap I_k| = |M \cap J_k^2|$, we get

$$\lambda_{a,b} + \lambda_{-,b} = y \in \{0, 1\},$$

i.e. there exists at most one $b_i \in M$.

If $c \notin M$, i.e. $y = 0$, then $\lambda_{a,b} = \lambda_{-,b} = 0$, i.e. $\Lambda_{a,-} \cup \Lambda_{-,-} = \langle 1, n \rangle$. Since $\Lambda_{a,-} \neq \langle 1, n \rangle \neq \Lambda_{-,-}$, there exist indices $i_1 \in \Lambda_{-,-}$ and $i_2 \in \Lambda_{a,-}$. Then

$$|M \cap I_{i_1}| = |M \cap I_{i_2}| + 1,$$

i.e. $M \hookrightarrow_= K$ cannot hold both in I_{i_1} and in I_{i_2}, i.e. $C \not\models M \hookrightarrow_= K$. Thus, we get $c \in M$, i.e. $y = 1$, i.e. there exists exactly one $b_i \in M$. Now we can distinguish between two cases:
   - $\lambda_{a,b} = 1$ and $\lambda_{-,b} = 0$,
   - $\lambda_{a,b} = 0$ and $\lambda_{-,b} = 1$.

Firstly, assume $\lambda_{a,b} = 1$ and $\lambda_{-,b} = 0$, and let $\Lambda_{a,b} = \{i_1\}$. If $\lambda_{a,-} \neq 0$, then consider $i_2 \in \Lambda_{a,-}$. Then

$$|M \cap I_{i_1}| = |M \cap I_{i_2}| + 1,$$

i.e. $M \hookrightarrow_= K$ cannot hold both in I_{i_1} and in I_{i_2}, i.e. $\mathcal{C} \not\models M \hookrightarrow_= K$. Thus, $\lambda_{a,-} = 0$, i.e. $M = \{a_{i_1}, b_{i_1}, c\}$. Then, it holds $i_1 = k$, since otherwise $M \hookrightarrow_= K \in \mathcal{C}'$. But then

$$|M \cap J_k^2| = 0 \neq 2 = |M \cap I_k|.$$

i.e. $M \hookrightarrow_= K$ cannot hold both in $I_{J_k^2}$ and in I_k, i.e. $\mathcal{C} \not\models M \hookrightarrow_= K$. Secondly, assume $\lambda_{a,b} = 0$ and $\lambda_{-,b} = 1$, and let $\Lambda_{-,b} = \{i_1\}$. If $\lambda_{a,-} \neq 0$, then consider $i_2 \in \Lambda_{a,-}$. Then

$$|M \cap I_{i_1}| = |M \cap I_{i_2}| + 2,$$

i.e. $M \hookrightarrow_= K$ cannot hold both in I_{i_1} and in I_{i_2}, i.e. $\mathcal{C} \not\models M \hookrightarrow_= K$. Thus, $\lambda_{a,-} = 0$, i.e. $M = \{b_{i_1}, c\}$. Then $\lambda_{-,-} \neq 0$, and we can consider $i_2 \in \Lambda_{-,-}$. Then

$$|M \cap I_{i_1}| = 2 \neq 1 = |M \cap I_{i_2}|.$$

i.e. $M \hookrightarrow_= K$ cannot hold both in I_{i_1} and in I_{i_2}, i.e. $\mathcal{C} \not\models M \hookrightarrow_= K$.

2. If $C = \{a_1, \ldots, a_n\} \hookrightarrow_= n-1$, then $\mathcal{C}'$ has got the following $2^n + 1$ additional models, which are not models of $\mathcal{C}$:

- $J_{\langle x_1, \ldots, x_n \rangle} = \{x_1, \ldots, x_n\} \cup \{c\}$, for all vectors $\langle x_1, \ldots, x_n \rangle \in \{a_1, b_1\} \times \ldots \times \{a_n, b_n\}$,
- $J = \{a_1, b_1, \ldots, a_n, b_n\}$.

Firstly, we want to show that the a_i and b_i occur in pairs in M, i.e.

$$\Lambda_{a,b} \cup \Lambda_{-,-} = \langle 1, n \rangle.$$

Assume, that there exists $i' \in \langle 1, n \rangle \setminus (\Lambda_{a,b} \cup \Lambda_{-,-})$, i.e. $i' \in \Lambda_{a,-} \cup \Lambda_{-,b}$. Then we consider the two tuples $t_{a_{i'}}$ and $t_{b_{i'}}$, where

$$t_{x_{i'}} = \langle x_1, \ldots, x_{i'-1}, x_{i'}, x_{i'+1}, \ldots, x_n \rangle,$$

such that $t_{a_{i'}}$ and $t_{b_{i'}}$ coincide on the elements $x_i \in \{a_i, b_i\}$, $i \in \langle 1, n \rangle \setminus \{i'\}$, and differ on $x_{i'}$. Obviously, it holds

$$|M \cap J_{t_{a_{i'}}}| = |M \cap J_{t_{b_{i'}}}| + 1,$$

if $i \in \Lambda_{a,-}$, and

$$|M \cap J_{t_{a_{i'}}}| = |M \cap J_{t_{b_{i'}}}| - 1,$$

if $i' \in \Lambda_{-,b}$. Thus, $M \hookrightarrow_= K$ cannot hold both in $J_{t_{a_{i'}}}$ and in $J_{t_{b_{i'}}}$, i.e. $\mathcal{C} \not\models M \hookrightarrow_= K$.

In the following we will consider the models J and I_i for an arbitrary $i \in \langle 1, n \rangle$, and we will make use of the fact that $|M \cap I_i| = |M \cap J|$ must hold. If $c \in M$, then it holds $|M \cap I_i| = \lambda_{a,b} + 1$ and $|M \cap J| = 2 \cdot \lambda_{a,b}$. We get $\lambda_{a,b} = 1$. Thus, $M = \{a_{i'}, b_{i'}, c\}$. $\mathcal{C}' \not\models M \hookrightarrow_= K$ implies $K = 2$. Thus, $M \hookrightarrow_= K \in \mathcal{C}'$.

If $c \notin M$, then it holds $|M \cap I_i| = \lambda_{a,b}$ and $|M \cap J| = 2 \cdot \lambda_{a,b}$. We get $\lambda_{a,b} = 0$. Thus, $M = \emptyset$, which implies $K = 0$. Thus, $M \hookrightarrow_= K$ is trivial. $\square$

Semantic Classifications of Queries to Relational Databases

José María Turull Torres

Massey University
Information Science Research Centre
Department of Information Systems
PO Box 756, Wellington, New Zealand
j.m.turull@massey.ac.nz

Abstract. We survey different *semantic* classifications of queries to relational databases, as well as the different systems of partial isomorphisms developed as algebraic (hence, semantic) characterizations of equality of theories for databases, in *FO* and in weaker logics. We introduce an abstract notion of *similarity* for databases, as equality in the theories for a given logic and we define different sub-classes of queries by requiring that the queries in a given sub-class should not distinguish between databases which are similar. Then we use this general strategy to present a new semantic classification, with the class of bounded variable logics with counting (C^k), as the target logics. One important consequence of the definition of the two semantic classifications of queries which we present here is their orthogonality with the TIME-SPACE hierarchy defined in Turing machine complexity, allowing the definition of finer complexity classes by intersecting orthogonal hierarchies.

1 Introduction

It is a quite general approach in Science to classify a collection of objects in as many ways as possible, as a means to obtain knowledge about the objects. The more classifications we can define in the given collection of objects to study, the more knowledge we get about those objects. In relational database theory, in the last twenty years a computability theory for queries, as well as a complexity theory, starting with the seminal work of A. Chandra and D. Harel ([10]), is being developed. As a result of this process we have gained in knowledge about the nature of query computation in relational databases in a great extent.

In the developement of these theories, several sorts of models of computation of queries have been devised: formal programming languages ([10]), abstract machines ([3], [2]), logics in the finite model approach (see [13]), Boolean circuits ([12], [31]). And most classifications of queries have been done in terms of those models of computation. We may regard these classifications as *syntactic*, because they are defined in terms of properties of the expression in some given formalism, which represents the query (a program, a machine, a formula, or a family of circuits, respectively). Complexity measures in machines, such as space, time,

L. Bertossi et al. (Eds.): Semantics in Databases, LNCS 2582, pp. 200–228, 2003.

and number of variables used in dynamic queries; the less expressive logic needed to express a query, in the descriptive complexity approach (see [26]); complexity measures in families of circuits, such as size and depth; all these properties have induced corresponding syntactic classifications in the class of *computable queries* of Chandra and Harel.

On the other hand, what we may regard as *semantic* classifications have not been so frequent. We consider as semantic the classifications defined in terms of properties of the queries as functions. Examples of properties which yield this sort of classifications are the 0–1 *Laws*, and the property of *monotonicity* of a query. But these properties are Boolean, they do not induce infinite classifications. The 0–1 Law is a way to formalize the intuitive notion of a class of databases being "big" or "small", by means of asymptotic probability (we define it briefly in Section 2).

In [37] and [39], however, a hierarchy of infinite classes of queries has been defined ($\mathcal{QCQ}^\omega$), quite related to the work of S. Abiteboul, C. Papadimitriou, M. Vardi and V. Vianu on relational machines ([3], [4], [2]), which was proved to be strict. Moreover, it turns out that this hierarchy is *orthogonal* with the TIME-SPACE hierarchy defined in Turing machine complexity. We regard this classification as semantic, since each layer in it, is defined in terms of the number of variables k such that the queries in the given layer preserve equality of first order theories in the k bounded variable logic, denoted as FO^k.

The way in which the hierarchy was defined, turns out to be a natural generalization of the way in which the class of computable queries was defined, since these queries are the recursive queries which preserve equality of FO theories, or, equivalently, which preserve isomorphisms.

Then, we can think of an abstract notion of *similarity* for databases, and we can define different sub-classes of queries by requiring that the queries in a given sub-class should not distinguish between databases which are similar with our given notion of similarity.

We propose such a general scheme here, by defining similarity as equality in the theories for a given logic. When the logic we choose as the target logic for such theory preservation, has some measure which we can use as a parameter, then we get a corresponding hierarchy. This is the case, among other logics, with the logics FO^k and C^k (FO^k with the counting quantifiers, with the semantics "there exists at least m different elements in the database such that...", for every natural m), where the number of variables (k) can be used as such a parameter.

By using systems of partial isomorphisms with different extension conditions as algebraic characterizations of equivalence, or equality of theories, in different logics like FO^k and C^k, we can define the notion of similarity in a pure *algebraic* way, so that we get a pure *semantic* definition of the different hierarchies of sub-classes of queries.

In this article we propose, in Section 4, a new and general framework to define semantic classifications of queries in terms of preservation of equality in $\mathcal{L}$ theories, for different logics $\mathcal{L}$, weaker than first order, which is naturally induced by the classification defined in [37] and [39]. Then we survey, in Section 5, the results from [37] and [39], considered in the general framework defined above, as semantic classifications of queries. We also survey the different systems of partial isomorphisms developed to characterize in an algebraic way equivalence, or equality of theories for databases, in FO and in the logics which are weaker than FO (either defined by bounded quantifier rank and/or bounded number of variables in FO, or defined with the addition of counting quantifiers to each of these fragments, or existential fragments of the different logics).

We also present, in Section 6, a new semantic classification, defined by using the general strategy mentioned above, with the class of bounded variable logics with counting (C^k), as the target logics. We compare the two classifications by using known results on asymptotic probability. And we present the fact, derived naturally by the well known relation between the two logics FO^k and C^k regarding expressibility, that all sub-classes in the hierarchy QCQ^ω are "small", while all sub-classes in the hierarchy defined in Section 6 are "big", in a sense which we will make precise in Section 6 using results on asymptotic probabilities. Roughly, we can say that for every $k \geq 2$ and for every computable query q there is a query q' in the layer $\mathcal{C}_{C^k}$ which is equivalent to q over *almost all* databases. And this is not true for any layer in QCQ^ω, and not even for the whole hierarchy.

Once we defined a given sub-class following our general framework, it is natural to look for a computation model which expresses or captures exactly the sub-class we have defined. This makes sense also for whatever semantic way we use to define a sub-class of the computable queries. Then we define a variation of the reflective relational machine of [2] by using the logic C^k to express dynamic queries, getting what we call the *reflective counting machine*. It is similar to the machine defined in [32] to characterize the expressibility of fixed point logics with counting terms, though the model defined there is based in the relational machine of [3], instead. And we use our model to characterize the different sub-classes of the hierarchy which is based on the counting logics.

One important consequence of the definition of the two semantic classifications of queries which we present here is their orthogonality with the TIME-SPACE hierarchy defined in Turing machine complexity. This fact induces as to think in the definition of finer complexity classes by intersecting two or more orthogonal hierarchies. And we think that these cases are not just part of a few exceptions, but the general case, due to the fact that the hierarchies have been defined in pure semantic terms.

In Section 7 we consider some recent results on the descriptive complexity of finite groups, due to A. Koponen and K. Luosto, in order to follow with these results the approach we discussed above. And we show in a straightforward way that three properties of finite groups lye in the intersection of the class NLOGSPACE with one layer of the hierarchy QCQ^ω.

In Section 2 we present the necessary technical background for the material which we cover in this article.

2 Preliminaries

Unless otherwise stated, in the present article we will follow the usual notation in finite model theory, as in [13]. We use $\subseteq$ to denote set inclusion, and $\subset$ to denote proper set inclusion. We use ω to denote the set of natural numbers.

We define a *relational database schema*, or simply *schema*, as a set of relation symbols with associated arities. We don't allow constraints in the schema, and we don't allow constant symbols either. If $\sigma = \langle R_1, \ldots, R_s \rangle$ is a schema with arities $r_1, \ldots, r_s$, respectively, a *database instance*, or simply *database* over the schema σ is a structure $I = \langle D^I, R_1^I, \ldots, R_s^I \rangle$ where D^I is a finite set, which contains exactly all the elements of the database, and for $1 \leq i \leq s$, R_i^I is a relation of arity r_i, i.e., $R_i^I \subseteq (D^I)^{r_i}$. We will often use $dom(I)$ instead of D^I. We define the *size* of the database I as the size of D^I, i.e., $|D^I|$. We will use $\simeq$ to denote isomorphism. A *k-tuple* over a database I, with $k \geq 1$, is a tuple of length k formed with elements from $dom(I)$. We will denote a k-tuple of I as $\bar{a}_k$.

We will denote as $\mathcal{B}^\sigma$ the class of all the databases of schema σ.

2.1 Computable Queries and Relational Machines

We will consider in the present paper only *total* queries. Let σ be a schema, let $r \geq 1$, and let R be a relation symbol of arity r. A *computable query of arity r and schema σ* ([10]), is a total recursive function $q^r : \mathcal{B}^\sigma \to \mathcal{B}^{\langle R \rangle}$, which preserves isomorphisms, and such that for every database I of schema σ, $dom(q(I)) \subseteq dom(I)$. A *computable Boolean query of schema σ*, is a total recursive function $q : \mathcal{B}^\sigma \to \{0, 1\}$, which preserves isomorphisms. Boolean queries may also be regarded as 0-ary queries. We denote the class of computable queries of schema σ as $\mathcal{CQ}^\sigma$, and $\mathcal{CQ} = \bigcup_\sigma \mathcal{CQ}^\sigma$.

Relational machines (RM) have been introduced in [3]. A RM is a one-tape Turing machine (TM) with the addition of a *relational store* (rs), formed by a possibly infinite set of relations whose arity is *bounded* by some integer. The *only* way to access the relations in the rs is through FO (first order logic) formulas in the finite control of the machine, and the input database as well as the output relation are in rs. Each of these FO formulas, is evaluated in rs according to the transition function of the machine, and the resulting relation can be assigned to some relation symbol of the appropiate arity in the rs.

The reflective relational machine (RRM) was developed in [2], as an extension of the RM. In the RRM, the FO queries are generated *during the computation* of the machine, and they are called *dynamic queries*. Each of these queries is written in a *query tape* and it is evaluated by the machine in one step. Another important difference w.r.t. RM is that in RRM the relations in the rs can be of *arbitrary arity*. As a consequence of it, there is no such concept as the arity of a machine in RRM. The *variable complexity* of a RRM is the maximum number of variables which may be used in the dynamic queries which are generated by the machine throughout any computation. We will denote as RRM^k, with $k \geq 1$, the sub-class of RRM with variable complexity k. Further, we deine $RRM^{O(1)} = \bigcup_{k \geq 1} RRM^k$.

2.2   Finite Model Theory and Databases

We refer the reader to [14] as a source book for mathematical logic, to [13] and [26] for finite model theory, and [1] for the relation between databases and finite model theory. We will use the notion of a *logic* in a general sense, without a formal definition, which would be complicated and unnecessary to present our work. As it is usual in this setting, we will regard a logic as a language, that is, as a set of formulas. We will only consider signatures, or vocabularies, which are purely *relational*, i.e., there are no constant symbols nor function symbols. We will assume always that the signature includes a symbol for equality.

We consider *only finite* structures. So that our structures will always be *finite relational structures*. Consequently, if $\mathcal{L}$ is a logic, the notions of *satisfaction*, denoted as $\models_\mathcal{L}$ and *equivalence* between sentences, denoted as $\equiv_\mathcal{L}$ will be related to only finite structures.

If $\mathcal{L}$ is a logic, and σ is a signature, we will denote as $\mathcal{L}^\sigma$ the class of formulas from $\mathcal{L}$ with signature σ. If I is a structure of signature σ, or σ-*structure*, we define the $\mathcal{L}$ *theory of* I, as $Th_\mathcal{L}(I) = \{\varphi \in \mathcal{L}^\sigma : I \models_\mathcal{L} \varphi\}$.

A *database schema* will be regarded as a *relational signature*, and a *database instance* of some schema σ, as a finite and relational σ-*structure*. Consequently, the class of all finite σ-structures is exactly what we denoted as $\mathcal{B}^\sigma$ at the beginning of this section.

If φ is a sentence in $\mathcal{L}^\sigma$, we define $MOD(\varphi) = \{I \in \mathcal{B}^\sigma : I \models \varphi\}$.

By $\varphi(x_1, \ldots, x_r)$ we denote a formula of some logic, whose free variables are *exactly* $\{x_1, \ldots, x_r\}$. Let $free(\varphi)$ be the set of free variables of the formula φ. If $\varphi(x_1, \ldots, x_k) \in \mathcal{L}^\sigma$, $I \in \mathcal{B}^\sigma$, $\bar{a}_k = (a_1, \ldots, a_k)$ is a k-tuple over I, let $I \models \varphi(x_1, \ldots, x_k)[a_1, \ldots, a_k]$ denote that φ is TRUE, when interpreted by I, under a valuation v, where for $1 \leq i \leq k$, $v(x_i) = a_i$. That is, we consider the tuple $\bar{a}_k$ as a *valuation*. We will use the same notation also when the set of free variables of the formula is *strictly* included in $\{x_1, \ldots, x_k\}$, even when the formula is a sentence.

Then we can consider the set of all such valuations, as follows:

$$\varphi^I = \{(a_1, \ldots, a_k) : a_1, \ldots, a_k \in dom(I) \land I \models \varphi(x_1, \ldots, x_k)[a_1, \ldots, a_k]\}$$

That is, φ^I is the relation defined by φ in the structure I, and its arity is given by the number of free variables in φ.

Formally, we say that a formula $\varphi(x_1, \ldots, x_k)$ of signature σ, *expresses* a query q of schema σ, if for every database I of schema σ, is $q(I) = \varphi^I$. Similarly, a sentence φ expresses a Boolean query q if for every database I of schema σ, is $q(I) = 1$ iff $I \models \varphi$. Note that when using logics as query languages only *total* queries are considered.

We will deal also with *extensions* of structures. If R is a relation of arity k in the domain of a structure I, we denote as $\langle I, R \rangle$ the τ-structure resulting by adding the relation R to I, where τ is obtained from σ by adding a relation symbol of arity k. Similarly, if $\bar{a}_k$ is a k-tuple over I, we denote as $\langle I, \bar{a}_k \rangle$ the τ-structure resulting by adding the k-tuple $\bar{a}_k$ to I, where τ is obtained from σ by adding k constant symbols $c_1, \ldots, c_k$, and where for $1 \leq i \leq k$, the constant symbol c_i of τ, is interpreted in I by the i-th component of $\bar{a}_k$. This is the only case where we allow constant symbols in a signature.

2.3   Logics Different than First Order

We denote as FO^k, for some integer $k \geq 1$, the fragment of First Order Logic where only formulas whose variables are in $\{x_1, \ldots, x_k\}$ are allowed. In this setting, FO^k is itself a logic. This logic is clearly *less expressive* than FO.

We denote as C^k, for some integer $k \geq 1$, the logic which we get by adding to FO^k *counting quantifiers*, i.e., all the existential quantifiers of the form $\exists^{\geq m} x$, for every $m \geq 1$. Informally, the semantics of $\exists^{\geq m} x(\varphi)$ is that there exists at least m *different* elements in the structure or database, which satisfy φ.

The *infinitary logic* $\mathcal{L}^k_{\infty,\omega}$, for every $k \geq 1$, has the usual first order rules for the formation of formulas and in addition is closed under infinitary conjunctions and disjunctions. Formulas in $\mathcal{L}^k_{\infty,\omega}$ are restricted to use at most k different variables. And $\mathcal{L}^\omega_{\infty,\omega} = \bigcup_k \mathcal{L}^k_{\infty,\omega}$.

In the same way as the logics $\mathcal{L}^k_{\infty,\omega}$ and $\mathcal{L}^\omega_{\infty,\omega}$ extend FO^k, we can define corresponding infinitary logics as extensions of C^k. Then we get the *infinitary logics with counting*, denoted as $C^k_{\infty,\omega}$ and $C^\omega_{\infty,\omega}$, respectively. These logics are defined in the same way as $\mathcal{L}^k_{\infty,\omega}$ and $\mathcal{L}^\omega_{\infty,\omega}$, with the addition of all the *counting quantifiers*, i.e., $C^k_{\infty,\omega}$ has the same formation rules as $\mathcal{L}^k_{\infty,\omega}$, with the addition of the following rule: if ψ is a formula in $C^k_{\infty,\omega}$, x is a variable, and $m \geq 1$, then $\exists^{\geq m} x(\psi)$ is also a formula in $C^k_{\infty,\omega}$. Then $C^\omega_{\infty,\omega} = \bigcup_{k \geq 1} C^k_{\infty,\omega}$.

The *quantifier rank* of a formula is the maximum nesting depth of quantifiers in it. We denote as $FO^{;m}$, for every $m \geq 0$, the logic which is a fragment of FO where only formulas of quantifier rank up to m are allowed, and we denote as $FO^{k;m}$, for every pair of integers $k \geq 1$, and $m \geq 0$, the logic which is a fragment of the logic FO^k where only FO^k formulas of quantifier rank up to m are allowed. In the same way we define the logics $C^{k;m}$. We also define restrictions in terms of the quantifier rank for the infinitary logics $\mathcal{L}^k_{\infty,\omega}$ and $C^k_{\infty,\omega}$, getting the logics which we denote as $\mathcal{L}^{k;m}_{\infty,\omega}$ and $C^{k;m}_{\infty,\omega}$, respectively.

2.4   Types

Given a database I and a k-tuple $\bar{a}_k$ over I, we want to consider *all* the properties of $\bar{a}_k$ in the database I, including the properties of every component of the tuple, and the properties of all the different sub-tuples of $\bar{a}_k$, so we use the notion of *type*. We will usually follow the notation as in [33].

Definition 2.1. *Let $\mathcal{L}$ be a logic. Let I be a database of some schema σ, let $\bar{a}_k = (a_1, \ldots, a_k)$ be a k-tuple over I. The $\mathcal{L}$ type of $\bar{a}_k$ in I, denoted $tp^{\mathcal{L}}_I(\bar{a}_k)$, is the set of formulas in $\mathcal{L}^\sigma$ with free variables among $\{x_1, \ldots, x_k\}$, such that every formula in the set is TRUE when interpreted by I for any valuation which assigns the i-th component of $\bar{a}_k$ to the variable x_i, for every $1 \leq i \leq k$. In symbols*

$$tp_I^{\mathcal{L}}(\bar{a}_k) = \{\varphi \in \mathcal{L}^{\sigma} : free(\varphi) \subseteq \{x_1, \ldots, x_k\} \wedge I \models \varphi[a_1, \ldots, a_k]\}$$

It is note worthy that, according this Definition, the $\mathcal{L}$ *theory* of the database I, i.e., $Th_{\mathcal{L}}(I)$, is included in the $\mathcal{L}$ type of *every* tuple over I. Considering the logic $\mathcal{L}$ as a query language, we may also regard an $\mathcal{L}$ type as a *set of queries*. Moreover, the *(infinitary) conjunction* of all the $\mathcal{L}$ formulas which form the $\mathcal{L}$ type of $\bar{a}_k$, is itself a query, though not necessarily expressible in $\mathcal{L}$.

According Definition 2.1, a type is an *infinite* set of formulas which is *consistent*, i.e., there is a structure and a valuation which make all the formulas in the set TRUE. Moreover, the set is *maximally consistent*, that is, whichever formula we add to the set, we loose the consistency of the set.

Then we can also think of a type of a tuple, *without having a particular database in mind*, just by adding properties (formulas with the appropiate free variables) to a set as long as it remains consistent. That is, we can first define a type for a given schema and a given length of tuple, and then ask whether in a given database there is a tuple of that length and of that type. Let us denote as $Tp^{\mathcal{L}}(\sigma, k)$, for some $k \geq 1$, the class of *all* $\mathcal{L}$ types for k-tuples over databases of schema σ. In symbols

$$Tp^{\mathcal{L}}(\sigma, k) = \{tp_I^{\mathcal{L}}(\bar{a}_k) : I \in \mathcal{B}^{\sigma} \wedge \bar{a}_k \in (dom(I))^k\}$$

That is, $Tp^{\mathcal{L}}(\sigma, k)$ is a class of properties, or a set of sets of formulas.

Definition 2.2. Let $\mathcal{L}$ be a logic. Let σ be a schema, let $k \geq 1$, and let $\alpha \in Tp^{\mathcal{L}}(\sigma, k)$ (i.e., α is the $\mathcal{L}$ type of some k-tuple over some database in $\mathcal{B}^{\sigma}$). Let I be a database of schema σ. We say that I realizes the type α if there is a k-tuple $\bar{a}_k$ over I whose $\mathcal{L}$ type is α. That is, if $tp_I^{\mathcal{L}}(\bar{a}_k) = \alpha$.

Let I be a database of some schema σ, and let $k > 0$. We denote as $Tp^{\mathcal{L}}(I, k)$ the class of all $\mathcal{L}$ types for k-tuples, which are realized in I. That is, it is the class of the properties of all the k-tuples over the database I, which can be expressed in $\mathcal{L}$. In symbols

$$Tp^{\mathcal{L}}(I, k) = \{tp_I^{\mathcal{L}}(\bar{a}_k) : \bar{a}_k \in (dom(I))^k\}$$

The next well known result means that when two databases realize the same FO^k, or C^k types for tuples, it does not mind which length of tuples we consider.

Lemma 2.1. *For every $k \geq 1$, for every schema σ, for every pair of databases I, J of schema σ, and for every $1 \leq l \leq k$, the following holds:*

$$- \ Tp^{FO^k}(I, k) = Tp^{FO^k}(J, k) \Longleftrightarrow Tp^{FO^k}(I, l) = Tp^{FO^k}(J, l)$$
$$- \ Tp^{C^k}(I, k) = Tp^{C^k}(J, k) \Longleftrightarrow Tp^{C^k}(I, l) = Tp^{C^k}(J, l)$$

The following is also a well known result, which, among other sources, can be found as Proposition 2.1.1 in [13].

Proposition 2.1. *and for every pair of (finite) databases I, J of schema σ, the following holds: $I \equiv_{FO} J \Longleftrightarrow I \simeq J$.*

Moreover, the FO theory of every database is equivalent to a single FO^{n+1} sentence, where n is the size of the database.

Although types are infinite sets of formulas, due to the next result of M. Otto, a *single* C^k formula is equivalent to the C^k type of a tuple over a given database. And this equivalence holds for all the databases of the same schema.

Proposition 2.2. *([32]) For every schema σ, for every database I of schema σ, for every $k \geq 1$, for every $1 \leq l \leq k$, and for every l-tuple $\bar{a}_l$ over I, there is a C^k formula $\chi \in tp_I^{C^k}(\bar{a}_l)$ such that for any database J of schema σ and for every l-tuple $\bar{b}_l$ over J*

$$J \models \chi[\bar{b}_l] \Longleftrightarrow tp_I^{C^k}(\bar{a}_l) = tp_J^{C^k}(\bar{b}_l)$$

Moreover, such a formula χ can be built inductively for a given database. If a C^k formula χ satisfies the condition of Proposition 2.2, we say that χ is an *isolating formula* for $tp_I^{C^k}(\bar{a}_l)$.

2.5 Asymptotic Probabilities

The 0–1 Law was first defined for FO, independently by Y. Glebskij et al [18] and R. Fagin [16]. See [13], among other sources. Let σ be a schema, let I be a database of schema σ, let $\mathcal{L}$ be a logic, and let φ be a sentence in $\mathcal{L}^\sigma$. We define $MOD_n(\varphi) = \{I \in \mathcal{B}^\sigma : dom(I) = \{1, \ldots, n\} \wedge I \models \varphi\}$.

Let's denote as $\mathcal{B}^{\sigma,n}$ the sub-class of databases of schema σ with domain $\{1, \ldots, n\}$. We define the following limit, which we call *asymptotic probability of* φ:

$$\mu_\varphi = lim_{n \to \infty}(|MOD_n(\varphi)|/|\mathcal{B}^{\sigma,n}|)$$

We say that a logic $\mathcal{L}$ has a 0–1 *Law* if, for every sentence $\varphi \in \mathcal{L}$, μ_φ is defined, and if it is either 0 or 1. The same notion can also be defined on classes of databases, or Boolean queries.

This means that the asymptotic probability of every property which can be expressed in the formalism (or which belongs to the given class) always exist, and is either 0 or 1 (see [30] and [13]). Among other logics, FO has this Law. It is well known that there are computable Boolean queries which do not have a 0–1 Law, like the *parity* of the size of a database. Note that regarding this query, the limit is not defined.

3   Semantic Characterization of Equivalence in Logics

In the classical approach to formal semantics in model theory, isomorphic structures are considered as indistinguishable from the perspective of any given logic.

From Proposition 2.1, it follows that if we consider FO restricted to *finite* structures or models, isomorphism is not only a sufficient condition for elementary equivalence, it is also a necessary condition. The result stated in Proposition 2.1 may be regarded as a *semantic* characterization of equivalence in FO. More precisely, it is an *algebraic* characterization.

Note, however, that for the general case isomorphism is no longer a necessary condition. That is, two non isomorphic structures which are infinite may be indistinguishable in FO.

3.1   Systems of Partial Isomorphisms

If, on the other hand, we consider logics whose expressive power is smaller than FO, isomorphism is, again, no longer a necessary condition. Isomorphism is a notion which is too strong in the context of logics which are less expressive than FO. Thus, we need some sort of algebraic relation between databases, weaker than isomorphism. The concept of *partial isomorphism*, together with different extension properties constitute such weaker notions.

In what follows we exhibit well known results which give us algebraic characterizations of equivalence in several logics which are weaker than FO. The different systems of partial isomorphisms are algebraic notions which constitute different relaxations of the notion of isomorphism.

In this section we will allow constant symbols in the schemas, unless otherwise stated. If σ is a schema, c is a constant symbol in σ, and I is a database of schema σ, we will denote as c^I the interpretation of the constant c in the database I.

Definition 3.1. *Let I, J be two databases of some schema σ. Let f : $dom(I) \longrightarrow dom(J)$ be a partial function which is injective. We say that f is a* partial isomorphism *from I to J if it satisfies the following properties:*

a): *For every constant symbol c in σ: $c^I \in dom(f)$, and $f(c^I) = c^J$.*

b): *For every relation symbol R of arity r in σ, and for every r-tuple $(a_1, \ldots, a_r)$ of elements in $dom(f)$: $(a_1, \ldots, a_r) \in R^I \iff (f(a_1), \ldots, f(a_r)) \in R^J$.*

We denote as $Part(I, J)$ the set of partial isomorphisms from I to J.

If f, g, are in $Part(I, J)$, for some pair of databases I, J, we say that f is an *extension* of g if the graph of f includes the graph of g. We denote it as $f \supseteq g$.

That is, a partial isomorphism is an isomorphism between two sub-databases. The following result is well known (see Remarks 1.2.2 in [13], among others).

Lemma 3.1. *Let I, J be two databases of some schema σ. Let $f \in Part(I, J)$, with $dom(f) = \{a_1, \ldots, a_r\}$, and let $b_i = f(a_i)$, for $1 \le i \le r$. Then, for every FO formula φ, with free variables $\{x_1, \ldots, x_r\}$, which is quantifier free, the following holds:*

$$I \models \varphi[a_1, \ldots, a_r] \iff J \models \varphi[b_1, \ldots, b_r]$$

Thus, the existence of a partial isomorphism between two databases implies their equivalence with respect to FO, as long as their free variables are assigned only elements in the domain and range of the partial isomorphism, respectively, but *only* regarding *quantifier free* formulas. It is straightforward to find examples of databases and corresponding formulas with non zero quantifier rank, for which the statement of the Lemma does not hold. Thus, the equivalence of two databases under formulas with quantifiers requires a stronger relation between the two databases. It turns out that it requires a sort of *extension* property for the partial isomorphisms. As we will see later in this section, this is also true regarding the use of systems of partial isomorphisms for the characterization of equivalence in other logics, different than FO.

3.2   Equivalence in First Order

Definition 3.2. *Let I, J be two databases of some schema σ. Let $m > 0$. A back and forth system of partial isomorphisms of length m between I and J is a family of sets of partial isomorphisms from $Part(I, J)$, denoted as $(P_i)_{i \leq m}$, with the following properties:*

a):   *For every $0 \leq i \leq m$ the set P_i is non empty.*

b):   *The family of sets of partial isomorphisms $(P_i)_{i \leq m}$ is ordered by inclusion, as follows: $P_m \subseteq \ldots \subseteq P_0$.*

c):   *Forth Property: For every $0 < i \leq m$, for every partial isomorphism $f \in P_i$, and for every element $a \in dom(I)$, there is a partial isomorphism $g \in P_{i-1}$, and an element $b \in dom(J)$, such that $g \supseteq f$ and $(a, b) \in g$.*

d):   *Back Property: For every $0 < i \leq m$, for every partial isomorphism $f \in P_i$, and for every element $b \in dom(J)$, there is a partial isomorphism $g \in P_{i-1}$, and an element $a \in dom(I)$, such that $g \supseteq f$ and $(a, b) \in g$.*

That is, for every $0 \leq i \leq m$, every partial isomorphism in P_i can be extended up to i times, with every possible element of the database I (as well as with every possible element of the database J), and in such a way that the extensions lye in $P_{i-1}, P_{i-2}, \ldots, P_0$, respectively.

With this notion, we get an algebraic characterization of equivalence in FO, due to R. Fraïssé ([17]) (a proof can also be found in Theorem 1.3.2 of [13]). Note that this characterization holds for the general case, i.e., considering databases which may be infinite.

Theorem 3.1. *([Fraïssé]) Let I, J be two databases, either finite or infinite, of some schema σ, and let $m \geq 0$. Then, the following are equivalent:*

a):   *There is a back and forth system of partial isomorphisms of length m between I and J.*

b):   *For every sentence φ in FO with quantifier rank up to m: $I \models \varphi \Leftrightarrow J \models \varphi$, i.e., $I \equiv_{FO;m} J$.*

Note that the notion of back and forth system of partial isomorphisms can be easily extended to be of length ω, i.e., to consist of a countably infinite family of sets of partial isomorphisms $((P_i)_{i \in \omega})$. In this way we can characterize equivalence in *full FO*.

Corollary 3.1. *([Fraïssé]) Let I, J be two databases, either finite or infinite, of some schema σ. Then, the following are equivalent:*

a): *There is a back and forth system of partial isomorphisms of length ω between I and J.*

b): *For every sentence φ in FO: $I \models \varphi \Leftrightarrow J \models \varphi$, i.e., $I \equiv_{FO} J$.*

Thus, regarding *finite* databases the existence of a back and forth system of partial isomorphisms of length ω between two databases is equivalent to isomorphism.

3.3   Equivalence in Bounded Variable Logics

Now we will consider the use of back and forth systems of partial isomorphisms to characterize equivalence in the logics with bounded number of variables (FO^k, see Section 2). We define the *size* of a partial isomorphism f as the number of ordered pairs in f, and we denote it as $|f|$, and for any two given databases I, J of some schema σ, and for any $k \geq 0$ we will denote the set of partial isomorphisms of size at most k between I and J as $Part^k(I, J)$. As we will see below, in this case the size of the partial isomorphisms is bounded by the number of variables which is allowed in the logic. Thus, considering Definition 3.1, it is convenient to consider schemas with *only* relation symbols. Hence, in what follows we will not allow constant symbols in the schemas.

Definition 3.3. *Let I, J be two databases of some schema σ which contains only relation symbols. Let $m > 0$, and $k \geq 1$. A k-bounded back and forth system of partial isomorphisms of length m between I and J is a family of sets of partial isomorphisms from $Part^k(I, J)$, denoted as $(P_i^k)_{i \leq m}$, with the following properties:*

a): *For every $0 \leq i \leq m$ the set P_i^k is non empty.*

b): *The family of sets of partial isomorphisms $(P_i^k)_{i \leq m}$ is ordered by inclusion, as follows: $P_m \subseteq \ldots \subseteq P_0$.*

c): *Forth Property: 1): For every $0 < i \leq m$, for every partial isomorphism $f \in P_i^k$ with $|f| < k$, and for every element $a \in dom(I)$, there is a partial isomorphism $g \in P_{i-1}$, and an element $b \in dom(J)$, such that $g \supseteq f$ and $(a, b) \in g$. 2): For every $0 < i \leq m$, for every partial isomorphism $f \in P_i^k$ with $|f| = k$, for every ordered pair $(a', b') \in f$, and for every element $a \in dom(I)$, there is a partial isomorphism $g \in P_{i-1}$, and an element $b \in dom(J)$, such that $g = f - \{(a', b')\} \cup \{(a, b)\}$.*

d): *Back Property: It is defined similarly (see Definition 3.2).*

Thus, in this kind of system, as all partial isomorphisms must be bounded in size, once a given partial isomorphism has been extended up to the limit it can be further "extended" by replacing one of its ordered pairs by the new pair. This is the reason why the forth and the back properties have been stated by cases in Definition 3.3.

With this variation of the back and forth system of partial isomorphisms, we get an algebraic characterization of equivalence in FO^k, for every $k > 0$, due to N. Immerman ([25]).

Theorem 3.2. *([25]) Let I, J be two databases of some schema σ which has only relation symbols, and let $m \geq 0$, and $k \geq 1$. Then, the following are equivalent:*

a): *There is a k-bounded back and forth system of partial isomorphisms of length m between I and J.*

b): *For every sentence φ in FO^k with quantifier rank up to m: $I \models \varphi \Leftrightarrow J \models \varphi$, i.e., $I \equiv_{FO^{k;m}} J$.*

c): *For every sentence φ in $\mathcal{L}^k_{\infty,\omega}$ with quantifier rank up to m: $I \models \varphi \Leftrightarrow J \models \varphi$, i.e., $I \equiv_{\mathcal{L}^{k;m}_{\infty,\omega}} J$.*

Note that the finitary logics and the corresponding infinitary logics have the *same* discerning power, though the expressive power of the infinitary logics is much stronger (see [13], [20] and [33]). The same is also true for the logics which are existential fragments of the corresponding bounded variable logics (see Theorem 3.3 and Corollary 3.3), and when we add counting quantifiers to the bounded variable logics (see Theorem 3.4 and Corollary 3.4). And it is also trivially true regarding full FO w.r.t. the infinitary logic $\mathcal{L}_{\infty,\omega}$, since we are considering only finite structures.

It might seem that the infinitary logics are not much related to the computation of queries to finite databases. However, this is far from being true. The infinite length of a formula compensates the lack of a recursion mechanism in classical predicate logics. With such a formula, we can "unfold" each recursion construction of a program, for any given input. Note that for every input database we have a fixed number of iterations in every recursion construction of the given program. Then we can express with a finite formula the computation of the program on that input database. Finally, we can take an infinitary disjunction with the computations of the program for every possible input database, and thus we get an infinitary formula (see [38]).

Corollary 3.2. *([25]) Let I, J be two databases of some schema σ which has only relation symbols, and let $k \geq 1$. Then, the following are equivalent:*

a): *There is a k-bounded back and forth system of partial isomorphisms of length ω between I and J.*

b): *For every sentence φ in FO^k: $I \models \varphi \Leftrightarrow J \models \varphi$, i.e., $I \equiv_{FO^k} J$.*

c): *For every sentence φ in $\mathcal{L}^k_{\infty,\omega}$: $I \models \varphi \Leftrightarrow J \models \varphi$, i.e., $I \equiv_{\mathcal{L}^k_{\infty,\omega}} J$.*

Now we will briefly consider the *existential* fragments of the bounded variable logics. Let us denote as $FO^k(\exists)$, for every $k \geq 1$ the logic which we informally define as the logic FO^k with the restriction that universal quantifiers are not allowed in the formulas, and existential quantifiers may not be negated. We define in a similar way the existential fragments of the infinitary bounded variable logics, which we denote as $\mathcal{L}^k_{\infty,\omega}(\exists)$, for every $k \geq 1$. We also define the corresponding fragments of bounded quantifier rank of these logics in the usual way getting, for every $m \geq 0$ and for every $k \geq 1$ the logics $FO^{k;m}(\exists)$ and $\mathcal{L}^{k;m}_{\infty,\omega}(\exists)$.

Regarding equivalence in these logics, the usual symmetric relation $\equiv_{\mathcal{L}}$, does not make sense, as the following well known fact shows:

Lemma 3.2. *Let I, J be two databases of some schema σ which has only relation symbols, and let $k \geq 1$.*

 a): $I \equiv_{FO^k(\exists)} J \Longleftrightarrow I \equiv_{FO^k} J.$

 b): $I \equiv_{\mathcal{L}^k_{\infty,\omega}(\exists)} J \Longleftrightarrow I \equiv_{\mathcal{L}^k_{\infty,\omega}} J.$

Then we define the relation $\preceq_{\mathcal{L}}$, instead. So, for every logic $\mathcal{L}$, for every pair of databases I, J of some schema σ, and for every $k \geq 1$, we write $I \preceq_{\mathcal{L}} J$ to denote that for every sentence φ in $\mathcal{L}$, $I \models \varphi \Rightarrow J \models \varphi$. Note that $I \preceq_{\mathcal{L}} J \wedge J \preceq_{\mathcal{L}} I \Longleftrightarrow I \simeq_{\mathcal{L}} J$.

P. Kolaitis and M. Vardi proved ([29], [30]) a characterization of the relation $\preceq_{\mathcal{L}}$, for the existential logics which we defined above, with systems of partial isomorphisms, by requiring *only* the forth property. These logics have been further studied, still in the setting of finite models, in [35]. Then we modify slightly Definition 3.3, and we define, for every $m \geq 0$, and for every $k \geq 1$, a *k-bounded forth system of partial isomorphisms of length m* in the same way as in Definition 3.3, except that the *back* property is not required. So, we can give now an algebraic characterization of $\preceq_{\mathcal{L}}$ for the existential logics which we are considering.

Theorem 3.3. *([29], [30]) Let I, J be two databases of some schema σ which has only relation symbols, and let $m \geq 0$, and $k \geq 1$. Then, the following are equivalent:*

 a): *There is a k-bounded forth system of partial isomorphisms of length m between I and J.*

 b): *For every sentence φ in $FO^k(\exists)$ with quantifier rank up to m: $I \models \varphi \Longrightarrow J \models \varphi$, i.e., $I \preceq_{FO^{k;m}(\exists)} J$.*

 c): *For every sentence φ in $\mathcal{L}^k_{\infty,\omega}(\exists)$ with quantifier rank up to m: $I \models \varphi \Longrightarrow J \models \varphi$, i.e., $I \preceq_{\mathcal{L}^{k;m}_{\infty,\omega}(\exists)} J$.*

Corollary 3.3. *([29], [30]) Let I, J be two databases of some schema σ which has only relation symbols, and let $k \geq 1$. Then, the following are equivalent:*

 a): *There is a k-bounded forth system of partial isomorphisms of length ω between I and J.*

 b): *For every sentence φ in $FO^k(\exists)$: $I \models \varphi \Longrightarrow J \models \varphi$, i.e., $I \preceq_{FO^k(\exists)} J$.*

 c): *For every sentence φ in $\mathcal{L}^k_{\infty,\omega}(\exists)$: $I \models \varphi \Longrightarrow J \models \varphi$, i.e., $I \preceq_{\mathcal{L}^k_{\infty,\omega}(\exists)} J$.*

3.4   Equivalence in Bounded Variable Logics with Counting

We will consider now equivalence in logics with bounded number of variables with the addition of counting quantifiers (C^k, see Section 2). As we did before, in what follows we will not allow constant symbols in the schemas.

Definition 3.4. *Let I, J be two databases of some schema σ which contains only relation symbols. Let $m > 0$, and $k \geq 1$. A k-bounded bijective back and forth system of partial isomorphisms of length m between I and J is a family of sets of partial isomorphisms from $Part^k(I, J)$, denoted as $(BP^k_i)_{i \leq m}$, with the following properties:*

a): *For every $0 \le i \le m$ the set BP_i^k is non empty.*

b): *The family of sets of partial isomorphisms $(BP_i^k)_{i \le m}$ is ordered by inclusion, as follows: $BP_m \subseteq \ldots \subseteq BP_0$.*

c): Bijective Forth Property: *For every $0 < i \le m$, and for every partial isomorphism $f \in BP_i^k$ there exists a bijection $h_f : dom(I) \to dom(J)$, such that for every partial isomorphism $f' \subseteq f$ and for every element $a \in dom(I)$, with $|f' \cup \{(a, h_f(a))\}| \le k$, there exists a partial isomorphism $g \in BP_{i-1}^k$, with $g = f' \cup \{(a, h_f(a))\}$.*

Note that the forth property in this system is much stronger. There must be a bijection h_f for each partial isomorphism f, such that for the extension of *every* sub-partial isomorphism of f with the same element, say a, in one database, the corresponding element in the other database must be always the same, i.e., $h_f(a)$. Note also that the back property is implied by this stronger forth property. Then we get a characterization of equivalence in $C^{k;m}$, for every $k \ge 1$ and $m \ge 0$, due to L. Hella ([22]).

Theorem 3.4. *([22]) Let I, J be two databases of some schema σ which has only relation symbols, and let $m \ge 0$, and $k \ge 1$. Then, the following are equivalent:*

a): *There is a k-bounded bijective back and forth system of partial isomorphisms of length m between I and J.*

b): *For every sentence φ in C^k with quantifier rank up to m: $I \models \varphi \Leftrightarrow J \models \varphi$, i.e., $I \equiv_{C^{k;m}} J$.*

c): *For every sentence φ in $C_{\infty,\omega}^k$ with quantifier rank up to m: $I \models \varphi \Leftrightarrow J \models \varphi$, i.e., $I \equiv_{C_{\infty,\omega}^{k;m}} J$.*

Corollary 3.4. *([22]) Let I, J be two databases of some schema σ which has only relation symbols, and let $k \ge 1$. Then, the following are equivalent:*

a): *There is a k-bounded bijective back and forth system of partial isomorphisms of length ω between I and J.*

b): *For every sentence φ in C^k: $I \models \varphi \Leftrightarrow J \models \varphi$, i.e., $I \equiv_{C^k} J$.*

c): *For every sentence φ in $I \equiv_{C_{\infty,\omega}^k} : I \models \varphi \Leftrightarrow J \models \varphi$, i.e., $I \equiv_{C_{\infty,\omega}^k} J$.*

Remark 3.1. For every back and forth system of partial isomorphisms which we covered in this section, a corresponding game theoretic characterization has been defined, which we will not consider in this article. For the original system of Fraïssé, it was due to A. Ehrenfeucht [15]. For the k-bounded back and forth system, it was originally defined by J. Barwise [6], and then reinvented by N. Immerman [25] using pebbles, and also by B. Poizat [34]. For the k-bounded forth system, the game theoretic characterization was done also by P. Kolaitis and M. Vardi together with the system of partial isomorphisms ([29], [30]). Finally, for the bijective system there are two different corresponding games; one due to N. Immerman and E. Lander [24], and the other was defined by L. Hella [22], who proved also that the two games are equivalent. The definition of all these games and the corresponding proofs for the different characterizations can be found in [13] and also in [26].

4   Similarity of Databases

When we consider queries to databases, it is quite natural to ask that whenever two databases represent the *same* real world situation, the answer to the query must be the *same*. This requisite has been formalized by A. Chandra and D. Harel in their formal notion of computable query (see Section 2) by asking that the result of a Boolean query must be equal for databases which are isomorphic (i.e., the evaluation of the query must yield the same truth value for both databases). Similarly, the r-ary relations which result form the evaluation of an r-ary query on two databases which are isomorphic, are required to be also isomorphic. Now, we can rephrase the consideration of two databases representing the same real world situation in a more general setting by using some notion of *similarity*. Thus, we can require for a query to be regarded as computable that it should yield *similar* results when it is evaluated on *similar* databases (besides, of course, of the requisite of recursiveness of the query). Clearly, for the class $\mathcal{CQ}$ similarity of databases (as well as of the results of an r-ary query) means isomorphism. And similarity of results for Boolean queries means equality of truth values.

We may define different sub-classes of computable queries by using different interpretations for the notion of similarity between databases. That is, the statement *similar results on similar databases* leads us to the definition of new classes of queries, whose intuitive appeal derives naturally from the particular notion of similarity which we choose to use. Then, using $\approx$ to denote *similarity* between databases, we make a first attempt to formalize our approach, as follows.

Definition 4.1. *Let σ be a database schema, then we define:*

$$\mathcal{C}_{\approx}^{\sigma} = \{f \in \mathcal{CQ}^{\sigma} \mid \forall I, J \in \mathcal{B}^{\sigma} : I \approx J \Longrightarrow f(I) \approx' f(J)\}$$

and we define $\mathcal{C}_{\approx} = \bigcup_{\sigma} \mathcal{C}_{\approx}^{\sigma}$.

Note that we have used $\approx'$ to denote similarity between the results of the evaluation of a query, since it is not necessarily the same notion as for databases. We will be more precise to this regard later in this section.

Clearly, the definition of the different classes $\mathcal{C}_{\approx}^{\sigma}$ is *semantic*, since it is stated in terms of a property which a query, as a function, must satisfy. We are not making use of any formalism or model of computation to define the classes.

The strategy which we propose in this article is to formalize the notion of similarity as *equivalence* in some given logic. For the next definition, we will consider only Boolean queries.

Definition 4.2. *Let $\mathcal{L}$ be a logic, let σ be a schema, and let $\mathcal{CQ}_b^{\sigma}$ denote the sub-class of computable queries of schema σ which are Boolean (in the same way, we will use $\mathcal{CQ}_b$ to denote the sub-class of Boolean computable queries of any schema). Then we define*

$$\mathcal{C}_{\mathcal{L}}^{\sigma} = \{f \in \mathcal{CQ}_b^{\sigma} \mid \forall I, J \in \mathcal{B}^{\sigma} : I \equiv_{\mathcal{L}} J \Longrightarrow f(I) = f(J)\}$$

and we define $\mathcal{C}_{\mathcal{L}} = \bigcup_{\sigma} \mathcal{C}_{\mathcal{L}}^{\sigma}$.

That is, two databases are *similar* with respect to a given logic $\mathcal{L}$, when all the Boolean queries which can be expressed in $\mathcal{L}$ evaluate to the same truth value in both databases.

4.1   Similarity of Results for Non Boolean Queries

According Definition 4.2, the notion of *similar results* (i.e., the notion denoted as $\approx'$ in Definition 4.1), regarding Boolean queries, is naturally formalized as equality of truth values. However, if we consider r-ary queries, we must formalize that notion in some other way. Actually, what we need to formalize when considering r-ary queries, is the intuitive notion of similarity between the *tuples* in the relations which result from the evaluation of a query on the two given databases. That is, we want to formalize the idea that the r-ary queries which form the class $\mathcal{C}_\mathcal{L}$, for a given logic $\mathcal{L}$, are those queries which, when computed on similar databases, evaluate to relations built up with *similar tuples* from the two databases.

For this sake we use the notion of *type* from model theory (see Section 2). For a given logic $\mathcal{L}$, the $\mathcal{L}$ type of a tuple in a database captures the set of properties of that tuple, as far as they can be expressed in the logic $\mathcal{L}$. Hence, we will formalize the notion of similar tuples as tuples of the same $\mathcal{L}$ type. By doing so, we may now complete our definition of the classes $\mathcal{C}_\mathcal{L}$, considering all queries, either Boolean or r-ary, as follows:

Definition 4.3. *Let $\mathcal{L}$ be a logic, and let σ be a schema. Then we define*

$$\mathcal{C}_\mathcal{L}^\sigma = \{f^r \in \mathcal{CQ}^\sigma \mid \forall I, J \in \mathcal{B}^\sigma :$$

$$Tp^\mathcal{L}(I, r) = Tp^\mathcal{L}(J, r) \Longrightarrow Tp^\mathcal{L}(\langle I, f(I)\rangle, r) = Tp^\mathcal{L}(\langle J, f(J)\rangle, r)\}$$

where $\langle I, f(I)\rangle$ and $\langle J, f(J)\rangle$ are databases of schema $\sigma \cup \{R\}$, with R being a relation symbol of arity r. And, as we did before, we define $\mathcal{C}_\mathcal{L} = \bigcup_\sigma \mathcal{C}_\mathcal{L}^\sigma$.

Note that the types of the tuples in the resulting relations, must be considered with respect to an extended schema, not just the schema of the databases. Then, the class $\mathcal{C}_\mathcal{L}$ is the class of computable queries which *preserve realization* of $\mathcal{L}$ types.

4.2   Preservation of Equality of Logical Theories

Recall that we may regard Boolean queries as 0-ary queries, and Boolean constants as relation symbols of arity 0. On the other hand, because of Definition 2.1, the $\mathcal{L}$ type of a 0-tuple in a database I is the $\mathcal{L}$ theory of I (i.e., $Th_\mathcal{L}(I)$). Thus, in the case of Boolean queries, Definition 4.3, takes the following form:

$$\mathcal{C}_\mathcal{L}^\sigma|_{r=0} = \{f^0 \in \mathcal{CQ}^\sigma \mid \forall I, J \in \mathcal{B}^\sigma : Th_\mathcal{L}(I) = Th_\mathcal{L}(J) \Longrightarrow f(I) = f(J)\}$$

Recall that $\equiv_{\mathcal{L}}$ means equality of $\mathcal{L}$ theories (see Section 2), so that this expression is equivalent to the one we used in Definition 4.2 for Boolean queries. Hence, Definition 4.3 does consider *all* computable queries, either Boolean or r-ary.

Thus, we got a general scheme for defining sub-classes of queries in a semantic way, based on an abstract notion of similarity.

For some logics, it is known that equality of type realization is equivalent to equality of theories. This is the case for the logics which we consider in the present article, i.e., for FO, FO^k, $FO^{k;m}$, C^k, and $C^{k;m}$. This is a well known result, which, among other sources, can be derived as a Corollary from Corollaries 1.34 and 1.40 in [33].

Proposition 4.1. *Let* $\mathcal{L} \in \{FO, FO^k, FO^{k;m}, C^k, C^{k;m}\}_{k \geq 1, m \geq 0}$. *For every* $1 \leq r \leq k$, *for every schema* σ, *and for every pair of databases* I, J *of schema* σ, *the following holds:* $I \equiv_{\mathcal{L}} J \Longleftrightarrow Tp^{\mathcal{L}}(I, r) = Tp^{\mathcal{L}}(J, r)$.

Note that by Proposition 2.1 and Proposition 4.1, $\mathcal{C}_{FO}^{\sigma} = \mathcal{C}\mathcal{Q}^{\sigma}$. Thus, if we consider logics which are weaker than FO as to expressive power, then the classes which we define in this way are proper sub-classes of $\mathcal{C}\mathcal{Q}$.

Because of Proposition 4.1, if $\mathcal{L} \in \{FO, FO^k, FO^{k;m}, C^k, C^{k;m}\}_{k \geq 1, m \geq 0}$, we may define $\mathcal{C}_{\mathcal{L}}^{\sigma}$ also as follows:

$$\mathcal{C}_{\mathcal{L}}^{\sigma} = \{f \in \mathcal{C}\mathcal{Q}^{\sigma} \mid \forall I, J \in \mathcal{B}^{\sigma} : I \equiv_{\mathcal{L}} J \Longrightarrow \langle I, f(I) \rangle \equiv_{\mathcal{L}} \langle J, f(J) \rangle\}$$

So, it turns out that for these logics, preservation of realization of $\mathcal{L}$ types is equivalent to preservation of equality of $\mathcal{L}$ theories. Hence, if it is the case that $\mathcal{L} \in \{FO, FO^k, FO^{k;m}, C^k, C^{k;m}\}_{k \geq 1, m \geq 0}$, we can consider the class $\mathcal{C}_{\mathcal{L}}$ as the class of computable queries which *preserve equality* of $\mathcal{L}$ theories. That is, whenever two databases satisfy the same sentences of $\mathcal{L}$, the resulting relations of the evaluation of the query on both databases will also satisfy the same sentences of $\mathcal{L}$ (considering as their signature the schema of the databases with the addition of a relation symbol with the arity of the query).

By using any of the logics FO^k, $FO^{k;m}$, C^k, or $C^{k;m}$ to define sub-classes $\mathcal{C}_{\mathcal{L}}$, we may use the number of variables (k), and the quantifier rank (m) as parameters, so that we get *hierarchies* in $\mathcal{C}\mathcal{Q}$, or *classifications* of computable queries. We will see in Section 5 and 6, that considering the logics FO^k and C^k, respectively, and letting the number of variables k vary, we get hierarchies into sub-classes of $\mathcal{C}\mathcal{Q}$ which are *strict* and which are *orthogonal* with the classical hierarchies of TIME and SPACE complexity defined in the area of Turing machine complexity.

Remark 4.1. Considering the results stated in Section 3, we can also define the different sub-classes $\mathcal{C}_{\mathcal{L}}$ with the notion of similarity formalized in a pure *algebraic* way. Thus, it turns out that we can get different classifications inside $\mathcal{C}\mathcal{Q}$, by relaxing the algebraic notion of isomorphism in different ways.

5   Similarity as Equivalence in Bounded Variable Logics

In [37] and [39] a hierarchy of sub-classes of computable queries denoted as QCQ^ω was studied. Every sub-class in the hierarchy, denoted as QCQ^k, can be considered as being defined following the general scheme given in Definition 4.3, and taking FO^k as the target logic $\mathcal{L}$. In that article, the sub-classes QCQ^k, which we will denote in our context as $\mathcal{C}_{FO^k}$, were defined as a semantic characterization of the class of reflective relational machines with variable complexity k, denoted as RRM^k, in order to study the increment in expressive power of that model of computation when working on databases which satisfy different properties related to homogeneity.

The way in which the classes QCQ^k were defined in [37] is to some extent implicit in [3] and [4].

In what follows we will survey the results from [37], under the perspective of the general framework which we discussed in Section 4.

Definition 5.1. *Let σ be a database schema, and let $k \geq 1$ and $k \geq r \geq 0$. Then we define*

$$\mathcal{C}_{FO^k}^\sigma = \{f^r \in CQ^\sigma \mid \forall I, J \in \mathcal{B}^\sigma :$$

$$Tp^{FO^k}(I,k) = Tp^{FO^k}(J,k) \Longrightarrow Tp^{FO^k}(\langle I, f(I)\rangle, k) = Tp^{FO^k}(\langle J, f(J)\rangle, k)\}$$

where $\langle I, f(I)\rangle$ and $\langle J, f(J)\rangle$ are databases of schema $\sigma \cup \{R\}$, with R being a relation symbol of arity r. Further, we define $\mathcal{C}_{FO^k} = \bigcup_\sigma \mathcal{C}_{FO^k}^\sigma$, and $QCQ^\omega = \bigcup_{k \geq 1} \mathcal{C}_{FO^k}$.

That is, a query is in the sub-class $\mathcal{C}_{FO^k}^\sigma$ if for every pair of databases of the corresponding schema, say σ, and for every FO^k type for k-tuples if either both databases have no k-tuple of that type, or both databases have a non empty subset of k-tuples of that type then the relations defined by the query in each database also agree in the same sense, considering the schema of the databases with the addition of a relation symbol with the arity of the query. Recall that by Proposition 4.1, the sub-class $\mathcal{C}_{FO^k}^\sigma$ can also be considered as the class of queries which preserve equality of FO^k theories.

Note that by Lemma 2.1, and Proposition 4.1, we do not need to consider the arity of the query (r) as the arity of the tuples (compare to Definition 4.3).

By Proposition 2.1 and Proposition 4.1, $\mathcal{C}_{FO}^\sigma = CQ^\sigma$, so that queries in the class CQ *preserve realization of FO types.*

Hence, queries in CQ range, to this regard, from those which need to consider every property of the database up to isomorphism (i.e., every FO property), to those for which it is enough to consider the FO^k properties of the database, for some fixed k. And the different sub-classes $\mathcal{C}_{FO^k}$ in the hierarchy QCQ^ω, correspond to different degrees of "precision", say, with which the queries in the given sub-class *need* to consider the database for them to be evaluated.

The next result states that the classes $\mathcal{C}_{FO^k}$ form a *strict* hierarchy inside CQ. In [2] it is shown that RRM with sub-linear variable complexity form a strict hierarchy, according to the function which bounds the number of variables in dynamic queries depending on the size of the input database. That result is related, to some extent, to the next result, by Proposition 5.2.

Proposition 5.1. *([37]) For every $k \geq 1$, $\mathcal{C}_{FO^{k+1}} \supset \mathcal{C}_{FO^k}$.*

As an example of a query which belongs to the sub-class $(\mathcal{C}_{FO^{k+1}} - \mathcal{C}_{FO^k})$, consider the query *"nodes in $(k + 1)$-cliques which are subgraphs of the input graph"*, defined in the schema of the graphs.

Remark 5.1. The hierarchy defined by the classes $\mathcal{C}_{FO^k}$ is not only *strict*, but it is *orthogonal* with respect to the hierarchy of complexity classes defined in terms of TIME and SPACE in Turing machines (like $LOGSPACE \subseteq PTIME \subseteq NP \subseteq PSPACE \subseteq EXPTIME$, see [7]). Note that *any* recursive predicate evaluated on the number of equivalence classes in the (equivalence) relation defined by equality of FO^k types in the set of k-tuples of a database, is in $\mathcal{C}_{FO^k}$. So that there is no complexity class defined in terms of TIME or SPACE in Turing machines which include $\mathcal{C}_{FO^k}$. And this is the case for every $k \geq 1$.

Now we give a *syntactic* characterization of the sub-classes $\mathcal{C}_{FO^k}$, by means of a computation model. To the author's knowledge this is the first semantic characterization of the machine RRM^k.

Theorem 5.1. *([37]) For every $k \geq 1$, the class of total queries which are computable by RRM^k machines is exactly the class $\mathcal{C}_{FO^k}$.*

As to the RM, as it is known that the whole class of these machines is equivalent, as to computational power, to the sub-class $RRM^{O(1)}$ (see [3]), then it turns out that $\mathcal{QCQ}^\omega = total(RM)$. So that we get also a semantic characterization for this model.

So, the next result follows immediately from Theorem 5.1.

Corollary 5.1. $\mathcal{QCQ}^\omega = total(RRM^{O(1)}) = total(RM)$.

Now we consider the relation between $\mathcal{QCQ}^\omega$ and the infinitary logic $\mathcal{L}^\omega_{\infty,\omega}$.

Let $\mathcal{L}^\omega_{\infty,\omega}|_{rec}$ denote the fragment of $\mathcal{L}^\omega_{\infty,\omega}$, of the sentences whose class of models is recursive. Then we have the following result from [4], regarding only *Boolean queries*.

Theorem 5.2. *([4]) $total(RM) = \mathcal{L}^\omega_{\infty,\omega}|_{rec}$, i.e., the class of total Boolean queries which are computable by relational machines is exactly the class of Boolean queries expressible by sentences in $\mathcal{L}^\omega_{\infty,\omega}|_{rec}$.*

In [40], and regarding RRM, instead of RM, this characterization was generalized to $\mathcal{L}^k_{\infty,\omega}|_{rec}$, for every $k \geq 1$.

Theorem 5.3. *([40]) For every $k \geq 1$, $total(RRM^k) = \mathcal{L}^k_{\infty,\omega}|_{rec}$.*

Then, by Theorem 5.1, we get the relation between $\mathcal{QCQ}^\omega$ and $\mathcal{L}^\omega_{\infty,\omega}$, which we were looking for.

Corollary 5.2. *([40]) For every $k \geq 1$, the class $\mathcal{C}_{FO^k}$, restricted to Boolean queries, is exactly the class of Boolean queries which are expressible by sentences in $\mathcal{L}^k_{\infty,\omega}|_{rec}$. Correspondingly, the whole hierarchy QCQ^ω, restricted to Boolean queries, is exactly the class of Boolean queries expressible by sentences in $\mathcal{L}^\omega_{\infty,\omega}|_{rec}$.*

As it was proved in [30] that the infinitary logic $\mathcal{L}^\omega_{\infty,\omega}$ and, hence, also the fragment $\mathcal{L}^\omega_{\infty,\omega}|_{rec}$, have a 0–1 Law (see Section 2), the next result follows immediately (compare it with the corresponding result regarding the sub-classes $\mathcal{C}_{C^k}$ in Lemma 6.2).

Proposition 5.2. *The classes QCQ^ω and $\mathcal{C}_{FO^k}$, for every $k \geq 1$, restricted to Boolean queries, have a 0–1 Law.*

And Proposition 5.2 gives us naturally the relation between the hierarchy QCQ^ω and the class CQ, since clearly the class CQ has *not* a 0–1 Law. We should point out, though, that this result can be also proved independently of the relation between QCQ^ω and the class $RRM^{O(1)}$.

Corollary 5.3. $QCQ^\omega \subset CQ$.

5.1 Preservation of Automorphism Types

We will now consider a sub-class which was defined in [37] and [39], denoted as QCQ, and which was proved to lye between QCQ^ω and CQ (see Proposition 5.3). The sub-class QCQ was studied in those articles as a lower bound as to the increment of expressive power of the model RRM^k when working with databases which satisfy a certain homogeneity property. This sub-class was defined also in a semantic way, and is similar to $\mathcal{C}_{FO^k}$, but it does not fit in the general framework which we defined in Section 4.

The intuitive idea behind the definition of the class QCQ is that queries which belong to it preserve, for any two databases of the corresponding schema, the realization of types in FO bounded to a number of variables which is *enough* for *both* databases to define tuples up to automorphism. Note that for different pairs of databases these queries will preserve FO types with possibly different bounds in the number of variables.

We will first recall some definitions from [37] which are related to the notion of *homogeneity* in classical model theory (see [11]).

For every $k \geq 1$, and for any two k-tuples $\bar{a}_k = (a_1, \ldots, a_k)$ and $\bar{b}_k = (b_1, \ldots, b_k)$ over a given database I, we will denote as $\equiv_\sim$ the (equivalence) relation defined by the existence of an automorphism in the database I mapping one k-tuple onto the other. That is, $\bar{a}_k \equiv_\sim \bar{b}_k$ iff there exists an automorphism f on I such that, for every $1 \leq i \leq k$ is $f(a_i) = b_i$.

Let $k \geq 1$. We say that a database I is *k-homogeneous* if for every pair of k-tuples $\bar{a}_k$ and $\bar{b}_k$ over I, if $\bar{a}_k \equiv_k \bar{b}_k$, then $\bar{a}_k \equiv_\sim \bar{b}_k$. Further, we define the database I as *strongly k-homogeneous* if for every $r \geq k$ it is r-homogeneous.

Definition 5.2. *For any schema* σ, *let us denote as* $arity(\sigma)$ *the maximum arity of a relation symbol in the schema. We define the class* $\mathcal{QCQ}$ *as the class of queries* $f \in \mathcal{CQ}$, *of some schema* σ *and of any arity, for which there exists an integer* $n \geq max\{arity(f), arity(\sigma)\}$, *such that for every pair of databases* I, J *in* $\mathcal{B}^{\sigma}$

$$Tp^{FO^h}(I, h) = Tp^{FO^h}(J, h) \Longrightarrow Tp^{FO^h}(\langle I, f(I)\rangle, h) = Tp^{FO^h}(\langle J, f(J)\rangle, h)$$

where $h = max\{n, min\{k : I \text{ and } J \text{ are strongly } k\text{-homogeneous}\}\}$. *As in Definition 5.1, regarding the last equivalence we consider the schema* $\sigma \cup \{f\}$, *where* f *is a relation symbol with the arity of the query.*

Note that for every pair of databases, as they are finite, there is a k such that both databases are strongly k-homogeneous. The next result shows that this class strictly includes the whole hierarchy $\mathcal{QCQ}^{\omega}$.

Proposition 5.3. *([37])* $\mathcal{CQ} \supset \mathcal{QCQ} \supset \mathcal{QCQ}^{\omega}$.

An example of a query f which is in $(\mathcal{QCQ} - \mathcal{QCQ}^{\omega})$, is as follows: define f as *the nodes belonging to the smaller clique* in the sub-class of graphs which are the disjoint union of two cliques whose sizes differ at most by 1. And define f to be empty if the two cliques are of the same size or if the graph is not in that sub-class (see [37]). On the other hand, the *parity* query on the size of a database, defined in the empty schema (i.e., the schema of the sets with equality) is not in the sub-class $\mathcal{QCQ}$, though certainly it is computable. Finally, note that all the Boolean queries which are recursive predicates in the number of equivalence classes in the relation $\equiv_{\approx}$, are also in $(\mathcal{QCQ} - \mathcal{QCQ}^{\omega})$.

6   Similarity as Equivalence in Bounded Variable Logics with Counting

In this section we introduce some new results related to a hierarchy of sub-classes which we define in terms of the general framework presented in Section 4, by using the logics C^k, for every $k \geq 1$, as the target logics $\mathcal{L}$.

One of the main causes of the weakness of FO^k as to expressibility of queries, is its inability to count beyond the bound given by k. For instance, note that we need $k + 1$ different variables to express that a node in a graph has output degree at least k. Hence, it seems quite natural to add to FO^k the capability to count beyond that bound, while still restricting the number of different variables which may be used in a formula. In this way we get the logic C^k (see Section 2), which turned out to be much more expressive than FO^k (see [13] and [26]).

Definition 6.1. *Let* σ *be a database schema, and let* $k \geq 1$ *and* $k \geq r \geq 0$. *Then we define*

$$\mathcal{C}_{C^k}^{\sigma} = \{f^r \in \mathcal{CQ}^{\sigma} \mid \forall I, J \in \mathcal{B}^{\sigma} :$$

$$Tp^{C^k}(I,k) = Tp^{C^k}(J,k) \Longrightarrow Tp^{C^k}(\langle I, f(I)\rangle, k) = Tp^{C^k}(\langle J, f(J)\rangle, k)\}$$

where $\langle I, f(I)\rangle$ and $\langle J, f(J)\rangle$ are databases of schema $\sigma \cup \{R\}$, with R being a relation symbol of arity r. We define, further $\mathcal{C}_{C^k} = \bigcup_\sigma \mathcal{C}^\sigma_{C^k}$, and $\mathcal{QCQ}^{C^\omega} = \bigcup_{k \geq 1} \mathcal{C}_{C^k}$.

First we give an important result from [9] which we will use in our proofs. Then, we show that the hierarchy defined by the sub-classes $\mathcal{C}^\sigma_{C^k}$, for $k \geq 1$, is strict.

Proposition 6.1. *([9]) For every $k \geq 1$, there are two non isomorphic graphs G_k, H_k, such that $G_k \equiv_{C^k} H_k$.*

Proposition 6.2. *For every $k \geq 1$, there is some $h > k$ such that $\mathcal{C}^\sigma_{C^h} \supset \mathcal{C}^\sigma_{C^k}$.*

Proof. The inclusion is trivial, and can be also easily obtained as a Corollary to Theorem 6.1. For the strict inclusion we will use the graphs G_k, H_k of Proposition 6.1. Note that by Proposition 2.1, for every pair of the graphs G_k, H_k there exists an integer $h > k$ such that the C^h types are FO types for both graphs. Let us write h as $h(k)$. Then, by Proposition 6.1, for every $k \geq 1$ in one of the graphs, say G_k, there are nodes whose $C^{h(k)}$ types are not realized in the other graph, H_k. Then we define, for every $k \geq 1$, the query f_k in the schema of the graphs, say σ, as the nodes of the input graph whose $C^{h(k)}$ types are not realized in the graph H_k. We will show first that $f_k \in \mathcal{C}^\sigma_{C^{h(k)}}$. For every pair of graphs I, J, whenever $I \equiv_{C^{h(k)}} J$, if they are in turn $C^{h(k)}$ equivalent to H_k, then the result of f_k will be the empty set for both graphs. If, on the other hand, $I \equiv_{C^{h(k)}} J$, but they are not $C^{h(k)}$ equivalent to H_k, then clearly the nodes in the result of f_k will have the same $C^{h(k)}$ types in both graphs. So, f_k preserves realization of $C^{h(k)}$ types, and hence $f_k \in \mathcal{C}^\sigma_{C^{h(k)}}$. Now, we will show that $f_k \notin \mathcal{C}^\sigma_{C^k}$. To see that, note that $f_k(H_k) = \emptyset$ by definition of f_k, but by Proposition 6.1, and by our asumption, $f_k(G_k) \neq \emptyset$, and $H_k \equiv_{C^k} G_k$. Thus, f_k does not preserve realization of C^k types, and hence $f_k \notin \mathcal{C}^\sigma_{C^k}$.

Proposition 6.3. $\mathcal{QCQ}^{C^\omega} \subset \mathcal{CQ}$.

Proof. The inclusion is trivial, since clearly every query in $\mathcal{QCQ}^{C^\omega}$ is computable. For the strict inclusion we will use again the graphs G_k, H_k of Proposition 6.1. We define a query f on the schema of the pairs of disjoint graphs, say σ, as the nodes in the first graph, whose FO type is not realized in the second graph. Clearly, f is computable and total. Now, running towards a contradiction, let us consider that $f \in \mathcal{QCQ}^{C^\omega}$. Then, for some $h \geq 1$, $f \in \mathcal{C}^\sigma_{C^h}$. For some order of the pairs of corresponding graphs as in Proposition 6.1, say (G_h, H_h), the result of f is non empty, by definition of f. If we consider now the pair (G_h, G_h), the result of f is empty. Since $(G_h, H_h) \equiv_{C^h} (G_h, G_h)$, it turns out that $f \notin \mathcal{C}^\sigma_{C^h}$, which is a contradiction. Thus, $f \notin \mathcal{QCQ}^{C^\omega}$.

6.1   A Reflective Machine for Logics with Counting

We will define now a model of computation to characterize the sub-classes $\mathcal{C}_{C^k}^\sigma$.

In [32], M. Otto defined a new model of computation of queries, inspired in the RM of [3], to characterize the expressive power of fixed point logic with the addition of *counting terms* (see [26]). The finite set of operations which the machine of Otto can perform includes the capability to count, where the parameter m of the counting quantifier $\exists^{\geq m}$ is dynamically defined.

Here, we define a machine which is similar to the model of Otto, but which is inspired in the RRM of [2], instead. We will not be concerned in the present paper with the comparison as to expressive power between the model of Otto and ours. However, it is straightforward to prove that the machine of Otto can be simulated in our model. The other direction should be further studied, though the simulation seems to be possible, as well.

Definition 6.2. *We define, for every $k \geq 1$, the* reflective counting machine *of variable complexity k, which we denote by RCM^k, as a* reflective relational *machine RRM^k, where dynamic queries are C^k formulas, instead of FO^k formulas. In all other aspects, our model works in exactly the same way as RRM^k. We define $RCM^{O(1)} = \bigcup_{k \geq 1} RCM^k$.*

We need first a technical result.

Definition 6.3. *Let φ be a formula of some logic $\mathcal{L}$, where the relation symbols $R_1, \ldots, R_k$, with respective arities $r_1, \ldots, r_k$, are used. Let $\varphi_1(x_{11}, \ldots, x_{1r_1})$, $\ldots$, $\varphi_k(x_{k1}, \ldots, x_{kr_k})$ be also formulas in $\mathcal{L}$. We say that $\hat{\varphi}$ is obtained from φ by* composition *with $\varphi_1, \ldots, \varphi_k$, if for every $1 \leq i \leq k$, every occurrence of an atomic formula of the form $R_i(z_1, \ldots, z_{r_i})$ in φ, is replaced by the formula φ_i in such a way that, for all $1 \leq j \leq r_i$, each occurrence of the free variable x_{ij} in φ_i is replaced by the variable z_j.*

Lemma 6.1. *Let $k \geq 1$, let σ be a schema, let I be a database of schema σ, and let $\mathcal{M}$ be a RCM^k which computes a query of schema σ. Then, the result of the computation of $\mathcal{M}$ on input I, is equivalent to the evaluation of a formula $\varphi_{\mathcal{M},I}$ in C^k on the database I. Moreover, $\varphi_{\mathcal{M},I}$ depends only on $\mathcal{M}$ and I.*

Proof. The only transitions of $\mathcal{M}$ which may affect the content of the output relation in the relational store of $\mathcal{M}$, are those transitions where a C^k formula is evaluated, and the relation defined by the evaluation is assigned to some relation symbol in the relational store. Informally, in every computation step where a C^k formula, say ψ, is evaluated, we can use composition as in Definition 6.3, replacing each relation symbol R which is used in ψ, and which is not in σ, by the last C^k formula whose resulting relation from its evaluation on the relational store was assigned to R. Then, by running an induction on the length of the computation of $\mathcal{M}$ on input I, and by applying composition as in Definition 6.3, it is straightforward to prove that we get finally one C^k formula $\varphi_{\mathcal{M},I}$ which, evaluated on I, defines the same output relation as the computation of $\mathcal{M}$ on input I. And clearly this formula depends only on $\mathcal{M}$ and I. Finally, note that C^k is closed under the operation of composition, as defined in Definition 6.2.

Now we can prove the characterization of the expressive power of the model RCM^k. The proof follows the same strategy as the proof of Theorem 5.1 in [37]. Hence, we will give here just a sketch of it.

Theorem 6.1. *For every $k \geq 1$, the class of total queries which are computable by RCM^k machines is exactly the class C_{C^k}.*

Proof. **a)** ($\subseteq$): Let f be a total r-ary query of some schema σ which is computable by a RCM^k M, for some $k \geq r$. And let I and J be two databases of schema σ such that $Tp^{C^k}(I, k) = Tp^{C^k}(J, k)$, or $I \equiv_{C^k} J$ (which is equivalent by Proposition 4.1). By Lemma 6.1, the computations of M on I and J, are equivalent to the evaluation of the C^k formulas $\varphi_{M,I}$ on the database I, and $\varphi_{M,J}$ on the database J, respectively. Thus, the computations of M on I and J preserve equality of C^k theories. So that $f \in C_{C^k}$.

b) ($\supseteq$): Let $f \in C_{C^k}$ be an r-ary query of schema σ for some $k \geq r$. We build a RCM^k machine $\mathcal{M}_f$, which will compute f. We use a countably infinite number of k-ary relation symbols in its relational store. With the input database I in its relational store, $\mathcal{M}_f$ will build in its TM tape an encoding of a database I', such that $Tp^{C^k}(I, k) = Tp^{C^k}(I', k)$. For this purpose, $\mathcal{M}_f$ will work as follows:

b1): $\mathcal{M}_f$ finds out the size of I, say n. Note that this can be done through an iteration, by varying m in the query $\exists^{\geq m} x(x = x) \wedge \neg \exists^{\geq m+1} x(x = x)$, which is in C^1. Then $\mathcal{M}_f$ builds an encoding of every possible database I' of schema σ and of size n in its TM tape, with domain $\{1, \ldots, n\}$.

b2): For every k-tuple s_i over I', $\mathcal{M}_f$ builds in its TM tape the isolating formula χ_{s_i} (as in Proposition 2.2) for the C^k type of s_i in the database I', and in this way we get isolating formulas for the types in $Tp^{C^k}(I', k)$.

b3): $\mathcal{M}_f$ evaluates as dynamic queries the formulas χ_{s_i}, for every i, which are in C^k, and assigns the results to the working k-ary relation symbols S_i in the relational store, respectively (note that these queries are evaluated on the input database I).

b4): If every relation S_i is non empty, and if the union of all of them is the set of k-tuples over I, then it means that $Tp^{C^k}(I, k) = Tp^{C^k}(I', k)$ and I' is the database we were looking for; otherwise, we try another database I' (note that the set of k-tuples over I can be generated quite easily, like through the dynamic query $\psi(x_1, \ldots, x_k) \equiv x_1 = x_1 \wedge \ldots \wedge x_k = x_k$).

b5): Now $\mathcal{M}_f$ computes $f(I')$ in its TM tape, which is possible because $f \in CQ$ and f is defined on I' because it is *total*, and then $\mathcal{M}_f$ expands the r-ary relation $f(I')$ to a k-ary relation $f^k(I')$, with cartesian product with $dom(I')$.

b6): $\mathcal{M}_f$ builds $f^k(I)$ in the relational store as the union of the relations S_i which correspond to the C^k types χ_{s_i} of the k-tuples s_i which form $f^k(I')$, and finally it reduces the k-ary relation $f^k(I)$ to an r-ary relation $f(I)$, by projecting out the components which were added in the previous step.

Corollary 6.1. $RCM^{O(1)} = QCQ^{C^\omega}$.

Remark 6.1. The considerations made in Remark 5.1 remain true regarding the hierachy of the classes $\mathcal{C}_{C^k}$. That is, this hierarchy is not only *strict*, but it is *orthogonal* with respect to the hierarchy of complexity classes defined in terms of TIME and SPACE in Turing machine complexity.

6.2   Comparison of the Two Hierarchies

Note that the two hierarchies defined by the sub-classes $\mathcal{C}_{FO^k}$ and $\mathcal{C}_{C^k}$, by using k as a parameter, have some similar properties. Both are strict and properly contained in $\mathcal{CQ}$, both are orthogonal with the Turing machine complexity hierarchy, and the two of them are characterized syntactically by machines which have a very similar definition. However, the expressive power of every logic C^k is much bigger than that of the corresponding logic FO^k. And it turns out that the sub-classes $\mathcal{C}_{FO^k}$ are "*very small*", while the sub-classes $\mathcal{C}_{C^k}$ are "*very big*", in a sense which we will precise next.

This fact can be clearly noted by using the notion of asymptotic probability (see Section 2). Recall that each sub-class $\mathcal{C}_{FO^k}$, as well as the whole hierarchy $\mathcal{QCQ}^\omega$, have a 0–1 Law (Proposition 5.2), and this implies a strong limitation as to expressive power. As we mentioned in Section 2, a query as simple as the parity query has not a 0–1 Law, and this means that this query does not belong even to the whole hierarchy $\mathcal{QCQ}^\omega$.

On the other hand, the parity query belongs to the first layer of the hierarchy $\mathcal{QCQ}^{C^\omega}$. Recall that in the second part of the proof of Theorem 6.1 we defined a machine RCM^k which in first place computed the size of the input database, and it was done by using a dynamic query in C^1, so that the *parity* query can be computed by a RCM^1 machine, and hence, it belongs to the sub-class $\mathcal{C}_{C^1}$. Then the next result follows immediately.

Lemma 6.2. *No sub-class $\mathcal{C}_{C^k}$, for any $k \geq 1$, nor the whole hierarchy $\mathcal{QCQ}^{C^\omega}$, have a 0–1 Law.*

The extent of the difference in the size of the corresponding sub-classes, as well as of the hierarchies, is clearly shown by the two following results. Proposition 6.4 can be found in [20], and is obtained as a corollary to the combination of a result by L. Babai, P. Erdös, and S. Selkow ([8]), and a result by N. Immerman and E. Lander ([24]). Proposition 6.5 can be also found in [20], and follows from results of P. Kolaitis and M. Vardi ([30]).

Proposition 6.4. *There is a class $\mathcal{C}$ of graphs with $\mu_\mathcal{C} = 1$ such that for all graphs $I, J \in \mathcal{C}$ we have $I \simeq J \Longleftrightarrow I \equiv_{C^2} J$. Moreover, for all $I \in \mathcal{C}$, and $a, b \in dom(I)$, there is an automorphism mapping a to b iff $tp_I^{C^2}(a) = tp_I^{C^2}(b)$.*

Note that the class $\mathcal{C}$ of Proposition 6.4 is C^2 homogeneous, moreover, it is also strongly C^2 homogeneous. So, the next corollary is immediate.

Corollary 6.2.

1. *Almost all graphs are strongly C^2 homogeneous.*

2. *Almost all graphs which are C^2 homogeneous are also strongly C^2 homogeneous.*

Note also that this result is quite relevant, not only in our context, but also in complexity theory, since the isomorphism problem is well known to be in NP, while M. Grohe proved that, for every $k \geq 1$, C^k equivalence is $PTIME$ complete, under quantifier free reductions ([19]). Examples of classes of well known graphs, though not having asymptotic probability 1, where C^k equivalence coincides with isomorphism, are the class of planar graphs ([21]), and the class of trees ([24]). On the other hand, the class of linear graphs is an example of a class where FO^2 equivalence coincides with isomorphism (see [13]).

Proposition 6.5. *Let $k \geq 1$. If C is a class of graphs such that for all graphs $I, J \in C$ we have $I \simeq J \Longleftrightarrow I \equiv_{FO^k} J$, then $\mu_C = 0$.*

Next and following [23] though using a slightly different perspective, we define the notion of equality of queries *almost everywhere*. Let σ be a schema, and let q, q' be two computable queries of schema σ. Let $\mu_{(q=q')}$ be as follows:

$$\mu_{(q=q')} = \lim_{n \to \infty} \frac{|\{I \in \mathcal{B}^\sigma : dom(I) = \{1, \ldots, n\} \wedge q(I) = q'(I)\}|}{|\{I \in \mathcal{B}^\sigma : dom(I) = \{1, \ldots, n\}\}|}$$

By Proposition 6.4 for *every* computable query q there is a query q' in $\mathcal{C}_{C^2}$ (and, hence in each layer $\mathcal{C}_{C^k}$, for $k \geq 2$) such that $\mu_{(q=q')} = 1$, i.e., such that q' coincides with q over *almost all* databases. On the other hand, by Proposition 6.5, this cannot be true for any layer in $\mathcal{QCQ}^\omega$, and not even for the whole hierarchy.

7   Relation with Complexity Theory

As we pointed out in Remarks 5.1 and 6.1, the hierarchies $\mathcal{QCQ}^\omega$ and $\mathcal{QCQ}^{C^\omega}$ are *orthogonal* with respect to the hierarchy of complexity classes defined in terms of TIME and SPACE in Turing machine complexity. Then, we can use these hierarchies to refine the TIME and SPACE complexity classes, by intersecting these classes with the different hierarchies $\mathcal{QCQ}^\omega$ and $\mathcal{QCQ}^{C^\omega}$. In this way, we get complexity classes which are much finer, with the obvious consequence of gaining in knowledge and understanding on the nature of the queries in the originally defined, and coarser, complexity classes. As we mentioned before, it is important to define as many different classifications of queries as we can, in order to improve our knowledge about them. And the orthogonality of different classifications allows as to intersect them, thus getting new classifications in turn.

We think that this is not a consequence of the target logics, FO^k and C^k, which we used to define $\mathcal{QCQ}^\omega$ and $\mathcal{QCQ}^{C^\omega}$, but of the way in which we defined them, following the general strategy proposed in Section 4.

To illustrate our point, we will mention in this section some results on the descriptive complexity of finite groups, recently found by A. Koponen and K. Luosto ([27] and [28]), which we may relate to the hierarchy $\mathcal{QCQ}^\omega$.

Proposition 7.1. *([28]) The following Boolean queries are expressible by sentences in $\mathcal{L}^4_{\infty,\omega}$, in the class of finite groups:*

1): *Given a group G, it is completely reducible and centreless.*

2): *Given a group G and a pair of elements a, b, the subgroup generated by a is a subgroup of the subgroup generated by b.*

3): *Given a group G and a pair of elements a, b, the two elements generate the same subgroup.*

4): *Given a group G and an element a, the subgroup generated by a is minimal normal.*

As all the Boolean queries considered in Proposition 7.1 are recursive, then it turns out that all of them can be expressed also by sentences in the fragment $\mathcal{L}^4_{\infty,\omega}|_{rec}$ (recall its definition after Corollary 5.1). On the other hand, it is known that the first three properties are expressible in FO with the addition of the *transitive closure* quantifier, which we denote by $(FO + TC)$. The logic thus obtained is known to be strictly included in the complexity class $NLOGSPACE$ (see [13]). Then, by Proposition 7.1, and Theorems 5.3 and 5.1, we have the next result.

Proposition 7.2.

a): *The first three Boolean queries of Proposition 7.1, defined in the class of finite groups, belong to the sub-class $(\mathcal{C}_{FO^4} \cap NLOGSPACE)$.*

b): *The fourth Boolean query of Proposition 7.1, defined in the class of finite groups, belongs to the sub-class $\mathcal{C}_{FO^4}$.*

Let $(FO + DTC)$ denote FO with the addition of the *deterministic transitive closure* quantifier. This logic is known to be strictly included in the complexity class $DLOGSPACE$ ([13]).

Proposition 7.3. *([27]) In the class of finite groups, the Boolean queries defined by the properties simplicity, nilpotency and solvability, are expressible by sentences in $(FO + DTC)$.*

Note that by Theorems 5.3 and 5.1, every Boolean query which is expressible by a sentence in $\mathcal{L}^\omega_{\infty,\omega}|_{rec}$ belongs to a sub-class $\mathcal{C}_{FO^k}$, for some $k \geq 1$. As the logic $(FO + DTC)$ is known to be less expressive than $\mathcal{L}^\omega_{\infty,\omega}|_{rec}$, then it turns out that the three properties of Proposition 7.3 will belong to some layers in the hierarchy $\mathcal{QCQ}^\omega$.

Finally, we give a result from [27] which implies that the isomorphism query defined in the class of finite abelian groups *does not* belong to the hierarchy $\mathcal{QCQ}^\omega$.

Proposition 7.4. *([27]) The isomorphism problem of finite abelian groups is not expressible by any sentence in $\mathcal{L}^\omega_{\infty,\omega}$.*

References

1. Abiteboul, S., Hull, R. and Vianu, V.: Foundations of Databases. Addison-Wesley (1994)
2. Abiteboul, S., Papadimitriou, C. and Vianu, V.: Reflective Relational Machines. Information and Computation **143** (1998) 110–136
3. Abiteboul, S. and Vianu, V.: Computing with first-order logic. Journal of Computer and System Sciences **50(2)** (1995) 309–335
4. Abiteboul, S., Vardi, M. and Vianu, V.: Computing with Infinitary Logic. Theoretical Computer Science **149(1)** (1995) 101–128
5. Abiteboul, S., Vardi, M. and Vianu, V.: Fixpoint Logics, Relational Machines, and Computational Complexity. Journal of ACM **44(1)** (1997) 30–56
6. Barwise, J.: On Moschovakis Closure Ordinals. Journal of Symbolic Logic **42** (1977) 292–296
7. Balcázar, J., Díaz, J., Gabarró, J.: Structural Complexity I, 2nd ed. Springer (1995)
8. Babai, L., Erdös, P. and Selkow, S.: Random Graph Isomorphism. SIAM Journal on Computing **9** (1980) 628–635
9. Cai, J. Y., Fürer, M. and Immerman, N.: An Optimal Lower Bound on the Number of Variables for Graph Identification. Combinatorica **12(4)** (1992) 389–410
10. Chandra, A. K. and Harel, D.: Computable Queries for Relational Data Bases. Journal of Computer and System Sciences **21(2)** (1980) 156–178
11. Chang, C. and Keisler, H.: Model Theory, 3rd ed. Elsevier North Holland (1992)
12. Denenberg, L., Gurevich, Y., Shelah, S.: Definability by Constant Depth Polinomial Size Circuits. Information and Control **70** (1986) 216–240
13. Ebbinghaus, H. and Flum, J.: Finite Model Theory, 2nd ed. Springer (1999)
14. Ebbinghaus, H., Flum, J., Thomas, W.: Mathematical Logic. Springer (1984)
15. Ehrenfeucht, A.: An Application of Games to the Completeness Problem for Formalized Theories. Fundamenta Mathematicae **49** (1961) 129–141
16. Fagin, R.: Probabilities on Finite Models. Journal of Symbolic Logic **41** (1976) 50–58
17. Fraïssé, R.: Sur Quelques Classifications des Systèmes de Relations (english summary). Université d'Alger, Publications Scientifiques, Série A **1** (1954) 35–182
18. Glebskij, Y. V., Kogan, D. I., Liogon'kij, M. I., Talanov, V. A.: Range and Degree of Realizability of Formulas in the Restricted Predicate Calculus. Cybernetics **5** (1969) 142–154
19. Grohe, M.: Equivalence in Finite Variable Logics is Complete for Polynomial Time. Proc. of 37th IEEE Symposium on Foundations of Computer Science (1996) 264–273
20. Grohe, M.: Finite Variable Logics in Descriptive Complexity Theory. Preliminary version (1998)
21. Grohe, M.: Fixed Point Logics on Planar Graphs. Proc. of 13th IEEE Symposium on Logic in Computer Science (1998) 6–15
22. Hella, L.: Logical Hierarchies in PTIME. Information and Computation **129(1)** (1996) 1–19
23. Hella, L., Kolaitis, P. and Luosto, K.: Almost Everywhere Equivalence of Logics in Finite Model Theory. The Bulletin of Symbolic Logic **2(4)** (1996) 422–443
24. Immerman, N. and Lander, E.: Describing Graphs: A First Order Approach to Graph Canonization. Complexity Theory Retrospective, A. Selman, ed. Springer (1990) 59–81

25. Immerman, N.: Upper and Lower Bounds for First Order Expressibility. Journal of Computer and System Sciences **25(1)** (1982) 76–98
26. Immerman, N.: Descriptive Complexity. Springer (1999)
27. Koponen, A. and Luosto, K.: Definability of Group Theoretic notions. Research Report **227** of the Department of Mathematics of the University of Helsinki (1999)
28. Koponen, A. and Luosto, K.: personal comunication (2000)
29. Kolaitis, P., Vardi, M.: On the Expressive Power of Datalog: Tools and a Case Study. Proceedings of the 9th ACM Symposium on Principles of Database Systems (1990) 61–71
30. Kolaitis, P. and Vardi, M.: Infinitary Logic and 0–1 Laws. Information and Computation **98** (1992) 258–294
31. Otto, M.: Symmetry and First Order: Explicitly Presentation-Invariant Circuits. Preliminary report (1994)
32. Otto, M.: The Expressive Power of Fixed Point Logic with Counting. Journal of Symbolic Logic **61, 1** (1996) 147–176
33. Otto, M.: Bounded Variable Logics and Counting. Springer (1997)
34. Poizat, B.: Deux ou trois choses que je sais de L_n. Journal of Symbolic Logic **47** (1982) 641–658
35. Rosen, E., Weinstein, S.: Preservation Theorems in Finite Model Theory. Proceedings of the International Workshop on Logic and Computational Complexity, Springer LNCS **960** (1994) 480–502
36. Turull Torres, J.M.: Untyped Queries, Untyped Reflective Machines and Conditional Quantifiers. Proceedings of the 2nd East European Symposium on Advances in Database and Information Systems, ADBIS'98, Springer, LNCS **1475** (1998) 35–45
37. Turull Torres, J.M.: Reflective Relational Machines Working on Homogeneous Databases. Proceedings of the First International Symposium on Foundations of Information and Knowledge Systems, FoIKS 2000, Springer, LNCS **1762** (2000) 288–303
38. Turull Torres, J.M.: On the Expressibility and the Computability of Untyped Queries. Annals of Pure and Applied Logic **108** (2001) 345–371
39. Turull Torres, J. M.: A Study of Homogeneity in Relational Databases. To appear in Annals of Mathematics and Artificial Intelligence (2001)
40. Turull Torres, J.M.: Relational Databases and Homogeneity in Logics with Counting. Draft (2001)

Author Index

Lecture Notes in Computer Science

For information about Vols. 1–2535

please contact your bookseller or Springer-Verlag

Vol. 2575: L.D. Zuck, P.C. Attie, A. Cortesi, S. Mukhopadhyay (Eds.), Verification, Model Checking, and Abstract Interpretation. Proceedings, 2003. XI, 325 pages. 2003.

Vol. 2576: S. Cimato, C. Galdi, G. Persiano (Eds.), Security in Communication Networks. Proceedings, 2002. IX, 365 pages. 2003.

Vol. 2578: F.A.P. Petitcolas (Ed.), Information Hiding. Proceedings, 2002. IX, 427 pages. 2003.

Vol. 2580: H. Erdogmus, T. Weng (Eds.), COTS-Based Software Systems. Proceedings, 2003. XVIII, 261 pages. 2003.

Vol. 2581: J.S. Sichman, F. Bousquet, P. Davidsson (Eds.), Multi-Agent-Based Simulation II. Proceedings, 2002. X, 195 pages. 2003. (Subseries LNAI).

Vol. 2582: L. Bertossi, G.O.H. Katona, K.-D. Schewe, B. Thalheim (Eds.), Semantics in Databases. Proceedings, 2001. IX, 229 pages. 2003.

Vol. 2583: S. Matwin, C. Sammut (Eds.), Inductive Logic Programming. Proceedings, 2002. X, 351 pages. 2003. (Subseries LNAI).

Vol. 2585: F. Giunchiglia, J. Odell, G. Weiß (Eds.), Agent-Oriented Software Engineering III. Proceedings, 2002. X, 229 pages. 2003.

Vol. 2586: M. Klusch, S. Bergamaschi, P. Edwards, P. Petta (Eds.), Intelligent Information Agents. VI, 275 pages. 2003. (Subseries LNAI).

Vol. 2587: P.J. Lee, C.H. Lim (Eds.), Information Security and Cryptology – ICISC 2002. Proceedings, 2002. XI, 536 pages. 2003.

Vol. 2588: A. Gelbukh (Ed.), Computational Linguistics and Intelligent Text Processing. Proceedings, 2003. XV, 648 pages. 2003.

Vol. 2589: E. Börger, A. Gargantini, E. Riccobene (Eds.), Abstract State Machines 2003. Proceedings, 2003. XI, 427 pages. 2003.

Vol. 2590: S. Bressan, A.B. Chaudhri, M.L. Lee, J.X. Yu, Z. Lacroix (Eds.), Efficiency and Effectiveness of XML Tools and Techniques and Data Integration over the Web. Proceedings, 2002. X, 259 pages. 2003.

Vol. 2591: M. Aksit, M. Mezini, R. Unland (Eds.), Objects, Components, Architectures, Services, and Applications for a Networked World. Proceedings, 2002. XI, 431 pages. 2003.

Vol. 2592: R. Kowalczyk, J.P. Müller, H. Tianfield, R. Unland (Eds.), Agent Technologies, Infrastructures, Tools, and Applications for E-Services. Proceedings, 2002. XVII, 371 pages. 2003. (Subseries LNAI).

Vol. 2593: A.B. Chaudhri, M. Jeckle, E. Rahm, R. Unland (Eds.), Web, Web-Services, and Database Systems. Proceedings, 2002. XI, 311 pages. 2003.

Vol. 2594: A. Asperti, B. Buchberger, J.H. Davenport (Eds.), Mathematical Knowledge Management. Proceedings, 2003. X, 225 pages. 2003.

Vol. 2595: K. Nyberg, H. Heys (Eds.), Selected Areas in Cryptography. Proceedings, 2002. XI, 405 pages. 2003.

Vol. 2597: G. Păun, G. Rozenberg, A. Salomaa, C. Zandron (Eds.), Membrane Computing. Proceedings, 2002. VIII, 423 pages. 2003.

Vol. 2598: R. Klein, H.-W. Six, L. Wegner (Eds.), Computer Science in Perspective. X, 357 pages. 2003.

Vol. 2599: E. Sherratt (Ed.), Telecommunications and beyond: The Broader Applicability of SDL and MSC. Proceedings, 2002. X, 253 pages. 2003.

Vol. 2600: S. Mendelson, A.J. Smola, Advanced Lectures on Machine Learning. Proceedings, 2002. IX, 259 pages. 2003. (Subseries LNAI).

Vol. 2601: M. Ajmone Marsan, G. Corazza, M. Listanti, A. Roveri (Eds.) Quality of Service in Multiservice IP Networks. Proceedings, 2003. XV, 759 pages. 2003.

Vol. 2602: C. Priami (Ed.), Computational Methods in Systems Biology. Proceedings, 2003. IX, 214 pages. 2003.

Vol. 2604: N. Guelfi, E. Astesiano, G. Reggio (Eds.), Scientific Engineering for Distributed Java Applications. Proceedings, 2002. X, 205 pages. 2003.

Vol. 2606: A.M. Tyrrell, P.C. Haddow, J. Torresen (Eds.), Evolvable Systems: From Biology to Hardware. Proceedings, 2003. XIV, 468 pages. 2003.

Vol. 2607: H. Alt, M. Habib (Eds.), STACS 2003. Proceedings, 2003. XVII, 700 pages. 2003.

Vol. 2609: M. Okada, B. Pierce, A. Scedrov, H. Tokuda, A. Yonezawa (Eds.), Software Security – Theories and Systems. Proceedings, 2002. XI, 471 pages. 2003.

Vol. 2611: S. Cagnoni, J.J. Romero Cardalda, D.W. Corne, J. Gottlieb, A. Guillot, E. Hart, C.G. Johnson, E. Marchiori, J.-A. Meyer, M. Middendorf, G.R. Raidl (Eds.), Applications of Evolutionary Computing. Proceedings, 2003. XXI, 708 pages. 2003.

Vol. 2612: M. Joye (Ed.), Topics in Cryptology – CT-RSA 2003. Proceedings, 2003. XI, 417 pages. 2003.

Vol. 2614: R. Laddaga, P. Robertson, H. Shrobe (Eds.), Self-Adaptive Software: Applications. Proceedings, 2001. VIII, 291 pages. 2003.

Vol. 2615: N. Carbonell, C. Stephanidis (Eds.), Universal Access. Proceedings, 2002. XIV, 534 pages. 2003.

Vol. 2616: T. Asano, R. Klette, C. Ronse (Eds.), Geometry, Morphology, and Computational Imaging. Proceedings, 2002. X, 437 pages. 2003.

Vol. 2618: P. Degano (Ed.), Programming Languages and Systems. Proceedings, 2003. XV, 415 pages. 2003.

Vol. 2619: H. Garavel, J. Hatcliff (Eds.), Tools and Algorithms for the Construction and Analysis of Systems. Proceedings, 2003. XVI, 604 pages. 2003.

Vol. 2620: A.D. Gordon (Ed.), Foundations of Software Science and Computational Structures. Proceedings, 2003. XII, 441 pages. 2003.

Vol. 2621: M. Pezzè (Ed.), Fundamental Approaches to Software Engineering. Proceedings, 2003. XIV, 403 pages. 2003.

Vol. 2622: G. Hedin (Ed.), Compiler Construction. Proceedings, 2003. XII, 335 pages. 2003.

Vol. 2623: O. Maler, A. Pnueli (Eds.), Hybrid Systems: Computation and Control. Proceedings, 2003. XII, 558 pages. 2003.

Vol. 2625: U. Meyer, P. Sanders, J. Sibeyn (Eds.), Algorithms for Memory Hierarchies. Proceedings, 2003. XVIII, 428 pages. 2003.

Vol. 2626: J.L. Crowley, J.H. Piater, M. Vincze, L. Paletta (Eds.), Computer Vision Systems. Proceedings, 2003. XIII, 546 pages. 2003.